# NITROARENES
## Occurrence, Metabolism, and Biological Impact

# ENVIRONMENTAL SCIENCE RESEARCH

*Series Editor:*

Herbert S. Rosenkranz
*Department of Environmental and Occupational Health*
*Graduate School of Public Health*
*University of Pittsburgh*
*130 DeSoto Street*
*Pittsburgh, Pennsylvania*

*Founding Editor:*
Alexander Hollaender

## Recent Volumes in this Series

A Continuation Order Plan is available for this series. A continuation order will
bring delivery of each new volume immediately upon publication. Volumes are
billed only upon actual shipment. For further information please contact the
publisher.

# NITROARENES
## Occurrence, Metabolism, and Biological Impact

**Edited by**
## Paul C. Howard
*Case Western Reserve University*
*Cleveland, Ohio*

## Stephen S. Hecht
*American Health Foundation*
*Valhalla, New York*

## and
## Frederick A. Beland
*National Center for Toxicological Research*
*Jefferson, Arkansas*

SPRINGER SCIENCE+BUSINESS MEDIA, LLC

Library of Congress Cataloging-in-Publication Data

International Conference on N-Substituted Aryl Compounds: Occurrence,
   Metabolism, and Biological Impact of Nitroarenes (4th : 1989 :
   Cleveland, Ohio)
      Nitroarenes : occurrence, metabolism, and biological impact /
   edited by Paul C. Howard, Stephen S. Hecht, and Frederick A. Beland.
         p.   cm. -- (Environmental science research ; v. 40)
      "Proceedings of the Fourth International Conference on N
   -Substituted Aryl Compounds: Occurrence, Metabolism, and Biological
   Impact of Nitroarenes, held July 15-19, 1989, in Cleveland, Ohio"-
   -T.p. verso.
      Includes bibliographical references and index.
      ISBN 978-1-4613-6694-2     ISBN 978-1-4615-3800-4 (eBook)
      DOI 10.1007/978-1-4615-3800-4
      1. Nitroaromatic compounds--Carcinogenicity--Congresses.
   2. Nitroaromatic compounds--Mutagenicity--Congresses.
   3. Nitroaromatic compounds--Metabolism--Congresses.   I. Howard,
   Paul C.   II. Hecht, Stephen S.   III. Beland, F. A. (Frederick A.)
   IV. Title.   V. Series.
   RC268.7.A74I59  1989
   616.99'4071--dc20                                        90-22875
                                                               CIP

Proceedings of the Fourth International Conference on
N-Substituted Aryl Compounds: Occurrence, Metabolism,
and Biological Impact of Nitroarenes, held July 15–19, 1989,
in Cleveland, Ohio

ISBN 978-1-4613-6694-2

© 1990  Springer Science+Business Media New York
Originally published by Plenum Press in 1990
Softcover reprint of the hardcover 1st edition 1990

PREFACE

Prior to 1979, consideration of the problem of the carcinogenicity of the aromatic amine class of chemicals took place primarily in poster sessions and symposia of annual meetings of the American Association for Cancer Research and analogous international associations.

In November 1979 the first meeting concerned with the aromatic amines was held in Rockville, Maryland under primary sponsorship of the National Cancer Institute. The proceedings from this meeting were published as Monograph 58 of the *Journal of the National Cancer Institute* in 1981.

The second meeting in this series, the Second International Conference on N-Substituted Aryl Compounds, was held in March/April of 1982 in Hot Springs, Arkansas. The National Cancer Institute and The National Center for Toxicological Research were the primary sponsors of this meeting. The proceedings were published as Volume 49 of the journal *Environmental Health Perspectives* in 1983.

The third meeting in this series was held in April of 1987 at the Dearborn Hyatt in Dearborn, Michigan. The principal sponsor of this meeting was the Meyer L. Prentis Comprehensive Cancer Center of Metropolitan Detroit. The proceedings, *Carcinogenic and Mutagenic Responses to Aromatic Amines and Nitroarenes*, were published in 1987 by Elsevier Press.

The fourth meeting was held in Cleveland, Ohio, on July 15-19, 1989. This meeting was also organized as a satellite of the Fifth International Conference on Environmental Mutagens. This meeting differed from the previous three meetings, in that the program was more focused on the nitrated polycyclic aromatic hydrocarbons than the aromatic amines. This was the first meeting devoting much of the program to the nitrated polycyclic aromatic hydrocarbons since the Fifth Chemical Industry Institute for Toxicology Conference on Toxicology in January of 1982 at Raleigh, North Carolina.

The primary interest of the Organizing and Program Committee was to bring together the leading researchers in the nitrated polycyclic aromatic hydrocarbons and related areas of aromatic amine research. With this as the goal, the conference was very effective. The program was indeed representative of international researchers in these areas and was highlighted by the keynote address by Takashi Sugimura, President of the National Cancer Center, Tokyo, Japan.

The editors would like to thank the members of the Organizing and Program Committee for their efforts in setting the speaking agenda: Frederick A. Beland, Jefferson, Arkansas; Robert P.P. Fuchs, Strasbourg, France; Stephen S. Hecht, Valhalla, New York; Paul C. Howard, Cleveland,

Ohio; Charles M. King, Detroit, Michigan; Dennis R. McCalla, Hamilton, Ontario, Canada; Gerald M. Mulder, Leiden, The Netherlands; Miriam C. Poirier, Bethesda, Maryland; Herbert S. Rosenkranz, Cleveland, Ohio; Dennis Schuetzle, Detroit, Michigan.

Our thanks is also offered to the financial sponsors of this conference:

THE FORD MOTOR COMPANY
THE U.S. ENVIRONMENTAL PROTECTION AGENCY
CASE WESTERN RESERVE UNIVERSITY
ICN BIOCHEMICALS, INC.

Paul C. Howard
Stephen S. Hecht
Frederick A. Beland

CONTENTS

METABOLISM

EVIDENCE FROM ANIMAL STUDIES FOR THE

CARCINOGENICITY OF INHALED DIESEL EXHAUST

Joe L. Mauderly, William C. Griffith, Rogene F. Henderson,
Robert K. Jones, and Roger O. McClellan

Inhalation Toxicology Research Institute
Lovelace Biomedical and Environmental Research Institute
P. O. Box 5890
Albuquerque, NM  87185

INTRODUCTION

Concerns in the mid 1970s for petroleum shortages and the mandating of
fuel efficiency standards for automobiles led to speculation that the use
of diesel engines in the U. S. light-duty fleet would increase substantially.  It had been known for 20 years that diesel soot carried a solvent-
extractable organic fraction containing numerous polycyclic aromatic
hydrocarbons, and that the extract was carcinogenic to mouse skin (Kotin et
al., 1955); thus, concern arose within the U. S. Environmental Protection
Agency (EPA) for the potential contribution of increased diesel emissions
to lung cancer in the U. S.  The EPA issued a precautionary notice in 1977
(U. S. EPA, 1977) that the extracts were mutagenic in the Ames Salmonella
assay, and numerous studies have subsequently provided detailed information
on the nature of soot-associated mutagenic activity (reviewed in Ishinishi
et al., 1986).

The EPA began studies of the toxicity of diesel exhaust in animals
exposed repeatedly by inhalation with a two-month exposure of cats and
rodents in 1977, followed by initiation of a two-year exposure of several
species in 1978 (Pepelko and Peirano, 1983).  General Motors began a chronic
inhalation study in 1980 (Kaplan et al., 1983), and soon thereafter, other
long-term inhalation studies were begun in the U. S., Germany, Japan, and
Switzerland.  The preliminary results of these studies were presented at a
symposium in 1986 (Ishinishi et al., 1986) and were reviewed by McClellan
in 1987 (McClellan, 1987).  The final results of most, but not all, of
these studies have now been published.

This paper reviews the results of the long-term inhalation studies and
the evidence they present for the pulmonary carcinogenicity of diesel
exhaust.  It also reviews the current issue of the specificity of the
carcinogenic response of rats for diesel exhaust and the research that is
now being conducted to resolve this issue.  The chemistry, dosimetry, and
genotoxicity of inhaled diesel exhaust are reviewed in other papers in this
volume, and are not discussed in detail in this paper.

## Rats

Eight studies of the carcinogenicity of inhaled diesel exhaust in rats have been reported (Table 1). The report of the EPA long-term study of rats and other species did not discuss carcinogenicity in the rat (Pepelko and Peirano, 1983). Five of the rat carcinogenicity assays yielded positive results, with four involving exposures of 30 mo or longer and one involving exposures for 24 mo. Two of the negative studies were 24 mo or less in duration and one involved exposures for 30 mo. Most of the studies involved interim sacrifices. To an unknown extent, therefore, some of the differences in tumor expression might be ascribed to differences among the sacrifice schedules and differences in the manner in which sacrificed rats were included in the populations at risk.

The report of the EPA study did not discuss carcinogenicity in the rat (Pepelko and Peirano, 1983). The subsequent studies were largely consistent in demonstrating increased incidences of lung cancer in rats exposed repeatedly to high concentrations of exhaust for approximately 24 mo or longer. Key parameters of these studies and the lung tumor incidences observed are listed in Table 1, in alphabetical order by laboratory.

A study funded by the Committee of Common Market Automobile Constructors was conducted at Battelle Laboratories in Geneva, Switzerland (Brightwell et al., 1986). Male and female F344 rats were exposed 16 hr/day, 5 days/wk for 24 mo to exhaust from a 1.5 L Volkswagen engine operated on an urban duty cycle, and survivors were held without exposure for up to an additional 6 mo. Rats were also exposed to the same concentrations of exhaust with the soot fraction removed by filtration. Significant increases in lung tumor incidences were observed in rats exposed to exhaust at 2.2 and 6.6 mg soot/$m^3$, but no increase was observed in rats exposed to filtered exhaust.

General Motors (GM) conducted a study at Southwest Research Institute (Kaplan et al., 1983; White et al., 1983), in which male F344 rats were exposed 20 hr/day, 7 days/wk for 15 mo to exhaust from GM 5.7 L engines operated at constant speed (40 mph) and load, and survivors were held without exposure for up to 8 additional months. No significant increase in lung tumor incidence was observed.

The Fraunhofer Institute conducted a study funded by Volkswagen (Heinrich et al., 1986), in which female Wistar rats were exposed 19 hr/day, 5 days/wk for up to 32 mo to exhaust from 1.6 L Volkswagen engines operated on an urban duty cycle. Rats were also exposed to the same concentration of exhaust with the soot fraction removed by filtration. A significant increase in lung tumors was observed in rats exposed to exhaust at 4.2 mg soot/$m^3$, but no increase was observed in rats exposed to filtered exhaust.

The Japan Automobile Research Institute (JARI) (Ishihara, 1988) conducted studies in which male and female F344 rats were exposed 16 hr/day, 6 days/wk for up to 30 mo to exhaust from either 1.8 L light-duty (LD) or 11.0 L heavy-duty (HD) engines operated at constant speed (LD = 50 kph, HD = 40 kph) and load. A significant increase in lung tumor incidence was observed only in the group exposed to HD exhaust 3.7 mg soot/$m^3$. The information on the composition of the two exposure atmospheres does not suggest a clear hypothesis for the reason for the differences in response to LD and HD exhaust. Most gas concentrations were similar at corresponding soot concentrations. Vapor-phase hydrocarbon concentrations were relatively higher for HD than for LD exhaust, but soot-associated benzo(a)pyrene and

Table 1. Summary of Pulmonary Carcinogenicity Observed in Long-Term Studies of Rats Exposed Repeatedly to Diesel Exhaust by Inhalation

| Laboratory | Rat Strain | Rats per Group | Hours per Day | Days per Week | Maximum Months | Soot mg/m$^3$ | Percent Lung Tumors |
|---|---|---|---|---|---|---|---|
| Battelle-Geneva[a] | F344 | 144 | 16 | 5 | 24(+6)[b] | 0 | 1.4 |
| | | | | | | 0.7 | 0.7 |
| | | | | | | 2.2 | 9.7[c] |
| | | | | | | 6.6 | 38.5 |
| General Motors[d] | F344 | 24 | 20 | 7 | 15(+8) | 0 | 0 |
| | | | | | | 0.25 | 1.7 |
| | | | | | | 0.75 | 5.0 |
| | | | | | | 1.5 | 1.7 |
| Fraunhofer Inst.[e] | Wistar | 95 | 19 | 5 | 32 | 0 | 0 |
| | | | | | | 4.2 | 15.8 |
| JARI[f] (Light-Duty) | F344 | 123 | 16 | 6 | 30 | 0 | 3.3 |
| | | | | | | 0.1 | 2.4 |
| | | | | | | 0.4 | 0.8 |
| | | | | | | 1.1 | 4.1 |
| | | | | | | 2.3 | 2.4 |
| (Heavy-Duty) | | | | | | 0 | 0.8 |
| | | | | | | 0.5 | 0.8 |
| | | | | | | 1.0 | 0 |
| | | | | | | 1.8 | 3.3 |
| | | | | | | 3.7 | 6.5 |
| Lovelace ITRI[g] | F344 | 220 | 7 | 5 | 30 | 0 | 0.9 |
| | | | | | | 0.35 | 1.3 |
| | | | | | | 3.5 | 3.6 |
| | | | | | | 7.1 | 12.8 |
| NIOSH[h] | F344 | 180 | 7 | 5 | 24 | 0 | 3.3 |
| | | | | | | 1.95 | 3.8 |
| Res. Inst. Tuberc.[i] | F344 | 22 | 8 | 7 | 24 | 0 | 4.5 |
| | | | | | | 4.9 | 42.1 |

[a]Brightwell et al. (1986).
[b]Values in parentheses = months of observation after exposure.
[c]Underline = significant difference from control incidence.
[d]Kaplan et al. (1983); White et al. (1983).
[e]Heinrich et al. (1986).
[f]Japan Automobile Research Institute, Ishihara (1988).
[g]Mauderly et al. (1987).
[h]Lewis et al. (1986). Tumor incidences presented orally, not given in text.
[i]Research Institute of Tuberculosis, Japan, Iwai et al. (1986).

1-nitropyrene concentrations were relatively lower for HD. In contrast to the tumor incidences, the extent of epithelial hyperplasia was reported to be greater in lungs of rats exposed to LD exhaust than in those exposed to HD exhaust.

The Lovelace Inhalation Toxicology Research Institute (ITRI) conducted a study funded by the U.S. Department of Energy (Mauderly et al., 1987) in which male and female F344 rats were exposed 7 hr/day, 5 days/wk for up to 30 mo to exhaust from 5.7 L GM engines operated on an urban duty cyle. Significant increases in lung tumor incidence were observed in rats exposed at 3.5 and 7.1 mg soot/$m^3$.

The National Institute of Occupational Safety and Health (NIOSH) (Lewis et al., 1986) conducted a study in which male and female F344 rats were exposed 7 hr/day, 5 days/wk for up to 24 mo to exhaust from a 7.0 L Caterpillar Model 3304 engine fitted with an exhaust water scrubber and operated on a variable-speed and load ("load-haul-dump") mine cycle. No significant increase in lung tumor incidence was observed.

The Japan Research Institute of Tuberculosis (Iwai et al., 1986) conducted a study in which female F344 rats were exposed 8 hr/day, 7 days/wk for up to 24 mo to exhaust from 2.4 L engines operated at constant speed (1000 rpm) and load. A significant increase in lung tumor incidence was observed in rats exposed at 4.9 mg soot/$m^3$.

The above results indicate that diesel exhaust is a pulmonary carcinogen in rats, and that the effect is primarily associated with the soot fraction. The data in Table 1 indicate that, in general, exposure of rats to diesel exhaust at high concentrations for 7 or more hr/day, 5 or more days/wk for 24 mo or longer significantly increases the incidence of lung tumors. The general exposure-response relationship is illustrated in Fig. 1A, in which lung tumor incidence is plotted against the weekly soot exposure concentration x time product (CxT) for all exposures producing significant increases in lung tumors, and for the highest exposure in each study failing to produce a significant increase in tumor incidence. One can note the overlap between the three lowest weekly exposures causing significant tumor induction (123-192 mg·hr·$m^{-3}$, ITRI, Battelle, JARI) and the highest weekly exposure failing to induce tumors (210 mg·hr·$m^{-3}$, GM). The importance of the total length of exposure is illustrated in Fig. 1B, in which lung tumor incidence is plotted against the cumulative soot exposure CxT for the total exposure. In this plot, the overlap disappears, and all total exposures of $15.9 \times 10^3$ mg·hr·$m^{-3}$ or greater are shown to cause significant lung tumor induction in rats.

Figure 1A demonstrates that, with the exception of the Iwai et al. study, the relationship between weekly soot exposure CxT and lung tumor incidence is approximately curvilinear. The reason for the apparently greater relative tumor response in the Iwai et al. study is unknown; the small number of rats in this study and the information presented in that report does not provide a basis for judging the significance of the apparent difference in response from the other studies. By comparing Figs. 1A and 1B, one can note that for exposures of 24 months or longer, tumor incidence appears to be better correlated with exposure intensity expressed as a weekly CxT than with the total cumulative exposure.

Two studies have examined the potential of diesel exhaust exposure for increasing the response of rats to known respiratory tract carcinogens. The Fraunhofer study described above (Heinrich et al., 1986) included Wistar rats given weekly subcutaneous injections of 250 or 500 mg dipentyl-nitrosamine (DPN)/kg body weight for the first 25 weeks of exhaust exposure. Both total exhaust and exhaust with the soot removed by filtration was used.

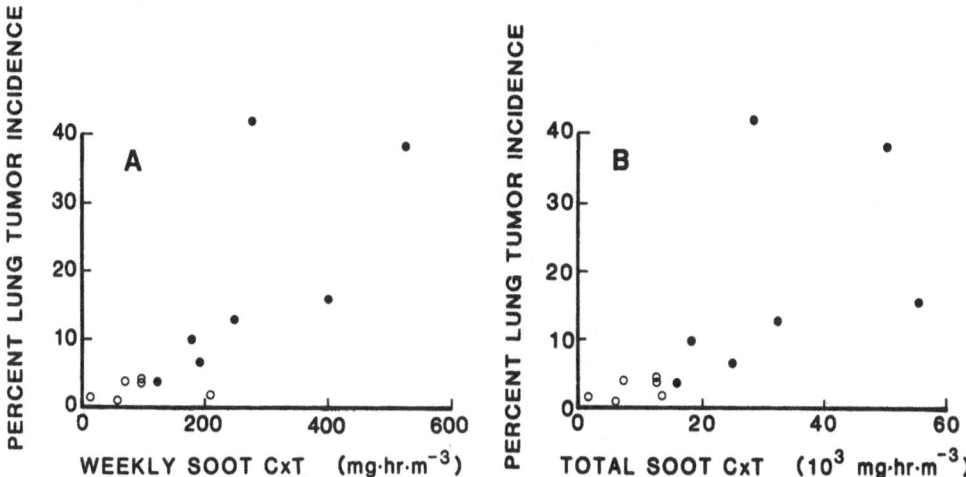

Fig. 1.  Relationship between diesel soot exposure concentration x time
(CxT) and lung tumor incidence in rats among experimental groups
from Table 1 having significant increases in lung tumor incidence
(filled circles) and groups having the highest exposures without
significant increases in tumor incidence (open circles). The
exposure CxT is expressed on a weekly basis in Fig. 1A and as the
total cumulative exposure CxT in Fig. 1B.


The DPN treatment markedly increased the baseline lung tumor incidence.
Exhaust exposure did not increase the incidence of total lung tumors in
DPN-treated rats, but did increase the incidence of squamous cell
carcinomas.  The effect of total exhaust was greater than that of filtered
exhaust.  The Saitama Medical School in Japan (Takemoto et al., 1986)
exposed untreated F344 rats and rats given 3 weekly intraperitoneal
injections of 1 g di-isopropanol-nitrosamine (DIPN)/kg body weight for 4
hr/day, 4 days/wk for 24 mo to exhaust at 2-4 mg soot/$m^3$ from a small (269
cc) diesel generator.  Although the small number of rats does not support a
judgement of the significance of the response, the results suggest that
exhaust exposure increased the lung tumor incidence above that resulting
from DIPN treatment alone.  The results of these studies suggest that
diesel exhaust may act as a cocarcinogen in rats.

Mice

     Five studies of the carcinogenicity of inhaled diesel exhaust in mice
have been conducted.  The results were variable, and do not clearly define
a carcinogenic response in mice.  The EPA study (Pepelko and Peirano, 1983)
included exposures of Strain A and Sencar mice 8 hr/day, 7 days/wk to
exhaust from a 3.2 L Nissan engine operated on an urban duty cycle.
Exposures of Strain A mice for approximately 7 mo caused an increase in the
number of lung tumors per mouse in females, but not males, exposed at 6 mg
soot/$m^3$, but a decrease in the number of lung tumors in both sexes exposed
at 12 mg soot/$m^3$.  An exposure of Sencar mice for 15 mo included an
unspecified period at 6 mg soot/$m^3$ followed by additional exposure at 12 mg
soot/$m^3$.  The exposure caused an increase in the lung tumor incidence in
females, but not males.

     The General Motors study described above (Kaplan et al., 1983) included
exposure of Strain A/J mice 20 hr/day, 7 days/wk for 8 mo to exhaust at
0.25, 0.75, and 1.5 mg soot/$m^3$.  The exposure did not increase the numbers
of lung tumors per mouse; indeed, the number of tumors tended to decrease
with increasing exposure concentration.

The Fraunhofer study described above (Heinrich et al., 1986) included exposure of female NMRI mice 19 hr/day, 5 days/wk for up to 28 mo to exhaust at 4.2 mg soot/$m^3$, and to the same concentration of exhaust with the soot removed by filtration. The lung tumor incidence was significantly increased in both groups exposed to exhaust (whole exhaust = 32%, filtered exhaust = 29%) compared to controls (13%). The increased tumor incidence was almost completely attributable to induction of adenocarcinomas; the incidence of adenomas was unchanged.

The Lovelace ITRI study described above (Mauderly et al., 1987) included exposure of CD-1 mice 7 hr/day, 5 days/wk for up to 24 mo to exhaust at 0.35, 3.5, and 7.1 mg soot/$m^3$. The final results are being analyzed; preliminary evaluation suggests no increase in lung tumor incidence.

The Saitama Medical School study described above (Takemoto et al., 1986) included C57BL/6N and ICR mice exposed 4 hr/day, 4 days/wk for up to 24 mo to exhaust at 2-4 mg soot/$m^3$. The exposure increased the incidence of lung tumors slightly, but the increase was insignificant. It is noteworthy that no lung tumors were found in 15 F344 rats exposed identically in a companion study and sacrificed between 18 and 24 mo of exposure.

The results of these studies are not conclusive with respect to the carcinogenicity of diesel exhaust in mice. Diesel exhaust was found to have little or no carcinogenic effect in the short-term Strain A mouse carcinogenicity model. Among Sencar mice, a strain developed for its sensitivity to carcinogens, females responded to 15 mo of exposure, but not males. Diesel exhaust at the highest exposure CxT used in mouse studies induced tumors in NMRI mice, but the effect occurred with or without the soot fraction. Tumors were apparently not induced by long-term exhaust exposure in the CD-1 mouse, but the weekly exposure CxT was only 62% of that experienced by the NMRI mice (249 vs 399 mg·hr·$m^{-3}$). The maximum possible weekly soot exposure CxT used for the C57BL and ICR mice was only 64 mg·hr·$m^{-3}$, which would not be expected to be carcinogenic in rats. It is unknown whether the trend toward increased lung tumor incidence in these strains would have become significant had greater exposures or larger group sizes been used.

One study examined the effect of diesel exhaust exposure on the response of mice to known carcinogens. The Fraunhofer study described above (Heinrich et al., 1986) included mice given 10 weekly intratracheal instillations of 50 µg benzo(a)pyrene (BaP) or dibenz(a,h)anthracene (DBA), or 20 weekly instillations of 100 µg BaP during exposure. The study also included mice given subcutaneous injections of 5 or 10 µg DBA 24 hr after birth, and subsequently exposed, as the other groups, 19 hr/day, 5 days/wk to exhaust at 4.2 mg soot/$m^3$. In no case did exhaust exposure increase the tumor response to the carcinogens; indeed, exhaust exposure tended to reduce the tumor response. These results suggest that diesel exhaust is not a cocarcinogen in mice.

## Hamsters

Four life-span studies of Syrian hamsters exposed by inhalation to diesel exhaust have been reported. No exposure-related increase in lung tumor incidence was observed in any of the studies.

The Battelle Pacific Northwest Laboratories (Cross et al., 1978) conducted a study funded by the National Institute of Environmental Health Sciences in which male hamsters were exposed 6 hr/day, 5 days/wk for life span (approximately 20 mo) to exhaust at 7.3 mg soot/$m^3$ from a 3-cylinder,

43 bhp engine operated on a variable speed and load cycle. No lung tumors were found in exhaust-exposed or control hamsters.

The Fraunhofer Institute (Heinrich et al., 1982) conducted a study in which female hamsters were exposed 7-8 hr/day, 5 days/wk for life span to exhaust at 3.9 mg soot/m$^3$ from a 2.4 L Daimler-Benz engine operated at constant load and speed (2400 rpm). No lung tumors were found in exhaust-exposed or control hamsters.

The Fraunhofer study described in the above sections (Heinrich et al., 1986) also included exposure of hamsters 19 hr/day, 5 days/wk for life span to exhaust at 4.2 mg soot/m$^3$. No lung tumors were observed in controls or in hamsters exposed only to exhaust.

The Battelle-Geneva study described above (Brightwell et al., 1986) also included exposure of hamsters 16 hr/day, 7 days/wk for up to 24 mo to exhaust at 0.7, 2.2, or 6.6 mg soot/m$^3$. No exposure-related increase in lung tumor incidence was observed.

The above results indicate that diesel exhaust is not a pulmonary carcinogen in the Syrian hamster.

The latter three studies also included hamsters treated with known carcinogens in addition to diesel exhaust exposure. The first Fraunhofer study included hamsters given subcutaneous injections of 1.5 or 4.5 mg diethylnitrosamine (DEN)/kg body weight before exhaust exposures began, and hamsters given intratracheal instillations of 0.1 or 0.3 mg DBA weekly for 20 weeks. The DEN treatment alone induced tracheal papillomas (13% and 45% incidence at 1.5 and 4.5 mg DEN/kg, respectively), but DBA alone did not induce tracheal or lung tumors. Exhaust exposure increased the incidence of tracheal papillomas in DEN-treated hamsters (from 45% to 70% at 4.5 mg DEN/kg), but tumors were not induced by the combination of exhaust and DBA.

Both the second Fraunhofer study and the Battelle-Geneva study included hamsters given subcutaneous injections of 4.5 mg DEN/kg body weight before exhaust exposures began. The Fraunhofer study also included hamsters given 20 weekly intratracheal instillations of 259 μg BaP. In both studies, exposure of the treated hamsters to exhaust did not increase the respiratory tract tumor incidence above that caused by the carcinogens alone. Since the suggestion of cocarcinogenesis in the first Fraunhofer study was not supported by subsequent findings, the present information suggests that diesel exhaust is not a cocarcinogen in Syrian hamsters.

## Summary: Carcinogenesis

Chronically inhaled diesel exhaust is a pulmonary carcinogen, and probably a cocarcinogen in the rat. Chronically inhaled diesel exhaust is of questionable pulmonary carcinogenicity in the mouse, and is not a pulmonary carcinogen in the Syrian hamster. Chronically inhaled diesel exhaust is not a pulmonary cocarcinogen in the mouse or Syrian hamster.

## MECHANISMS OF EXHAUST-INDUCED CARCINOGENESIS IN THE RAT

## Relevance to Risk Assessment

The finding that diesel exhaust, inhaled chronically at high concentrations, is a pulmonary carcinogen in the rat is of considerable interest, despite the failure of the predicted increase in diesel engine use in the U. S. light-duty fleet. Both early and current risk estimates suggest little need for concern for increases in lung cancer among the general U. S. population, even if a modest increase in light-duty diesel use were to occur (McClellan, 1986; McClellan et al., 1989). However, diesel engines are used more extensively in automobiles in Europe than in the U. S., and potential health effects among the general population remain a concern in several countries. The International Agency for Research on Cancer recently reviewed the toxicological and epidemiological evidence for the carcinogenicity of diesel exhaust and has classified it as carcinogenic to laboratory animals and probably carcinogenic to people (IARC, 1989).

The present focus of concern for the carcinogenicity of diesel exhaust in the U. S. is on occupational groups with high exposures. This concern was intensified by the finding of a positive association between estimated diesel exhaust exposure and lung cancer in railroad workers (Garshick et al., 1987; Garshick et al., 1988). Although this concern encompasses railroad workers, truck drivers, and heavy equipment operators, there is special concern for improving the estimates of risk to workers in dieselized coal mines (NIOSH, 1988). The present standard of 2 mg respirable dust/$m^3$ might not be sufficiently protective if a substantial portion of the dust were diesel soot.

As attempts are underway to estimate health risks to workers and to set standards for diesel exhaust exposures, it is especially important at this time to be able to place the results from rats in their proper context. It is therefore very important to resolve issues concerning the mechanisms of the exhaust-induced carcinogenicity in rats, as a basis for judging the extent to which the findings are applicable to man.

## The Case for Carcinogenesis by Chemical Genotoxicity

Present information suggests that, at least in part, diesel exhaust induces lung tumors in rats via the genotoxicity of the soot-associated organic compounds. First, the soot fraction is required to produce the effect in rats. Second, the soot-associated organic compounds are known to be mutagenic in bacterial and mammalian cells and to be carcinogenic to mouse skin. Third, it was found in the ITRI carcinogenesis study that lung DNA adduct levels were increased after 30 mo of exhaust exposure (Wong et al., 1986). These findings suggest that the carcinogenicity in rats might be extrapolated to estimates of health risk to man, as is done for other chemical carcinogens.

The rat studies suggest that diesel exhaust might exert carcinogenicity via an "initiation-promotion" pathway. If the formation of lung DNA adducts signals an "initiating" event, the response of the lung to the progressive accumulation of soot certainly provides "promoting" factors. Chronic exhaust exposures producing lung tumors in the rat also cause chronic, active inflammation in the rat lung, accompanied by cytotoxicity, epithelial proliferation, epithelial metaplasia, and focal fibrosis with attendant impairment of lung function (Heinrich et al., 1986; Henderson et al., 1988; Mauderly et al, 1988). The increased epithelial cell division would increase the probability that genetic alterations resulting from chemical-induced DNA damage might be propogated and expressed as transformed preneoplastic or neoplastic cell populations.

An initiation-promotion-progression sequence is supported by the finding that, although lung DNA adduct levels are increased after a few weeks of exposure (Bond et al., 1989) and inflammatory and proliferative responses are present from the early stages of exposure (Henderson et al., 1988), the lung tumors are a late-occurring phenomenon. In the ITRI carcinogenicity study (Mauderly et al., 1987), more than 80% of the lung tumors were observed after 24 mo of exposure. Moreover, in the GM study (White et al., 1983), the lung tumor incidence was not increased in rats exposed for only 15 mo, although the weekly exposure CxT exceeded that proven carcinogenic in more prolonged exposures. This occurred despite the observation of the GM rats for an additional 8 mo after exposure.

The importance of the contribution of the persistent inflammatory and proliferative activity is further supported by the recent finding (Bond et al., 1989) that an exposure level not carcinogenic in the ITRI carcinogenesis study, 0.35 mg soot/m$^3$ (Mauderly et al., 1987), increased rat lung DNA adducts to levels similar to those caused by exposures which did increase the tumor incidence (3.5 and 7.1 mg soot/m$^3$). The absence of inflammatory and proliferative responses at this exposure level in the ITRI study probably contributed to the lack of carcinogenesis. In addition, recent information on the time-course of exhaust exposure-related lung DNA adduct formation and repair suggests that a steady-state adduct level would be established during the first few months of chronic exposures (Bond et al., 1989).

### The Case for an Epigenetic Mechanism

There is concern that the lung tumors might not be induced in rats by chemical genotoxicity, and that the mechanism by which they are induced might not occur in man. Long-term inhalation exposures of rats to particles with little or no bioavailable mutagenic organic content have produced lung tumors of types similar to those produced by diesel exhaust. Examples presented by the following studies are illustrated in Fig. 2, in which lung tumor incidence is plotted vs the weekly particle exposure CxT. The highest CxT groups from the Battelle-Geneva, Fraunhofer, and ITRI diesel exhaust studies are included for comparison.

Two of the studies involved exposures to alpha-quartz (Min-U-Sil, Pennsylvania Glass and Sand). Holland et al. (1985) exposed F344 rats 6 hr/day, 4 days/wk for 19 mo to 12 mg quartz/m$^3$ (weekly CxT = 288 mg·hr·m$^{-3}$) and observed lung tumors in 29% of the rats. Dagle et al. (1985) exposed F344 rats 6 hr/day, 5 days/wk for 24 mo to 50 mg quartz/m$^3$ (weekly CxT = 1500 mg·hr·m$^{-3}$) and observed lung tumors in 11% of the rats.

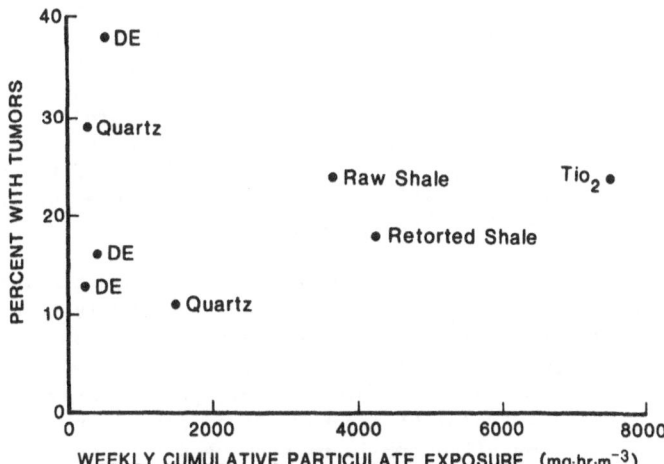

Fig. 2.  Relationship between weekly particle exposure CxT and lung tumor incidence among rats exposed to carcinogenic concentrations of diesel exhaust (DE), quartz, raw and retorted oil shale dust, and titanium dioxide. The sources of data are described in the text.

Holland et al. (1985) also exposed F344 rats to raw and retorted Paraho oil shale dust which contained approximately 10% quartz. The kerogen in raw shale must be extracted by high-temperature retorting; little, if any, organic material would be released in vivo. Rats were exposed 6 hr/day, 4 days/wk to 152 mg raw shale/$m^3$ or 176 retorted shale/$m^3$ (weekly CxT = 3648 and 4224 mg·hr·$m^{-3}$, respectively). The lung tumor incidences in the rats exposed to raw and retorted shale were 24% and 18%, respectively.

Lee et al. (1985) exposed CD rats chronically to titanium dioxide, a material sometimes used as a "nontoxic" or "negative" control particle. Rats were exposed 6 hr/day, 5 days/wk for 24 mo to 250 mg titanium dioxide/$m^3$ (weekly CxT = 7500 mg·hr·$m^{-3}$), which produced a lung tumor incidence of 24%.

A common feature of all the exposures in Fig. 2 was the progressive accumulation of particles in the lung accompanied by progressive pneumo-coniotic responses, including chronic, active inflammation, epithelial hyperplasia and metaplasia, and focal fibrosis leading to scar formation. Present information suggests that any long-term, high-level dust exposure which overwhelms particle clearance pathways, by overloading clearance capacity or damaging clearance mechanisms, can lead to this common, relatively nonspecific pneumoconiotic response in rats. The mechanisms by which dust "overloading" might occur and the implications of this phenomenon for interpreting toxicological studies were discussed by Morrow (1988). It is speculated that this response can also include tumor formation, and if so, that the tumors result from epigenetic mechanisms and may have little specificity for the chemical composition of the particles (Vostal, 1986; Morrow, 1988).

The extent to which chemical genotoxicity and epigenetic mechanisms contributed to the lung tumors in the rats in the diesel exhaust studies is unknown. It was demonstrated in the ITRI and Fraunhofer carcinogenicity studies that particle clearance was retarded (presumably due to overloading of clearance pathways) by the exposures (Heinrich et al., 1986; Wolff et al., 1987). This finding and the pneumoconiotic responses observed at carcinogenic exposure levels in all studies suggests that epigenetic mechanisms may have played a role. Although the information presented does not support a rigorous analysis, Fig. 2 suggests that the slope of the exposure-response curve might be steeper for diesel exhaust than for the other materials. With the exception of quartz, which is particularly toxic to alveolar macrophages, the exposure CxTs required to produce tumors by the inorganic particles were orders of magnitude greater than those required for diesel exhaust. This suggests that the genotoxicity of the soot-associated organic compounds might have contributed to the carcinogenicity of exhaust, perhaps in addition to epigenetic effects.

Since exposures of humans to diesel exhaust at concentrations and durations sufficient to cause lung overloading are unlikely to occur, the contribution of epigenetic mechanisms to exhaust-induced lung tumors in rats is a critical issue in the use of the rat data for predicting the risk of exhaust-induced cancer in man. This issue is particularly important because of the need to extrapolate exposure-dose-response relationships derived from observations of tumors in rats exposed to high concentrations of exhaust to cancer incidences expected in humans exposed to concentrations not shown to be carcinogenic in rats.

CURRENT RESEARCH DIRECTIONS

Current research on the carcinogenicity of diesel exhaust is focused in two areas: 1) studies at the molecular level of the genotoxicity of

10

soot-associated organic compounds; and 2) comparisons of the carcinogenicity of diesel exhaust and carbon particles with no mutagenic organic content. Examples of the first area include studies of the relationship between diesel exhaust exposure and DNA adduct formation, such as those described by Bond et al. (1989), and the evaluation of the genotoxicity of nitroarenes, as described in other papers in this volume.

Comparisons of the pulmonary carcinogenicities in the rat of diesel exhaust and carbon black are underway at the Lovelace ITRI and Fraunhofer Institute. In both laboratories, rats are being exposed chronically to exhaust at concentrations shown previously to cause lung cancer, and to the same concentration of carbon black of similar particle size, but having no solvent-extractable mutagenic organic content. In the ITRI study, funded jointly by the Health Effects Institute (supported jointly by EPA and the U. S. automotive industry) and the Department of Energy, F344 rats are being exposed 16 hr/day, 5 days/wk for 24 mo to diesel exhaust or carbon black at 2.5 and 6.5 mg particles/$m^3$. In the Fraunhofer study, also funded jointly by the German government (Bundesministerium fur Forschung und Technologie) and the German automotive industry (Forschungsgruppe Automobil-Technik), Wistar rats are being exposed 19 hr/day, 5 days/wk to diesel exhaust or carbon black at 7.5 mg particles/$m^3$ (additional exhaust, but not carbon, concentrations are included). Both studies include measurements of the amounts of particles accumulated in the lung and the effect of exposure on the clearance of tracer particles. The ITRI study is also examining the relationships among exposure, lung DNA adduct formation, and tumor occurence. Neither study has been underway long enough to produce data on lung tumor incidences at this time.

The research underway should improve our understanding of the usefulness of carcinogenicity data from rats in predicting carcinogenicity in man. The molecular studies should improve our understanding of the role of metabolites of soot-associated organic compounds in DNA injury, and whether or not the injury is of a type likely to play a role in tumor induction. The comparisons of the carcinogenicities of diesel exhaust and carbon black should provide estimates of the relative portions of the tumorigenic activity contributed by genotoxicity and epigenetic mechanisms.

## ACKNOWLEDGEMENTS

This report was prepared and the research at ITRI was conducted with support from the Office of Health and Environmental Research, U. S. Department of Energy (DOE), under Contract No. DE-AC04-76EV10103, in facilities fully accredited by the American Association for Accreditation of Laboratory Animal Care. The current study at ITRI of the carcino-genicities of diesel exhaust and carbon black are jointly funded by the DOE and the Health Effects Institute (HEI) under DOE Funds-In-Agreement No. DE-FI04-88AL52257 and HEI Research Agreement No. 88-2.

## REFERENCES

Bond, J. A., Harkema, J. R., Henderson, R. F., Mauderly, J. L., McClellan, R. O., and Wolff, R. K., 1989, Molecular dosimetry of inhaled diesel exhaust, in: "Assessment of Inhalation Hazards: Integration and Extrapolation Using Diverse Data," U. Mohr, ed., Springer-Verlag, Berlin (in press).

Brightwell, J., Fouillet, X., Cassano-Zopi, A.-L., Gatz, R., and Duchosal, F., 1986, Neoplastic and functional changes in rodents after chronic inhalation of engine exhaust emissions, in: "Carcinogenic and

Mutagenic Effects of Diesel Engine Exhaust," N. Ishinishi, A. Koizumi, R. McClellan, and W. Stöber, eds., Elsevier, Amsterdam.

Cross, F. T., Palmer, R. F., Filipy, R. E., Busch, R. H., and Stuart, B. O., 1978, "Study of the Combined Effects of Smoking and Inhalation of Uranium Ore Dust, Radon Daughters, and Diesel Oil Exhaust Fumes in Hamsters and Dogs", Report No. PNL-2744: UC-48, Pacific Northwest Laboratory, Richland.

Dagle, G. E., Wehner, A. P., Clark, M. L., and Buschbom, R. L., 1985, Chronic inhalation exposure of rats to quartz, in: "Silica, Silicosis and Cancer," D. Goldsmith, D. Winn, and C. Shy, eds., Praeger, New York.

Garshick, E., Schenker, M. B., Munoz, A., Segal, M., Smith, T. J., Woskie, S. R., Hammond, S. K., and Speizer, F. E., 1987, A case-control study of lung cancer and diesel exhaust exposure in railroad workers, Am. Rev. Respir. Dis., 135:1242.

Garshick, E., Schenker, M. B., Munoz, A., Segal, M., Smith, T. J., Woskie, S. R., Hammond, S. K., and Speizer, F., 1988, A retrospective cohort study of lung cancer and diesel exhaust exposure in railroad workers, Am. Rev. Respir. Dis., 137:820.

Heinrich, U., Peters, L., Funcke, W., Pott, F., Mohr, U., and Stöber, W., 1982, Investigation of toxic and carcinogenic effects of diesel exhaust in long-term inhalation exposure of rodents, in: "Toxicological Effects of Emissions From Diesel Engines," J. Lewtas, ed., Elsevier, New York.

Heinrich, U., Muhle, H., Takenaka, S., Ernst, H., Fuhst, R., Mohr, U., Pott, F., and Stöber, W., 1986, Chronic effects on the respiratory tract of hamsters, mice, and rats after long-term inhalation of high concentrations of filtered and unfiltered diesel engine emissions, J. Appl. Toxicol., 6:383.

Henderson, R. F., Pickrell, J. A., Jones, R. K., Sun, J. D., Benson, J. M., Mauderly, J. L., and McClellan, R. O., 1988, Response of rodents to inhaled diluted diesel exhaust: Biochemical and cytological changes in bronchoalveolar lavage fluid and in lung tissue, Fundam. Appl. Toxicol., 11:546.

Holland, L. M., Wilson, J. S., Tillery, M. I., and Smith, D. M., 1985, Lung cancer in rats exposed to fibrogenic dusts, in: "Silica, Silicosis and Cancer," D. Goldsmith, D. Winn, and C. Shy, eds., Praeger, New York.

IARC, 1989, "Engine Exhaust and Nitro-Aromatic Compounds", Monographs on the Evaluation of the Carcinogenic Risk of Chemicals to Humans, International Agency for Research on Cancer, Lyon (in press).

Ishihara, T., ed., 1988, "Diesel exhaust and health risks," Final Report of HERP Studies, Health Effects Research Program, Tsukuba, Japan.

Ishinishi, N., Koizumi, A., McClellan, R. O., and Stöber, W., eds., 1986, "Carcinogenic and Mutagenic Effects of Diesel Engine Exhaust," Elsevier, Amsterdam.

Iwai, K., Udagawa, T., Yamagishi, M., and Yamada, H., 1986, Long-term inhalation studies of diesel exhaust on F344 SPF rats: Incidence of lung cancer and lymphoma, in: "Carcinogenic and Mutagenic Effects of Diesel Engine Exhaust," N. Ishinishi, A. Koizumi, R. McClellan, and W. Stöber, eds., Elsevier, Amsterdam.

Kaplan, H. L., Springer, K. J., and MacKenzie, W. F., "Studies of Potential Health Effects of Long-Term Exposure to Diesel Exhaust Emissions", Final Report No. SWRI 01-0750-103: SFRE 1239, Southwest Research Institute, San Antonio.

Kotin, P., Falk, H. L., Thomas, M., 1955, Aromatic hydrocarbons III: Presence in the particulate phase of diesel engine exhausts and the carcinogenicity of exhaust extracts, Arch. Ind. Health, 11:113.

Lee, K. P., Trochimowicz, H. J., and Reinhardt, C. F., 1985, Pulmonary response of rats exposed to titanium dioxide by inhalation for two years, Toxicol. Appl. Pharmacol., 79:179.

Lewis, T. R., Green, F. H. Y., Moorman, W. J., Burg, J. A. R., and Lynch, D. W., 1986, A chronic inhalation toxicity study of diesel emissions and coal dust, alone or combined, in: "Carcinogenic and Mutagenic Effects of Diesel Engine Exhaust," N. Ishinishi, A. Koizumi, R. McClellan, and W. Stöber, eds., Elsevier, Amsterdam.

Mauderly, J. L., Jones, R. K., Griffith, W. C., Henderson, R. F., and McClellan, R. O., 1987, Diesel exhaust is a pulmonary carcinogen in rats exposed chronically by inhalation, Fundam. Appl. Toxicol., 9:208.

Mauderly, J. L. Gillett. N. A., Henderson, R. F., Jones. R. K., and McClellan, R. O., 1988, Relationships of lung structural and functional changes to accumulation of diesel exhaust particles, Ann. Occup. Hyg., 32:659.

McClellan, R. O., 1986, Health effects of diesel exhaust: A case study in risk assessment, Am. Ind. Hyg. Assoc. J., 47:1.

McClellan, R.O., 1987, Health effects of exposure to diesel exhaust particles, Ann. Rev. Pharmacol. Toxicol., 27:279.

McClellan, R. O., Cuddihy, R. G., Griffith, W. C., and Mauderly, J. L., 1989, Integrating diverse data sets to assess the risks of airborne pollutants, in: "Assessment of Inhalation Hazards: Integration and Extrapolation Using Diverse Data," U. Mohr, ed., Springer-Verlag, Berlin (in press).

Morrow, P. E., 1988, Possible mechanisms to explain dust overloading of the lungs, Fundam. Appl. Toxicol., 10:369.

NIOSH, 1988, "Current Intelligence Bulletin 50: Carcinogenic effects of exposure to diesel exhaust," Publication No. 88-116, U. S. Department of Health and Human Services, Public Health Service, Centers for Disease Control, National Institute for Occupational Safety and Health, Division of Standards Development and Technology Transfer.

Pepelko, W. E., and Peirano, W. B., 1983, Health effects of exposure to diesel engine emissions, J. Am. Coll. Toxicol., 2:253.

Takemoto, K., Yoshimura, H., and Katayama, H., 1986, Effects of chronic inhalation exposure to diesel exhaust on the development of lung tumors in di-isopropanol-nitrosamine-treated F344 rats and newborn C57BL and ICR mice, in: "Carcinogenic and Mutagenic Effects of Diesel Engine Exhaust," N. Ishinishi, A. Koizumi, R. McClellan, and W. Stöber, eds., Elsevier, Amsterdam.

U. S. Environmental Protection Agency, 1977, "Precautionary Notice on Laboratory Handling of Exhaust Products From Diesel Engines", USEPA, Office of Research and Development, Washington, DC.

Vostal, J. J., 1986, Factors limiting the evidence for chemical carcino-genicity of diesel emissions in long-term inhalation studies, in: "Carcinogenic and Mutagenic Effects of Diesel Engine Exhaust," N. Ishinishi, A. Koizumi, R. McClellan, and W. Stöber, eds., Elsevier, Amsterdam.

White, H., Vostal, J. J., Kaplan, H. L., and MacKenzie, W. F., 1983, A long-term inhalation study evaluates the pulmonary effects of diesel emissions, J. Appl. Toxicol., 3:332.

Wolff, R. K., Henderson, R. F., Snipes. M. B., Griffith, W. C., Mauderly, J. L., Cuddihy, R. G., and McClellan, R. O., 1987, Alterations in particle accumulation and clearance in lungs of rats exposed chronically to diesel exhaust, Fundam. Appl. Toxicol., 9:154.

Wong, D., Mitchell, C. E., Wolff, R. K., Mauderly, J. L., and Jeffrey, A. M., 1986, Identification of DNA damage as a result of exposure of rats to diesel engine exhaust, Carcinogenesis, 7:1595.

INTERPRETATION OF CARCINOGENICITY AND EFFECTIVE DOSE IN CHRONIC EXPOSURES

OF RATS TO HIGH DIESEL EXHAUST CONCENTRATIONS

Werner Stöber

Lovelace Inhalation Toxicology Research Institute
P.O. Box 5890, Albuquerque, New Mexico
on leave from
Fraunhofer-Institute of Toxicology and Aerosol Research
Hannover, West Germany

INTRODUCTION

Experimental carcinogenicity of combustion engine exhaust was first described by Kotin et al. (1954, 1955) more than 30 years ago. However, recent concern focussed particularly on diesel engine exhaust, because the diesel soot particles in the exhaust may act as built-in condensation nuclei for the collection of a variety of polycyclic aromatic hydrocarbons (PAHs), some of which are supposed to induce tumors. Over the last ten years, various chronic exposure studies with animals have been conducted to gather evidence for the suspected carcinogenicity of inhaled diesel exhaust. These studies were just reviewed by Mauderly and coworkers (this volume). In summary, the results of several investigators have shown that diesel exhaust is a pulmonary carcinogen in rats whenever these animals are chronically exposed to high diesel exhaust concentrations, and diesel soot is actually the indispensible component in the exhaust emission necessary to induce the tumors (Heinrich et al., 1986).

In addition to these findings, recent epidemiological studies seem to give weak but statistically significant support to the hypothesis that long-time US railroad workers who were employed on diesel locomotives were exposed to excessive concentrations of diesel exhaust and, therefore, developed lung cancer after twenty or more years of service (Garshick et al., 1987, 1988).

After assessing these results in the light of the excessive mutagenicity of extracts of diesel soot in the Ames test (Lewtas et al., 1978), national agencies in the US (NIOSH, 1988) and West Germany (DFG-Senatskommission, 1988) as well as an agency of the World Health Organization (IARC, 1988) finally decided last year to label officially diesel exhaust a potential or probable human carcinogen.

This formal decision will satisfy the needs of national regulatory panels for a health-related justification of their present and future decisions on emission standards for diesel-powered vehicles. However, the scientific community cannot yet leave the results of the long-term inhalation studies with rats as they stand right now. These studies, as pointed out by Mauderly and coworkers (loc. cit.), leave room for different

mechanistic interpretations, and it appears that they may have raised more
questions than they did answer.

Altogether, the mutagenicity of the diesel soot extracts, the DNA
adduct formation in rats after diesel exhaust inhalation (Wong et al.,
1986) and the tumor induction in rats after long-term high concentration
diesel exhaust inhalation apparently suggest a chemical carcinogenesis of
the rat tumors.  However, the long-term exposure studies revealed an
unexpected response of the rats to insoluble particles.  Contrary to
expectation, the chronic exposures did not reach a steady state for the
alveolar particulate load.  Even when the deposition rate was relatively
low, the experimental data indicated that, after about one year of chronic
exposure, the alveolar clearance will suffer from a partial impairment
which causes a continuous increase of the alveolar load during the
remainder of the animals' life span.  In case of high deposition rates, the
alveolar clearance mechanism appears to be overwhelmed immediately, which
leads to an almost linear increase of the alveolar particle burden over the
lifetime of the rats.  This overload phenomenon, which may be specific for
rats, and its toxicological implications, were recently discussed by Morrow
(1988).  Chronic inhalation studies under comparable overload conditions
with supposedly inert particles like titanium dioxide (Lee et al., 1985) or
coal dust (Martin et al., 1977) resulted in tumor inductions in the lung as
found after the diesel soot exposures.  Therefore, Vostal (1986) claimed
that tumor induction in rats under overload conditions is not specific for
the chemical nature of the deposited particles and represents an epigenetic
mechanism which would not be relevant to human exposures.

This paper discusses the toxicokinetic differences of the two potential
tumor induction mechanisms and a retention model for the unusual behavior
of the alveolar clearance in rats which may be applicable to other mammals.


DOSE IN CHEMICAL CARCINOGENESIS OF DIESEL TUMORS IN RATS

There is sufficient experimental evidence to justify the assumption
that the observed tumor induction in rats after lifetime exposure to diesel
exhaust is based on a chemical carcinogenesis.  In this case, the biochemical mechanism is supposed to involve certain diesel exhaust components
consisting of chemical compounds of proven experimental carcinogenicity.
Most frequently discussed are polycyclic aromatic hydrocarbons or certain
derivatives like nitroarenes which seem to occur in diesel exhaust more
frequently than in other combustion engine exhaust emissions.  These compounds have low vapor pressures and, together with other noncarcinogenic
compounds of similar structure, they show an affinity to carbonaceous
surfaces.  Therefore, the diesel soot particles will serve for these
compounds as condensation nuclei in the engine exhaust system.  Diesel soot
was frequently analyzed for extractable organic material.  The extracts may
represent up to 40% or more of the total weight of the soot (Clark, 1982).
The carcinogenic moiety is most likely a small fraction of the extract.

It is interesting to note that the carcinogenic PAH fractions of
diesel soot used in a rat study (Heinrich et al., 1982) were generally
lower than the corresponding fractions found in urban particulate air
pollution.  Table 1 lists 13 typical polycyclic hydrocarbons and compares
their extractable fractions in urban particulate air pollution as measured
in several German cities with the corresponding extracted fractions from
diesel soot used in the rat study.  The data indicate that all of the
carcinogenic PAHs determined in this study occurred in urban areas in
higher concentrations per particle mass than on the diesel soot investigated.  If adsorbed or condensed polycyclic aromatic hydrocarbons are
indeed the carcinogenic agents in all of the experimental rat studies, then

Table 1.  PAH-Content of Particulate Urban Air Pollution and Diesel
         Soot Used in an Animal Study (Heinrich et al., 1982)

| Polycyclic Aromatic Hydrocarbon | | Sampling Place | | | | Exposure Diesel Soot |
|---|---|---|---|---|---|---|
| | | Industrial | | | Rural | |
| | | Dortmund | Bochum | Essen | Borken | |
| FLU | (−) | 26.1 | 21.5 | 38.5 | 17.0 | 134.6 |
| PYR | (−) | 21.9 | 19.6 | 35.9 | 13.1 | 65.8 |
| BaF | (−) | 7.1 | 6.1 | 12.2 | 2.4 | 5.4 |
| BaA | (+) | 36.7 | 35.6 | 51.0 | 17.8 | 5.3 |
| CHR | (+) | 80.9 | 80.2 | 101.6 | 46.2 | 25.7 |
| BFL | (+) | 138.7 | 141.3 | 165.4 | 94.1 | 22.2 |
| BeP | (−) | 59.4 | 58.3 | 69.5 | 42.0 | 14.1 |
| BaP | (+) | 27.2 | 27.0 | 42.0 | 14.7 | 7.0 |
| PER | (−) | 6.3 | 5.9 | 7.3 | 3.9 | − |
| IP | (+) | 36.6 | 37.2 | 42.7 | 23.5 | 13.4 |
| DBA | (+) | 14.6 | 11.2 | 13.5 | 7.2 | − |
| BghiP | (−) | 50.7 | 44.5 | 56.5 | 34.8 | 21.4 |
| COR | (−) | 24.0 | 19.2 | 26.9 | 18.2 | 12.5 |

(−) noncarcinogenic
(+)  carcinogenic  } in noninhalation animal test models

FLU = Fluoranthene
PYR = Pyrene
BaF = Benzo[a]pyrene
BaA = Benz[a]anthracene
CHR = Chrysene
BFL = Benzofluoranthene
BeP = Benzo[e]pyrene

BaP = Benzo[a]pyrene
PER = Perylene
IP = Indeno[1,2,3-cd]pyrene
DBA = Dibenzanthracene
BghiP = Benzo[ghi]perylene
COR = Coronene

it seems to be legitimate to conclude that the carcinogenic impact on the
rats from inhaled urban particulate matter could be more impressive than
from diesel soot.  Unfortunately, this conclusion is hard to check
experimentally, because it is next to impossible to collect sufficient
amounts of particulate urban air pollution for a lifetime high-dose
inhalation study on laboratory rats.

Usually, the soot particles and their condensation layer of organics
form irregularly shaped aggregates, but the aerodynamic size of these
aggregates is still below 1 mm (Cheng et al., 1984).  This size makes the
soot aerosol completely respirable.  Thus, when inhaled, the soot will
deposit to a large extent all over the upper airways and in the alveolar
region of the lungs.  Chemical carcinogenesis would then imply that the
organic surface layer on the soot particles will more or less rapidly
dissolve into the body fluid, and, after cellular resorption, the
carcinogenic compounds could be metabolized and transformed into ultimate
carcinogens.

The effective dose for this mechanism is not easy to assess.
Obviously, it depends strongly on the velocity of several competing
processes.  The clearance mechanism of the tracheobronchial tract removes
deposited particles from the bronchial tree with half-times of the order of
hours.  If the adsorbed carcinogens dissolve from the soot surface layer on
a comparable time scale, then a substantial part of the soot deposit in the
tracheobronchial tract and all of the soot deposited in the alveolar region
may be an adequate measure of the effective dose for the tumor induction.

The effective dose may then possibly be in proportion to the accumulated exposure time and, as long as the metabolic apparatus is not oversaturated, to the soot exposure concentration.

However, if the release of the carcinogenic components from the soot surface layer has half-times of the order of several days, then the tracheobronchial deposition is practically immaterial, and the effective dose is related to the deposition in the alveolar region alone. This is due to slow macrophage-mediated pulmonary clearance in the rat which has a half-time of some 40 or more days. Still, there may be a proportionality of the effective dose to the exposure concentration at a lower ratio, because, under usual experimental conditions, alveolar deposition is proportional to the exposure concentration.

But the situation will change significantly, when the dissolution rate is substantially slower than the macrophage-mediated clearance rate of the alveolar deposit. In this case, the retained alveolar deposit determines the dissolution rate, and the effective dose would be proportional to the time integral of the retained alveolar deposit, which of course, is no longer proportional to the exposure concentration. Assuming chemical carcinogenesis, this is a situation where the alveolar particle burden represents a depot of the carcinogenic agent which is released at a slow but constant and continuous rate to the hosting organism.

Since the dose-response relationship for the tumor induction in rats is known to be nonlinear, it is rather difficult to decide between these different possibilities of effective dose.


DOSE IN EPIGENETIC CARCINOGENESIS OF DIESEL TUMORS IN RATS

While the effective dose in chemical carcinogenesis of diesel tumors is somewhat ambiguous and depends on the release rate of the carcinogenic agent from the deposited soot, there is no such problem when an epigenetic mechanism of carcinogenesis is assumed. In this case, the presence of foreign particulate matter in sufficient quantities is the only requirement so that the retained lung burden is an adequate measure of effective dose. In all likelihood, a no-effect threshold may be assumed, but, except for the latency period, no time dependency has to be considered.

Under normal circumstances, an epigenetic carcinogenesis could probably play a role only under severe exposure conditions, because, in general, the alveolar clearance mechanism should establish a steady state for the lung burden which remains below the epigenetic no-effect threshold. However, this is apparently not the case in rats. After a time period of regular behavior during chronic exposures, there seems to be an impairment of alveolar clearance, and, for sufficient deposition rates, it may result in alveolar deposits exceeding the postulated threshold.

Figure 1 shows lung burden patterns of chronic diesel exposures for three different exposure concentrations. The experimental data are taken from the Lovelace chronic diesel inhalation study (Mauderly et al., 1987). While, for the first 300 days of the study, the lung burdens appear to approach a steady state value, there is an unforeseen steady increase of the lung burdens during the second half of the study period so that, in terms of the hypothesis of epigenetic carcinogenesis, the two higher exposure concentrations actually caused a lung burden increase beyond the epigenetic no-effect threshold and led to tumor inductions.

Investigators favoring this interpretation argue that the type of carcinogenesis which causes tumors by overloading the pulmonary region of

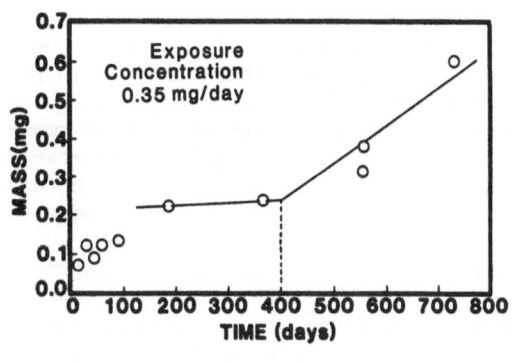

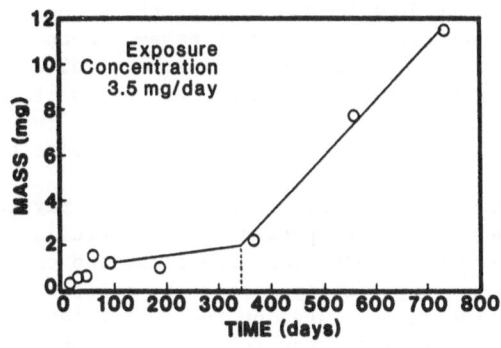

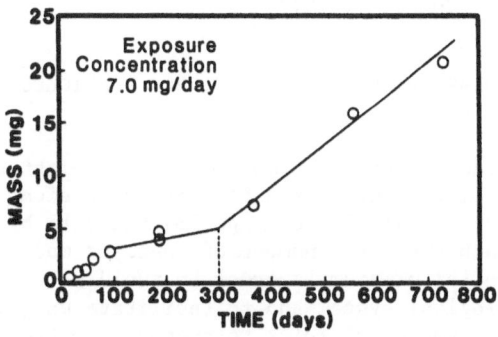

Fig. 1.  Estimates of sequestration onset in chronic diesel exhaust
inhalation studies (Mauderly et al., 1986).

the lung is the only one so far experimentally observed in diesel exposure
studies. This would have no relevance to human exposures, and there would
be no sound basis for a linear extrapolation of the obtained tumor
induction rates to exposure levels experienced by man.


## ALVEOLAR RETENTION OF DIESEL SOOT IN RATS

Apparently, the retained lung burden in the rat is a key parameter in
assessing the effective dose for an epigenetic mechanism of carcinogenesis,
but the lung burden may be also a significant parameter for a form of
chemical mechanism of carcinogenesis where the carcinogenic component is
slowly but continuously released from an undepletable depot in the lung.
Therefore, it is essential for pharmacokinetic assessments to consider
particulate alveolar retention as a potential parameter for an effective
dose and to investigate the unexpected failure of the alveolar clearance
mechanism in rats as observed by several investigators.

Experimental data from various laboratories (Strom et al., 1988, 1989;
Muhle et al., 1988; Wolff et al., 1987) show that, for sufficiently high
exposure concentrations, the alveolar clearance mechanism of the rat is
unable to maintain a steady state situation, where the clearance rate
compensates the deposition rate. Instead, the clearance mechanism fails
after some time and the lung burden will start increasing continuously.
Under extreme exposure conditions, the clearance mechanism will fail
immediately. Furthermore, subchronic exposure studies revealed that, once
the clearance system has failed, it does not regain its original strength,
even in long post-exposure periods up to one year. In these studies, parts
of the lung burden appeared to be practically irreversibly deposited.

As Morrow (1988) pointed out, there must be a physical limit to what
an alveolar macrophage can incorporate by phagocytosis, and he explained
the impairment of the macrophage-mediated clearance of the rat by postulat-
ing that the macrophages gradually lose their mobility with increasing
particulate macrophage load. Based on these postulates, a model of
alveolar retention of insoluble particles in the rats was proposed (Stöber
et al., 1989) which also made use of the concept of macrophage sequestration introduced by Soderholm (1981).


## SIMULATION OF ALVEOLAR RETENTION OF INSOLUBLE PARTICLES BY COMPARTMENTAL MODELING

The recent alveolar retention model features eight compartments.
Figure 2 shows the flow diagram of a simplified version. The design is not
supposed to achieve a simulation of experimental data by a conventional
mathematical approach where the number of employed model parameters is
reduced to an essential minimum in order to avoid potential ambiguities of
the complex physiological system and to facilitate unique numerical
solutions. Instead, the model tries to account for the dominant role of
alveolar macrophages in removing foreign particles from the pulmonary
region of the lungs. It links this role to some physiological characteristics of the macrophages. The phagocytic activity by which alveolar macrophages incorporate particle deposits on the alveolar surface, and the
carrier function of the macrophages as they eventually move with their
particulate burden onto the mucociliary escalator of the bronchial tree are
two typical model features. The influence of these characteristics is seen
as curtailed by the finite life span and a declining mobility of the
macrophages.

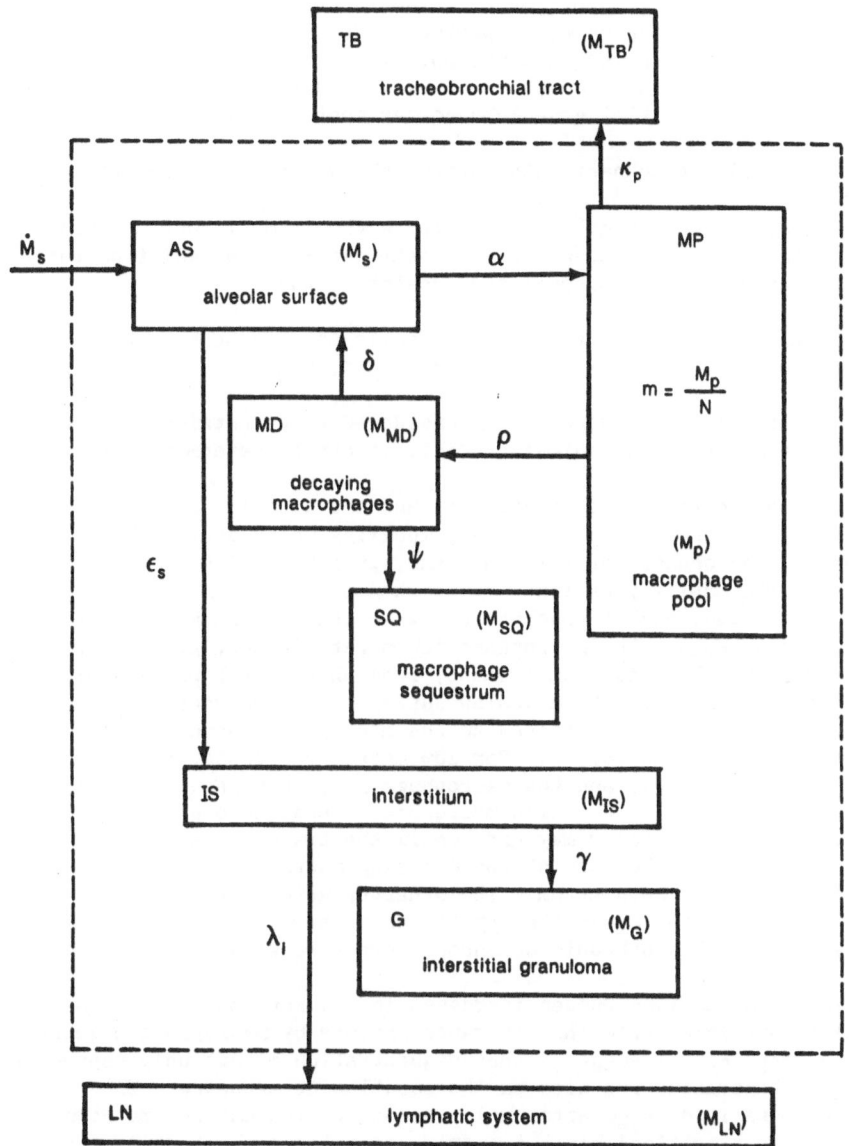

Fig. 2. Simplified model of alveolar retention by rats (Stöber et al., 1989).

Six of the compartments constitute the alveolar region  They are shown inside the dashed-line rectangle.  Outside of this rectangle are two recipient compartments for particles removed from the alveolar region.

Within the alveolar region are

1) the alveolar surface, AS, with a particulate mass deposit, $M_s$;
2) the macrophage pool, MP, with an incorporated mass, $M_p$;
3) the macrophage decay compartment, MD, with a mass, $M_{MD}$;
4) the macrophage sequestrum, SQ, with a sequestered deposit, $M_{SQ}$;
5) the interstitial compartment, IS, with a mass deposit, $M_{IS}$; and
6) the interstitial granuloma compartment, G, with a deposit, $M_G$.

The receiving compartments outside the alveolar region are

7) the tracheobronchial tract, TB, with a time-integrated transient mass, $M_{TB}$, representing the total particle removal from the alveolar region via the mucociliary escalator; and

8) the lymphatic system, LN, primarily the tracheobronchial lymph nodes, with an accumulated mass load, $M_{LN}$.

The lymph node deposit is considered to be outside the lung, because in experimental animal studies, it can be determined separately.

The phagocytosis rate factor, $\alpha$, and the alveolar clearance rate factor, $\kappa_p$, depend on the macrophage mobility which declines with increasing macrophage particle load and, possibly, with the age of the animals.  The mobility decline begins when a critical macrophage pool load, $M_{cf}$, is exceeded, and it reaches zero when the macrophages attain their maximum load, $M_{max}$.  The macrophage death rate factor, $\rho$, delivers dead macrophages to the decay compartment, from where their particulate load may be transferred back to the alveolar surface by a transfer rate factor, $\delta$. The load may also be transferred to the macrophage sequestrum by a sequestration rate factor, $\psi$.  For low macrophage loads, the sequestration rate vanishes ($\psi = 0$), and the macrophage loads are returned exclusively to the alveolar surface.  For high macrophage loads, however, the sequestration rate factor may grow until the transfer rate factor, $\delta \to 0$, disappears so that the load of the decaying macrophages is exclusively transferred to the sequestrum.  The other transfer rate factors, $\varepsilon_s$, $\lambda_i$ and $\gamma$, are constant, and $M_s$ is the deposition rate which is supposed to be constant for each subchronic or chronic exposure study.

The observed lung burden increases in old rats may be due to a with-drawal of particles from the clearance process by macrophage sequestration. In this case, Fig. 1 suggests that sequestration is not only dependent on the macrophage load, but also on the duration of exposure.  Thus, besides a critical mass load, $M_{ch}$, after which sequestration may become significant, there may be a critical time, $t_c$, prior to which sequestration may not become effective.  There may be also a later point in time, $t_{SQ}$, after which the macrophage load alone determines sequestration.  A detailed description of the modified model is given elsewhere (Stöber et al., to be published).

With these model parameters, simulations were attempted of the experimental data of the Lovelace chronic diesel exposure study as shown in Fig. 1 and the subchronic diesel study by Strom and coworkers (1988).  The objective of the simulations was to find a single set of parameters which, after accounting for the different exposure conditions, would represent all the experimental results of the two independent studies.

Table 2 gives the final set of parameters obtained by trial and error, and Figs. 3 to 9 show the experimental data and the final simulation patterns in graphical form. Besides the alveolar lung burden data, Figs. 6 to 9 give also experimental and simulation data on the lung-associated lymph node burdens.

Apparently, the representation of the experimental results by the simulation data is quite satisfactory. The selected lifetime of the alveolar macrophages, $\tau_L = 1/\rho = 7$ days, is in close agreement with the value suggested by Bowden (1983), and the phagocytosis half-time, $\tau_0 = 1/\alpha_{max} = 0.25$ days = 6 h, corresponds to results by Lehnert and Morrow (1985). The alveolar clearance rate, $\kappa_{pmax}$, corresponds to a half-time of 52.5 days, which is close to values found for rats exposed to other particulates. No comparative data are available for the other parameters. The exposure parameters at the bottom of Table 2, in particular the deposition rates, $\dot{M}_s$, properly reflect the different exposure conditions, and all other model parameters, with the exception of the time dependence of the macrophage sequestration indicated by $t_c$ and $t_{SQ}$, are the same for both studies.

In terms of the model design, the macrophage sequestration is suppressed for the first nine month in the chronic diesel exposures, while the very high exposure of the subchronic exposures leads to sequestration from the very beginning. Mathematically, this could be taken care of by assuming a dependency of $t_c$ on the deposition rate. However, the necessity of such remedial measure may just indicate that the model gives only an incomplete and approximate description of the sequestration phenomenon. Further adjustments should not be made before the model has been tested as

Table 2. Model Parameters for Simulating Experimental Data of a Subchronic and a Chronic Diesel Exhaust Inhalation Study

| Case Parameters | | Diesel Exhaust Exposure | | |
| --- | --- | --- | --- | --- |
| | | Subchronic (Strom et al., 1988) | Chronic (Wolff et al., 1987) | |
| $\tau_L$ | [days] | 7 | 7 | |
| $\tau_0$ | [days] | 0.25 | 0.25 | |
| $F_r$ | -- | 0.05 | 0.05 | |
| $\kappa_{pmax}$ | [day$^{-1}$] | 0.0132 | 0.0132 | |
| $\varepsilon_s$ | [day$^{-1}$] | 0.06 | 0.06 | |
| $S_r$ | -- | 1 | 1 | |
| $\lambda_i$ | [day$^{-1}$] | 0.001 | 0.001 | |
| $\gamma$ | [day$^{-1}$] | 0 | 0 | |
| $M_{max}$ | [mg] | 4.5 | 4.5 | |
| $M_{cf}/M_{max}$ | -- | 0.667 | 0.667 | |
| $M_{ch}/M_{max}$ | -- | 0 | 0 | |
| Q | -- | 4 | 4 | |
| $t_c$ | [days] | 0 | 274 | |
| $t_{SQ}$ | [days] | 500 | 365 | |
| Ms | [mg/day] | 0.275 | 0.0026  0.026  0.052 | |
| $T_{exp}$ | [days] | 7  21  42  84 | 735 | |
| $T_{end}$ | [days] | 465 | 735 | |

$F_r$ = Ratio of decay time to lifetime of alveolar macrophages
$S_r$ = Maximum fraction of decay being sequestered
Q = Exponential parameter for sequestration increase by macrophage load

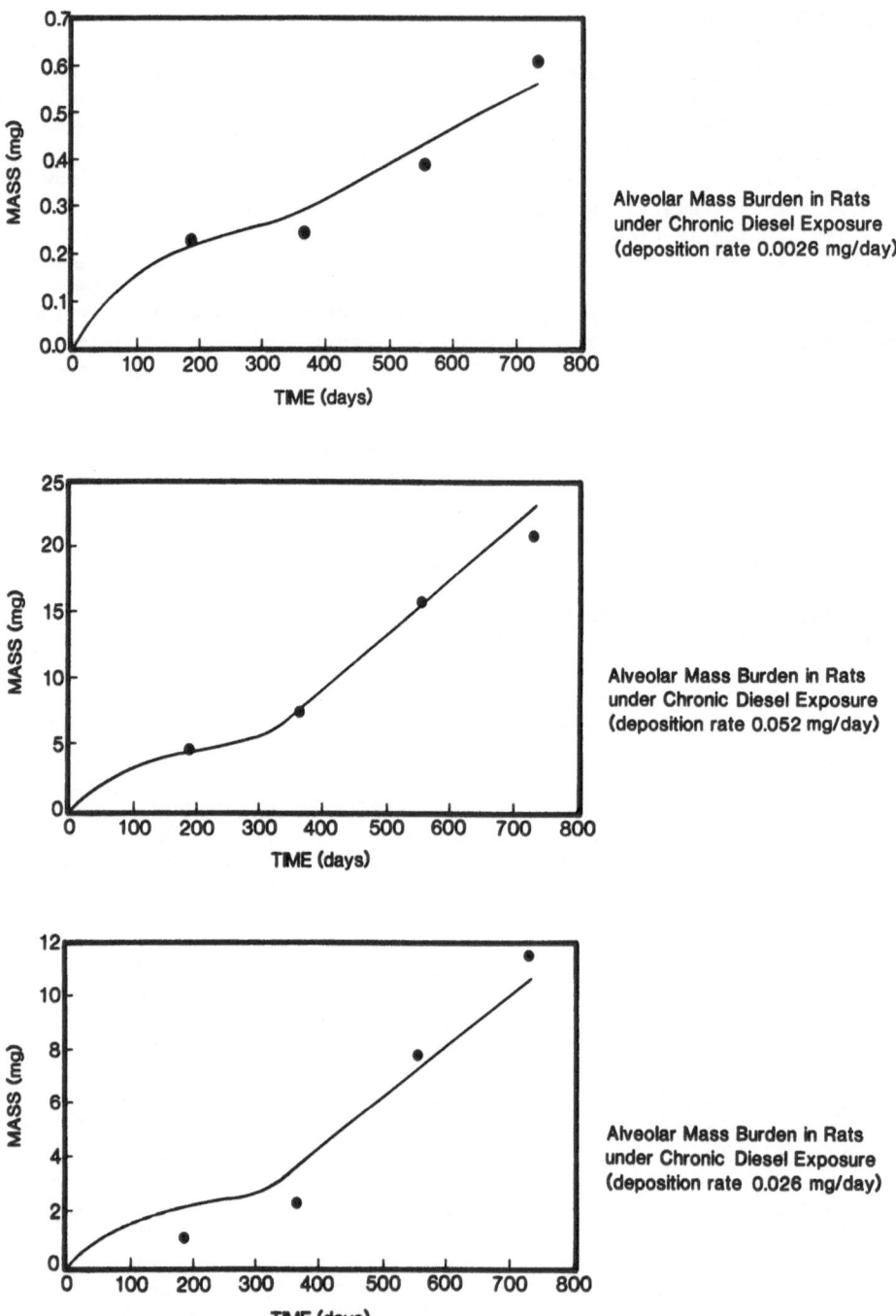

Alveolar Mass Burden in Rats
under Chronic Diesel Exposure
(deposition rate 0.0026 mg/day)

Alveolar Mass Burden in Rats
under Chronic Diesel Exposure
(deposition rate 0.052 mg/day)

Alveolar Mass Burden in Rats
under Chronic Diesel Exposure
(deposition rate 0.026 mg/day)

Figs. 3 to 5.  Model simulation of experimental
data of chronic diesel exhaust
inhalation studies (Mauderly et
al., 1986).

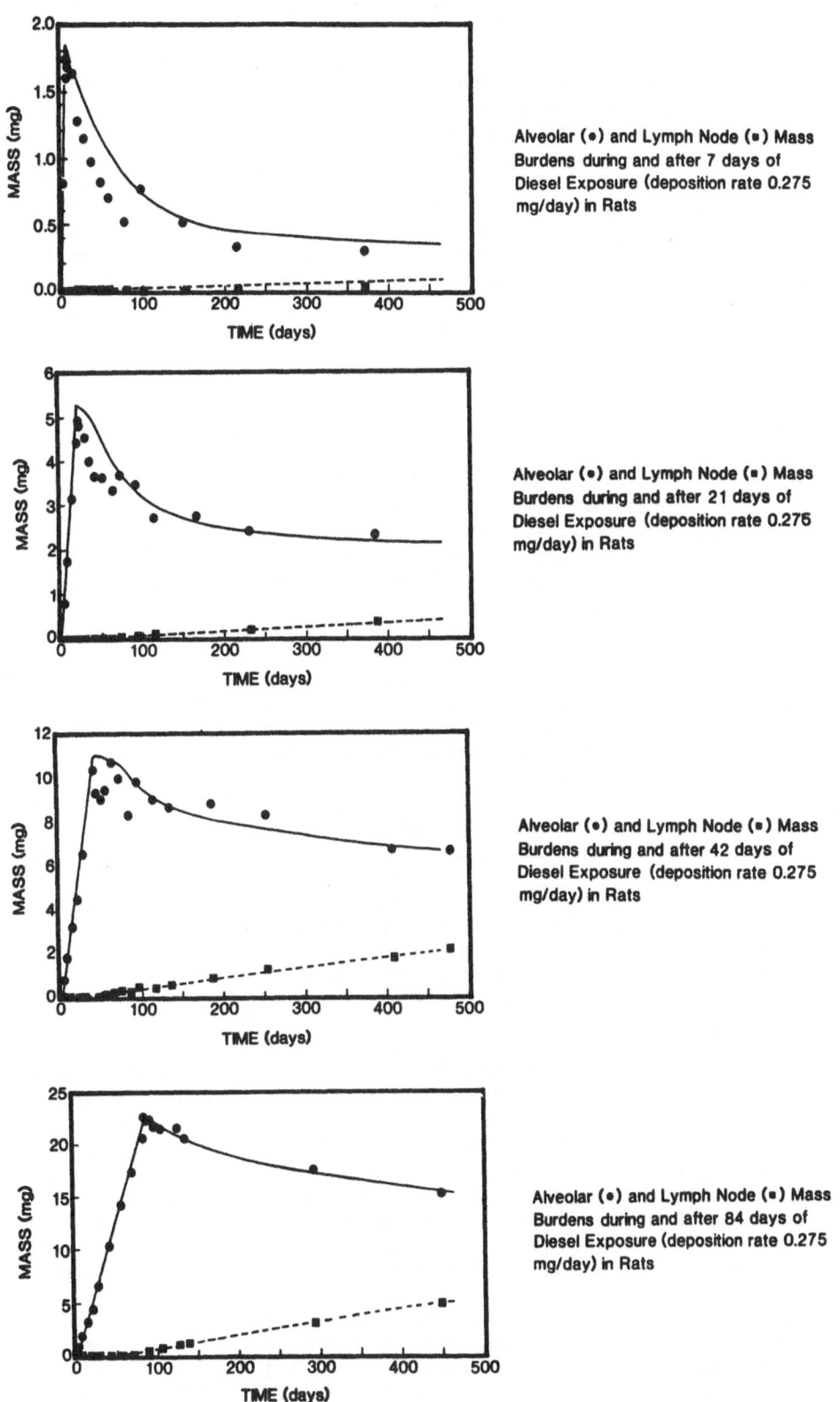

Alveolar (●) and Lymph Node (■) Mass Burdens during and after 7 days of Diesel Exposure (deposition rate 0.275 mg/day) in Rats

Alveolar (●) and Lymph Node (■) Mass Burdens during and after 21 days of Diesel Exposure (deposition rate 0.275 mg/day) in Rats

Alveolar (●) and Lymph Node (■) Mass Burdens during and after 42 days of Diesel Exposure (deposition rate 0.275 mg/day) in Rats

Alveolar (●) and Lymph Node (■) Mass Burdens during and after 84 days of Diesel Exposure (deposition rate 0.275 mg/day) in Rats

Figs. 6 to 9. Model simulation of experimental data of subchronic diesel exhaust inhalation studies (Strom et al., 1988).

to how well it simulates other, additional experimental data. There is also room left for some more detailed theoretical, physiology-oriented approaches.

The proper quantitative description of macrophage sequestration is apparently the key to a reliable prediction of the alveolar particulate lung burden as well as other lung compartments burdens, like the load of the lung-associated lymph nodes or the macrophage pool. The present model seems to be a successful step in the right direction, but, as a predicting device, the model may need some final improvement. Until then, there will be quantitative difficulties, whenever the build-up of an alveolar lung burden involves macrophage sequestration. Unfortunately, this seems to be the case for rats even when the exposure concentration is as low as 0.0026 mg/day (Fig. 2).

REFERENCES

Bowden, D. H., 1983, Cell turnover in the lung, Am. Rev. Respir. Dis., 128:S46.
Cheng, Y. S., Yeh, H. C., Mauderly, J. L., and Mokler, B. V., 1984, Characterization of diesel exhaust in a chronic inhalation study, Am. Ind. Hyg. Assoc. J., 45:547.
Clark, C. R., 1982, Mutagenicity of diesel exhaust particle extracts, in: "DOE Research and Development Report," LMF-96, National Technical Information Service, Springfield.
DFG-Senatskommission, 1988, Maximum concentrations at the workplace and biological tolerance values for working materials, Deutsche Forschungsgemeinschaft, VCH-Verlagsgesellschaft mbH, Weinheim.
Garshick, E., Schenker, M. B., Munoz, A., Segal, M., Smith, T. J., Woskie, S. R., Hammond, S. K., and Speizer, F. E., 1987, A case-control study of lung cancer and diesel exhaust exposure in railroad workers, Am. Rev. Respir. Dis., 135:1242.
Garshick, E., Schenker, M. B., Munoz, A., Segal, M., Smith, T. J., Woskie, S. R., Hammond, S. K., and Speizer, F. E., 1988, A retrospective cohort study of lung cancer and diesel exhaust exposure in railroad workers, Am. Rev. Respir. Dis., 137:820.
Heinrich, U., Peters, L., Funcke, W., Pott, F., Mohr, U., and Stöber, W., 1982, Investigation of toxic and carcinogenic effects of diesel exhaust in long-term inhalation exposure of rodents, in: "Toxicological Effects of Emissions from Diesel Engines," J. Lewtas, ed., pp. 225, Elsevier Biomedical, New York.
Heinrich, U., Muhle, H., Takenaka, S., Ernst, H., Fuhst, R., Mohr, U., Pott, F. and Stöber, W., 1986, Chronic effects on the respiratory tract of hamsters, mice and rats after long-term inhalation of high concentrations of filtered and unfiltered diesel engine emissions, J. Appl. Toxicol., 6:383.
IARC, 1989, "Engine Exhaust and Nitro-aromatic Compounds," Monographs on the evaluation of the carcinogenic risk of chemicals to humans, International Agency for Research on Cancer, Lyon (in press).
Kotin, P., Falk, H. L., and Thomas, M., 1954, Aromatic hydrocarbons. II. Presence in the particulate phase of gasoline-engine exhaust and the carcinogenicity of exhaust extracts, A.M.A. Arch. Ind. Hyg. Occup. Med., 9:164.
Kotin, P., Falk, H. L., and Thomas, M., 1955, Aromatic hydrocarbons. III. Presence in the particulate phase of diesel-engine exhaust and the carcinogenicity of exhaust extracts, Arch. Ind. Health, 11:113.
Lee, K. P., Trochimowicz, H. J., and Reinhardt, C. F., 1985, Pulmonary response of rats exposed to titanium dioxide by inhalation for two years, Toxicol. Appl. Pharmacol., 79:179.

Lehnert, B. E., and Morrow, P. E., 1985, Association of [59]iron oxide with alveolar macrophages during alveolar clearance, Exp. Lung Res. 9:1.

Lewtas, H. J., Bradow, R., Jungers, R., Claxton, L., Zweidinger, R., Tejada, S., Bumgarner, J., Duffield, F., Waters, M. D., Simmon, V. F., Hare, C., Rodriguez, C., and Snow, L., 1978, Application of bioassay to the characterization of diesel particle emissions, in: "Application of Short-term Bioassays in the Fractionation and Analysis of Complex Environmental Mixtures," M. D. Waters, S. Nesnow, H. J. Lewtas, S. S. Sandhu, and L. Claxton, eds., pp. 381, Plenum Press, New York.

Martin, J. C., Daniel, H., and Le Bouffant, L., 1977, Short- and long-term experimental study of the toxicity of coal-mine dust and some of its constituents, in: "Inhaled Particle IV," W. H. Walton, ed., Vol. 1, pp. 361, Pergamon, Oxford.

Mauderly, J. L., Jones, R. K., Griffith, W. C., Henderson, R. F., and McClellan, R. O., 1987, Diesel exhaust is a pulmonary carcinogen in rats exposed chronically by inhalation, Fundam. Appl. Toxicol., 9:208.

Mauderly, J. L., Griffith, W. C., Henderson, R. S., Jones, R. K., and McClellan, R. O., 1989, Evidence from animal studies for the carcinogenicity of inhaled diesel exhaust, this volume.

Morrow, P. E., 1988, Possible mechanisms to explain dust overloading of the lungs, Fundam. Appl. Toxicol., 10:369.

Muhle, H., Bellmann, B., Creutzenberg, O., Stöber, W., Kilpper, R., MacKenzie, J., Morrow, P., and Mermelstein, R., 1988, Pulmonary deposition, clearance and retention of test toner, $TiO_2$ and quartz during a long-term inhalation study in rats, The Toxicologist, 8:272.

NIOSH, 1988, Carcinogenic effects of exposure to diesel exhaust, Current Intelligence Bulletin 50.

Soderholm, S. C., 1981, Compartmental analysis of diesel particle kinetics in the respiratory system of exposed animals. Oral Presentation at EPA Diesel Emissions Symposium, Raleigh, NC, October 5-7.

Stöber, W., Morrow, P. E., and Hoover, M. D., Compartmental modeling of the long-term retention of insoluble particles deposited in the alveolar region of the lung, Fundam. Appl. Toxicol., in press.

Stöber, W., Morrow, P. E., Morawietz, G., and Koch, W., Simulation of experimental data of alveolar retention of inhaled insoluble particles by using a physiology-oriented compartmental kinetics model (to be published).

Strom, K. A., Chan, T. L., and Johnson, J. T., 1988, Pulmonary retention of inhaled submicron particles in rats: diesel exhaust exposures and lung retention model, Ann. Occup. Hyg., 32:645.

Vostal, J. J., 1986, Factors limiting the evidence for chemical carcinogenicity of diesel emissions in long-term inhalation experiments, in: "Carcinogenicity and Mutagenicity of Diesel Engine Exhaust, N. Ishinishi, A. Koizumi, R. O. McClellan, and W. Stöber, eds., pp. 381, Elsevier, Amsterdam.

Wolff, R. K., Henderson, R. F., Snipes, M. B., Griffith, W. C., Mauderly, J. L., Cuddihy, R. G., and McClellan, R. O., 1987, Alterations in particle accumulation and clearance in lungs of rats chronically exposed to diesel exhaust, Fundam. Appl. Toxicol., 9:154.

Wong, D., Mitchell, C. E., Wolff, R. K., Mauderly, J. L., and Jeffrey, A. M., 1986, Identification of DNA damage as result of exposure of rats to diesel engine exhaust, Carcinogenesis 7:1595.

# CARCINOGENICITY OF DINITROARENES IN RAT LUNG

Hiroshi Tokiwa,[1] Kazumi Horikawa,[1] Nobuyuki Sera,[1]
Keisuke Izumi,[2] Masanori Iwagawa,[2] Hisashi Otsuka,[2]
Yoshinari Ohnishi,[3] Akio Nakashima,[4] and Khoichi Nakashima[5]

[1]Department of Health Science, Fukuoka Environmental Research
Center, Fukuoka 818-01, Japan
[2]Department of Pathology, School of Medicine, The University
of Tokushima, Tokushima 770, Japan
[3]Department of Bacteriology, School of Medicine, The University
of Tokushima, Tokushima 770, Japan
[4]Department of Internal Medicine, Saiseikai Shimonoseki General
Hospital, Shimonoseki, Japan
[5]Department of Oral Medicine, School of Dentistry, Loma Linda
University, Loma Linda, California 92350

## INTRODUCTION

Of the dinitroarenes, dinitropyrene and dinitrofluoranthene are
classified as extraordinary mutagens, and small amounts of them are emitted
by incomplete combustion of fossil fuel; these compounds are mostly
coexistent with mononitroarenes in particles from diesel emission, kerosene
heater and city gas emission, and airborne pollutants (see review by
Tokiwa and Ohnishi, 1986).

Carcinogenicity of dinitropyrene for rats and mice after subcutaneous
injection has been reported by several workers (Ohgaki et al., 1984, 1985;
Tokiwa et al., 1984; Otofuji et al., 1987). In an earlier study, we found
that two dinitrofluoranthenes, derived from 3-nitrofluoranthene, were also
carcinogenic for rats by subcutaneous injection (Tokiwa et al., 1987).
Thus, dinitropyrene and dinitrofluoranthene are important carcinogen among
nitroarenes. In the course of this study, we detected 1,3-dinitropyrene
and 1-nitropyrene in lung tissue in a patient, a nonsmoker, with lung
cancer and pulmonary silicosis. In this case, there is no evidence that
the lung cancer was associated with these nitroarenes. However, it is
interesting that particles containing some nitroarenes were deposited in
the lung.

In the present study, we performed carcinogenicity tests in rat lung
for 1,6-dinitropyrene (DNP), 3,7- and 3,9-dinitrofluoranthene (DNF), and
benzo(a)pyrene (BaP) as a control, to see if tumors were induced by lower
doses of the chemicals. The aim of the study was to investigate the
association of nitroarenes with lung cancer.

## MATERIALS AND METHODS

### Chemicals

1,6-DNP was synthesized by nitration of pyrene (Tokyo Kasei Co. Tokyo,

Japan) by the procedure of Vollman et al. (1937) and purified to 99.8% (Iwagawa et al., 1989). 3,7- and 3,9-DNF were synthesized according to modifications of the procedures reported by Charlesworth and Lithown (1969), and Zinchenko and Burmistrov (1975) as described in a previous report (Nakagawa et al., 1987). To a solution of 3-nitrofluoranthene (NF) in 70 ml of nitric acid maintained at 50-55°C was added dropwise 17.5 ml of fuming nitric acid. After nitration was allowed to proceed for 30 min, the reaction mixture was chilled immediately and diluted with water. These material was applied to an alumina column (WOELM, neutral 5 x 30 cm) and eluted stepwise by hexane-benzene mixed, in order, at a ratio of 4 to 3, 1 to 1 and 3 to 4. Five yellow bands were eluted separately from the column. These fractions were purified further by high pressure liquid chromatography (h.p.l.c.) on a Unisil Q C18 column (Gaskuro Kogyo, Inc., Japan). The purity was 99.8% for 3,7-DNF, and 99.9% for 3,9-DNF. 3-NF (RK Chemical Co. Ltd.) was purified by h.p.l.c. as described previously (Nakagawa et al., 1987). BaP was purchased from Sigma Chemical Co.

## Animals

F344/NS1c male rats, used in the 1,6-DNP injection experiment, were purchased from Shizuoka Agricultural Cooperative for Experimental Animals, Hamamatsu, Japan. They were treated with the chemical when they were 11 weeks old and weighed 250 g. Six-week-old F344/DuCrj male rats used in the 3,7- and 3,9-DNF injection experiments were purchased from Charles River Japan Inc., Atsugi. The chemicals were implanted by the method of Stanton et al. (1972), as a suspension in a mixture of equal volumes of beeswax (Nakarai Chemicals, Kyoto, Japan) and tricaprylin (Sigma Chemical Co.). 1,6-DNP was given at doses of 0.003, 0.01, 0.03, 0.1 and 0.15 mg, and BaP at doses of 0.03, 0.1, 0.3 and 1.0 mg. 3,7-DNF and 3-NF were given at a single dose of 0.2 mg and 1.0 mg, respectively, and 3,9-DNF at doses of 0.05, 0.1 and 0.2 mg. All control rats were given only the beeswax-tricaprylin mixture. The rats were anesthetized by intramuscular injection of ketamine hydrochloride (100 mg/kg body weight, Sankyo Co., Ltd., Tokyo), left lateral thoracotomy was performed and 0.05 ml of beeswax-tricaprylin mixture containing the test chemical was injected directly into the left lung. After this procedure, the rats were observed for up to 104 weeks for 1,6-DNP, and up to 70 weeks for 3,7- and 3,9-DNF, and 3-NF. Some of the rats given lower doses (0.05 and 0.1 mg/rat) of 3,9-DNF are still under study.

## Identification of mutagens/carcinogens from a human lung specimen

Lung material (about 1.6 g) after surgery was immediately extracted with dichloromethane. Subsequently, the neutral fraction was obtained by the liquid-liquid separation procedure, and then chromatographed on a column of silica gel in the order of hexane, hexane/benzene, benzene, benzene/methanol and methanol as effluents according to the procedure described previously (Tokiwa et al., 1986). To identify the chemicals, analysis by h.p.l.c. and gas chromatography-mass spectrometry (GC/MS system, DB-5 capillary gas chromatograph interfaced to a Finnigan MAT ITD-80 mass spectrometer) was performed by the procedure described separately (Tokiwa et al., 1989).

## RESULTS

## Carcinogenicity of 1,6-DNP in the rat lung

Table 1 shows the data on lung lesions induced by intrapulmonary implantation of 1,6-DNP or BaP. At autopsy, the injection site was identified by the location of the beeswax pellet except in rats with large lung

Table 1. Lung lesions induced by intrapulmonary implantation of 1,6-DNP or BaP.

| Chemical | Dose mg | Dose nmol | No. affected of rats | No. of rats with lung lesions Granulomatous lesions | Squamous metaplasia | Squamous cell carcinoma | Undifferentiated neoplasm | Total with tumors (%) | Mean weight of lung tumor (g ± S.D.) | No. of rats with metastases (%) |
|---|---|---|---|---|---|---|---|---|---|---|
| 1,6-DNP | 0.003 | 10.3 | 39 | 39 | 0 | 0 | 0 | 0 | | |
| | 0.01 | 34.2 | 30 | 25 | 1 | 0 | 4 | 4 (13) | 10.6 ± 8.6 | 1 (25) |
| | 0.03 | 103 | 31 | 18 | 0 | 2 | 11 | 13 (42) | 13.1 ± 7.2 | 4 (31) |
| | 0.1 | 342 | 26 | 4 | 0 | 3 | 19 | 22 (85) | 14.3 ± 10.6 | 8 (36) |
| | 0.15 | 513 | 9 | 3 | 0 | 2 | 4 | 6 (67) | 10.8 ± 7.4 | 1 (17) |
| BaP | 0.03 | 119 | 29 | 28 | 0 | 0 | 1 | 1 (3) | 11.3 | 0 (0) |
| | 0.1 | 396 | 30 | 21 | 2 | 6 | 1 | 7 (23) | 15.6 ± 11.7 | 0 (0) |
| | 0.3 | 1189 | 29 | 4 | 3 | 20 | 2 | 22 (76) | 18.1 ± 9.4 | 6 (27) |
| | 1.0 | 3964 | 13 | 4 | 0 | 9 | 0 | 9 (69) | 12.4 ± 9.6 | 2 (22) |
| Control | 0 | 0 | 40 | 40 | 0 | 0 | 0 | 0 (0) | | |

tumors. Macroscopically, the tumors were grayish-white, solid and hard, often with irregular foci of hemorrhage and necrosis. The largest tumor, found in a rat treated with 0.1 mg of 1,6-DNP that died in week 64, weighed 41.6 g. The mean weights of lung tumors in all groups treated with 1,6-DNP and with BaP were 13.2 +9.0 g and 16.2 + 9.8 g, respectively. Histologically, the lung lesions induced by the two chemicals were classified as simple granulomatous lesions, keratinizing squamous metaplasias, squamous cell carcinomas, and undifferentiated neoplasms. Most of the tumors induced by 1,6-DNP were undifferentiated neoplasms and only seven tumors were squamous cell carcinomas (Table 1). In contrast, most of the tumors induced by BaP were squamous cell carcinomas, and undifferentiated neoplasms were found in only four rats given the lower doses of BaP. On the basis of these data, at the same doses of 1,6-DNP and BaP, the incidence of lung cancer was much higher after implantation of 1,6-DNP than of BaP : with doses of 0.03 mg equivalent to 103 and 119 nmol 1,6-DNP and BaP, respectively, the incidence was 42% and 3%, and with doses of 0.1 mg of these compounds, the incidence was 85% and 23%, respectively.

Lung tumors induced by a single injection of 3-NF, and 3,7- and 3,9-DNF

Table 2 shows the incidence of lung tumors caused by 3-NF, and 3,7- and 3,9-DNF. These data are the results by the 70th week after injection of the chemicals into the lungs of rats. The first tumors were found in rats given 0.2 mg of 3,9-DNF and sacrificed on day 257 after injection, and tumors were induced in 18 (85.7%) of 21 rats by day 450. The weight of the tumors induced ranged from 0.6 to 23.3 g. In the case of rats given 0.2 mg of 3,7-DNF, lung tumors were induced in 12 (54.5%) of 22 rats by day 514. The mean weight of the tumors induced was 14.3 + 16.5 g. In rats given 1.0 mg of 3-NF, tumors were induced in only 2 of 20 animals, showing it to be a weak carcinogen. The carcinogenic test in rats treated with lower doses of 3,9-DNF is still in progress: Four of the 10 rats given

Table 2. Lung tumors induced by a single injection of 3-NF, 3,7- and 3,9-DNF.
(Results observed until 70 weeks after injection of chemical)

| Compound | Dose | | No. of rats evaluated | No. of rats with tumors (%) | Mean weight of lung tumor (g + S.D.) |
| | mg/rat | nmol /rat | | | |
|---|---|---|---|---|---|
| 3,9-DNF | 0.2 | 680 | 21 | 18 (85.7) | 7.3 + 10.1 |
| | 0.1[a] | 340 | 10[a] | 4[a] | |
| | 0.05[a] | 170 | 10[a] | | |
| 3,7-DNF | 0.2 | 680 | 22 | 12 (54.5) | 14.3 + 16.5 |
| 3-NF | 1.0 | 3100 | 20 | 2 (10.0) | |
| Control | 0 | 0 | 20 | 0 | |

[a]The animals are still under study.

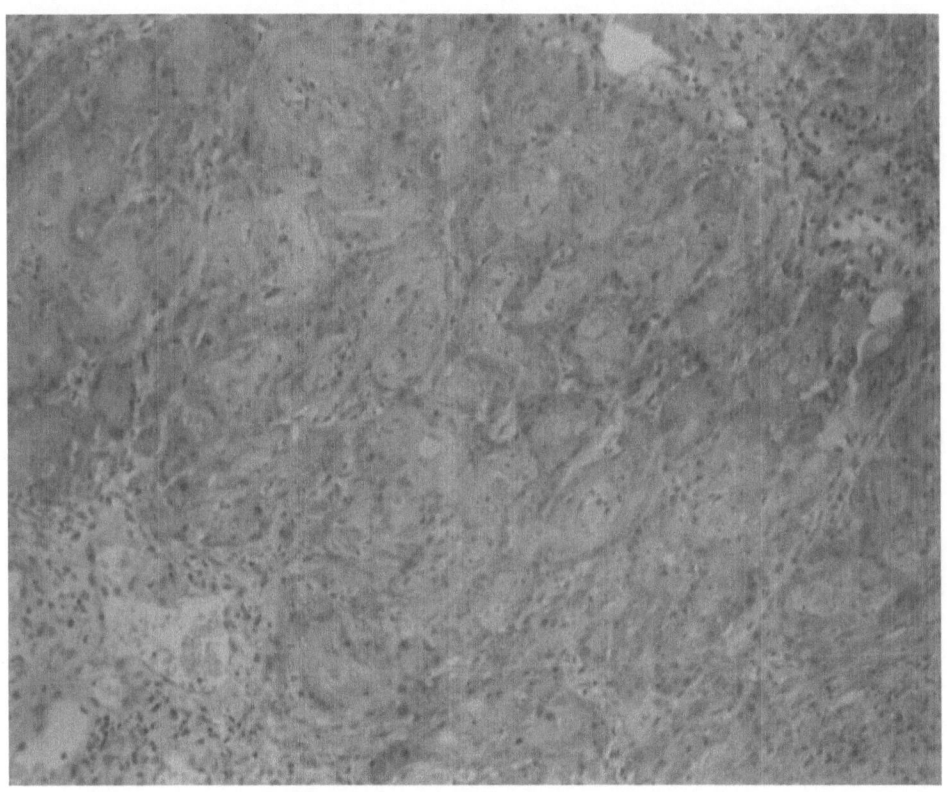

Fig.1. Well-differentiated squamous cell carcinoma induced by 3,7-DNF.
Bar = 100 µm x 120

0.1 mg developed tumors by 66 weeks after injection, but no tumors were
observed in those given 0.05 mg by 70 weeks. Tumors induced by 3,7-DNF
were histologically squamous cell carcinomas (Fig.1). The cumulative incidence of lung carcinomas is illustrated in Fig.2 ; a definite difference
was found in time of appearance of the tumors between 3,9-DNF and 3,7-DNF.

Detection of nitropyrene in human lung complicated with pulmonary silicosis

A part of the lung of a 64-year-old man, a nonsmoker and farmer, with
pulmonary silicosis was obtained. A deposit of dust like particles was
found in several places in his lung, and a specimen obtained at the oper-
ation revealed lung cancer which was diagnosed histologically as
keratinizing squamous cell carcinoma, well differentiated. Neither his past
history and occupation nor chest-X-ray film suggested pneumoconiosis.
Therefore, the putative mutagens/carcinogens in the particles were examined
chemically to determine the mechanism of the carcinogenesis. Table 3 shows
the results of the test for mutagenicity of each fraction for strain TA98.
The neutral fraction induced 12.2 revertants/ug in the absence of S9 mix
definitely indicating direct-acting mutagenicity. After silica gel column
chromatography, the benzene and benzene/methanol fractions each showed
strong direct-acting mutagenicity for the strain (Table 3). A mixture of
the two fractions was analyzed by h.p.l.c. on a Unisil Q C18 column and

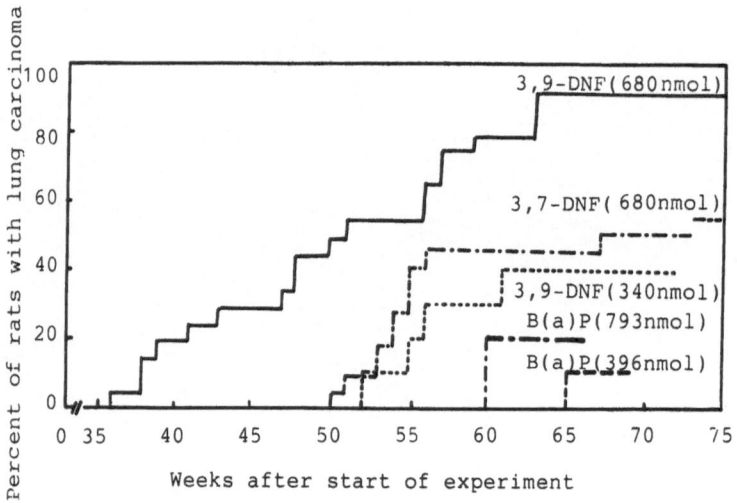

Fig.2. The cumulative incidence of lung carcinomas.
(Rats given BaP are still under study)

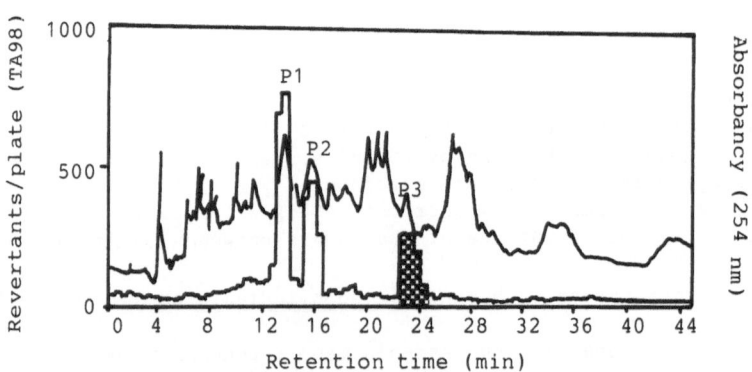

Fig.3. The h.p.l.c. analysis of particles deposited in a specimen of
a human lung bearing cancer.
Two white columns (P1 and P2) were found to have direct-acting
mutagenicity and the dark column to have indirect-acting
mutagenicity.

Table 3. Fractionation and mutagenicity of human lung specimen

| Fraction | Weight (mg) | TA98 (revertants/$\mu g$) | |
|---|---|---|---|
| | | -S9 | +S9 |
| Crude material | 1,517 | | |
| Dichloromethane extract | 244 | 1.12 | 0.19 |
| Neutral fraction | 18.3 | 12.2 | 2.12 |
| Silica gel column chromatography | | | |
| Hexane | 9.2 | 0.31 | 0.12 |
| Hexane/benzene | 3.5 | 0.65 | 0.21 |
| Benzene | 1.1 | 100 | 17.0 |
| Benzene/methanol | 1.4 | 52.6 | 7.1 |
| Methanol | 1.6 | 5.3 | 0.82 |

Table 4. Carcinogens detected in tissues of lungs bearing cancer.

| Chemical | Crude extract ($\mu g/g$) | Concentration in organ (ng/g) |
|---|---|---|
| 1-Nitropyrene | 1.2 | 7.5 |
| 1,3-Dinitropyrene | 3.3 | 6.5 |
| Chrysene | 1.8 | 1.8 |

three active peaks were produced (Fig. 3). Each fraction was analyzed by the GC/MS system. As shown in Table 4, five kinds of mutagens/carcinogens were identified; they were 1-NP, 1,3-DNP, chrysene, and unknown (MW 329). On the basis of the results of GC/MS, the concentrations of these compounds in the lung were 7.5 for 1-NP, 6.5 for 1,3-DNP and 1.8 ng for chrysene per gram of lung (Table 4).

DISCUSSION

It was reported by Maeda et al. (1986) that 1,6-DNP caused significant induction of tumors (mostly squamous cell carcinoma) in rat lung by intra-pulmonary administration at a single dose of 0.15 mg. The first purposes of this experiment, therefore, was to evaluate the carcinogenicity of 1,6-DNP at lower doses in the rat lung, and to see if the incidence of

tumors is dose-dependent. The second was to evaluate the carcinogenicity
of 3,7- and 3,9-DNF, having the same extent of mutagenicity as 1,6-DNP, in
the rat lung by the same route of administration. In the case of rats
treated with 1,6-DNP, the incidence of tumors in the lung was significant
at a small dose of 0.01 mg (34.3 nmol). In contrast, the significant incidence of tumors caused by BaP was at a dose of 0.1 mg (392 nmol), corresponding to a 10-fold dose of 1,6-DNP. On the other hand, both DNFs are
also expected to have similar carcinogenic potency to that of 1,6-DNP.
3,9-DNF induced tumors in 18 (85.7%) of 21 animals at a dose of 0.2 mg
(680 nmol). However, the carcinogenicity of 3,9-DNF seems to be more potent
than that of 3,7-DNF. We have already shown that in the results of subcutaneous injection of both DNFs in rats, carcinogenicity of 3,9-DNF
differed from that of 3,7-DNF in terms of potency and histological characteristics (Tokiwa et al., 1987). It is concluded that 3-NF is a weak
carcinogen in rat lung compared with both DNFs.

It is interesting that 1,3-DNP, 1-NP, and chrysene were detected in
the lung of a human with lung cancer complicated by pulmonary silicosis.
The diagnosis of pulmonary silicosis was made from lung specimens by means
of X-ray microanalysis; a characteristic silica peak was detected. Many
particles like dust were observed to be deposited in the lung and several
other places in the respiratory tract. Because of the patients nonsmoking
and occupational history, most of these particles were assumed to depend
on environmental agents. In the lung specimens, no metabolites of 1,3-DNP
and 1-NP could be detected by the GC/MS system. This may be due to the
use of only a limited amount of material. Thus, we could not define the
relationship between nitroarenes and the induction of lung cancer to the
evidence of the origin of the particles deposited. This is a case report
on nitroarenes detected in the lung of a human bearing cancer. We have to
focus on the behavior and biological impact of nitroarenes present in the
environment.

REFERENCES

Charlesworth, E.H. and Lithown, C.U., 1969, Fluoranthene studies, IV.
    Nitration of 2-nitro-, 3-nitro- and 2-acetamide-fluoranthene, Can.
    J. Chem., 47: 1595.
Iwagawa, M., Maeda, T., Izumi, K., Otsuka, H., Nishifuji, K., Ohnishi, Y.
    and Aoki, S., 1989, Dose-response studies on the carcinogenecities
    of 1,6-dinitropyrene and benzo(a)pyrene in the lung of F344 rats,
    Carcinogenesis, 10: 1285.
Maeda, T., Izumi, K., Otsuka, H.,Manabe, Y., Kinouchi, T. and Ohnishi, Y.,
    1986, Induction of squamous cell carcinoma in the rat lung by 1,6-
    dinitropyrene, J. Natl. Cancer Inst., 76: 693.
Nakagawa, R., Horikawa, K., Sera, N., Kodera, Y., and Tokiwa, H., 1987,
    Dinitrofluoranthene: induction, identification, and gene mutation,
    Mutation Res., 191:85.
Ohgaki, H., Hasegawa, H., Kato, T., Negishi, C., Sato, S. and Sugimura, T.,
    1985, Absence of carcinogenicity of 1-nitropyrene, correction of
    previous results, and new demonstration of carcinogenicity of 1,6-
    dinitropyrene in rats, Cancer Lett., 25:239.
Ohgaki, H., Negishi, C., Wakabayashi, K., Kusama, K., Sato, S. and
    Sugimura, T., 1984, Induction of sarcomas in rats by subcutaneous
    injection of dinitropyrenes, Carcinogenesis, 5:583.
Otofuji, T., Horikawa, K., Maeda, T., Sano, N., Izumi, K., Otsuka, H.,
    and Tokiwa, H., 1987, Tumorigenicity test of 1,3- and 1,8-dinitro-
    pyrene in BALB/c mice, J. Natl. Cancer Inst., 79:185.
Stanton, M.F., Miller, E., Wrench, C. and Blackwell, R., 1972,
    Experimental induction of epidermoid carcinoma in the lungs of rats
    by cigarette smoke condensate, J. Natl. Cancer Inst., 49:867.
Tokiwa, H., and Ohnishi, Y., 1986, Mutagenicity and carcinogenicity of

nitroarenes and their sources in the environment, CRC Critical Rev. Toxicol., 17:23.

Tokiwa, H., Otofuji, T., Horikawa, K., Kitamori, S., Otsuka, H., Manabe, Y., Kinouchi, T. and Ohnishi, Y., 1984, 1,6-Dinitropyrene: Mutagenicity in Salmonella and carcinogenicity in BALB/c mice, J. Natl. Cancer Inst., 73:1359.

Tokiwa, H., Otofuji, T., Horikawa, K., Sera, N., Nakagawa, R., Maeda, T., Sano, N., Izumi, K., and Otsuka, H., 1987, Induction of subcutaneous tumors in rats by 3,7- and 3,9-dinitrofluoranthene, Carcinogenesis, 8:1919.

Tokiwa, H., Otofuji, T., Nakagawa, R., Horikawa, K., Maeda, T., Sano, N., Izumi, K., and Otsuka, H., 1986, Dinitro derivatives of pyrene and fluoranthene in diesel emission particulates and their tumorigenicity in mice and rats, in: Carcinogenic and Mutagenic Effects of Diesel Engine Exhaust (N. Ishinishi et al., eds.), Elsevier Science Publishers B.V., p 253.

Tokiwa, H., Sera, N., Kai, M., Horikawa, K., and Ohnishi, Y., 1989, The role of nitroarenes in the mutagenicity of airborne particles indoors and outdoors, in: International Conference on Genetic Toxicology of Complex Mixtures, Plenum Publishing Corp., in press.

Vollmann, H., Becker, H., and Corell, M., 1937, Beitrage zur Kenntnis des Pyrenes und Senier Derivate, Justus Leibig Ann, Chem., 531:1.

Zinchenko, V.M., and Burmistrov, S.I., 1975, Synthesis of isomeric polynitrofluoranthenes, Zh. Org. Khim., 11:823.

# IN VIVO METABOLISM AND GENOTOXIC EFFECTS OF THE AIR POLLUTANT AND MARKER FOR NITRO-PAH'S, 2-NITROFLUORENE

L. Möller, J. Rafter, S. Törnquist, L. Eriksson, B. Beije, R. Toftgård, T. Midvedt, M. Corrie and J-Å. Gustafsson

Departments of Medical Nutrition, Medical Microbial Ecology and Pathology, Karolinska Institute and Department of Genetic and Cellular Toxicology, University of Stockholm, Stockholm, Sweden

## ABSTRACT

During incomplete combustion of organic matter there is a formation of polycyclic aromatic hydrocarbons (PAH) which can react with oxides of nitrogen, with the formation of nitro-PAH's as a result, a reaction which is catalyzed by a low pH. 2-Nitrofluorene (NF), a marker for nitro-PAH, is in vivo metabolized via two different routes. After inhalation there is a formation of potent mutagenic metabolites, ON-NF's, which are distributed in the body. After oral administration, NF is reduced to the amine, a reaction mediated by the intestinal microflora, and further acetylated to 2-acetylaminofluorene (AAF), a potent carcinogen. Further ringhydroxylation of AAF leads to detoxification and excretion.

Induction of cytochrome P450 c,d affects the metabolism in that more OH-NF's are formed. As a consequence, more mutagenic metabolites are found in the circulation. The liver excretes OH-NF's as, in terms of mutagenicity, totally harmless glucuronide conjugates. When these conjugates are excreted via the bile, intestinal beta-glucuronidase can liberate direct-acting mutagens in the intestine. Thus, inhalation of NF can lead to formation of potent mutagens in the intestine.

NF induces DNA-repair, in vivo, and is an initiatior and a weak promotor, measured as formation of preneoplastic lesions in the liver. Risk estimates, by two different methods, indicate that nitro-PAH's extrapolated from the marker NF, can expose humans to a cancer risk on a non-neglectable level.

## INTRODUCTION

Incomplete combustion is a major problem in terms of pollution. Examples include emissions from energy-production, vehicles, smoking and industrial processes.

The biological effects, resulting from incomplete combustion, can be divided into effects on the health or the ecosystem. Both effects can be

*Nitroarenes*, Edited by P. C. Howard *et al.*
Plenum Press, New York, 1990

acute or on a long term basis. The different biological responses can be related to each other because the same substance in the emissions can give rise to several reactions in the organism or the ecosystem. One example of this, are the nitrated polycyclic aromatic hydrocarbons (nitro-PAH's). For the formation of nitro-PAH's, incompletely combusted organic material (PAH) and oxidized nitrogen ($NO_x$) are necessary. A low pH ($SO_2$, $NO_x$) catalyzes the reaction. $NO_x$ is one important combustion product responsible for acidification of the environment, acute health effects [1] as well as formation of nitro-PAH's [2] which are strong genotoxic agents in mammalian systems [3-7]. Since the formation of nitro-PAH's is catalyzed by a low pH, the reaction product, $NO_x$, catalyzes its own reaction with PAH's in the formation of nitro-PAH's.

Nitro-PAH's are found in emissions from diesel driven [8] as well as petrol driven [9] vehicles, in the exhaust from kerosene heaters [10], in urban air [11-14], in river sediments [15] and in certain food products [16, 17]. Nitro-PAH's can be formed during the process of combustion or as a result of photochemical reactions of PAH's [18] or amino-PAH's [19]. Nitro-PAH formation has also been reported to occur in the water phase with nitrite as a donor of the nitro-group [20].

Nitro-PAH's are a group of at least 200 different substances. Many of them are mutagens [21-24] and the most potent mutagenic substances known of today (dinitropyrenes) are found in this group [25]. A number of the nitro-PAH's are also carcinogenic to laboratory animals [26-27].

2-Nitrofluorene (NF) is one of the more common nitro-PAH's, found in the environment [28-32] together with 1-nitropyrene (NP). Normally NP is the dominating species [29] although that is not always the case [30]. NF has been suggested to be a model substance for nitro-PAH's in the gas and particle phase, while NP is regarded to be a model substance for nitro-PAH's in the particle phase [35]. NF is a mutagen [24], as well as a carcinogen [27] in laboratory animals.

NF has been studied in detail in our laboratory from an analytical point of view [36], and with regard to metabolism [37-39], lung effects [40-41] and genotoxic effects [42-44]. A review of NF's prevalence and biological effects has also been published [34].


**RESULTS AND DISCUSSION**

Humans can be exposed to 2-nitrofluorene, and nitro-PAH's in general, mainly via two routes. The direct exposure is via inhalation when air, with NF in the gas phase as well as absorbed on the surface of particles, enter into the inhalatory system. Large particles will enter to the gastro intestinal tract after deposition in the upper part of the inhalatory system or as a consequence of ciliary transport up from the lungs. In other words, the inhalated dose expose the lungs as well as the gastro intestinal tract.

The indirect exposure will occur when deposition of particles occur on vegetables, or other products of agricultural origin, for human consumption. This way of contamination, with the addition of what can be added during food-processing, results in a dose to the gastro intestinal tract. In addition there also exist other, possible minor routes, like contaminated water and via water living organisms used as food. However, also in this case the endpoint is the gastro intestinal tract. These are the reasons why NF has been orally or intra tracheally administered in the following studies.

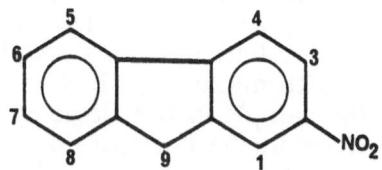

Figure 1. Chemical structure of 2-nitrofluorene (NF)

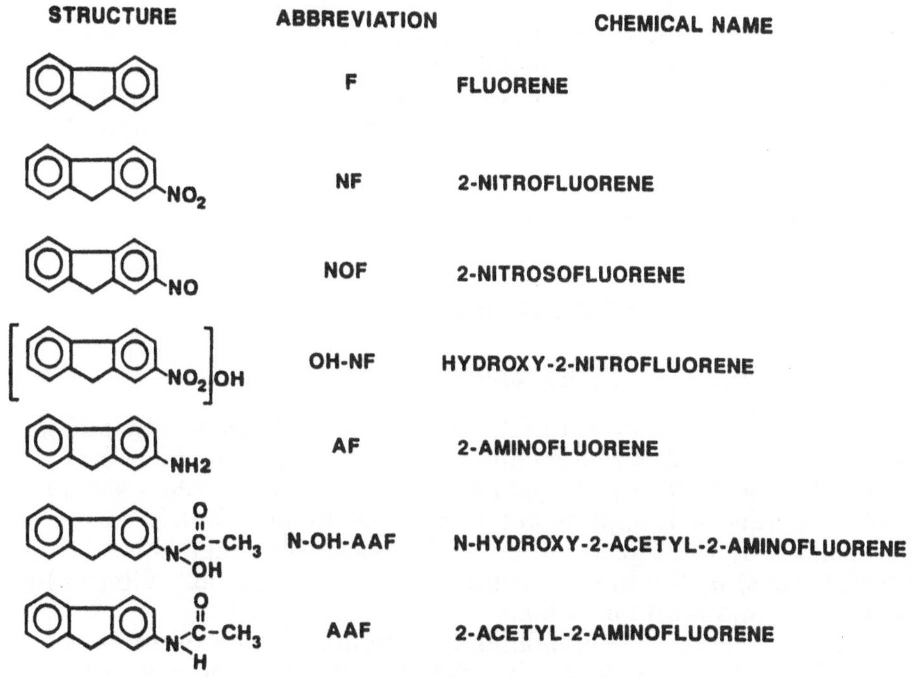

| STRUCTURE | ABBREVIATION | CHEMICAL NAME |
|---|---|---|
|  | F | FLUORENE |
|  | NF | 2-NITROFLUORENE |
|  | NOF | 2-NITROSOFLUORENE |
|  | OH-NF | HYDROXY-2-NITROFLUORENE |
|  | AF | 2-AMINOFLUORENE |
|  | N-OH-AAF | N-HYDROXY-2-ACETYL-2-AMINOFLUORENE |
|  | AAF | 2-ACETYL-2-AMINOFLUORENE |

Figure 2. Chemical structures of NF-related substances

LC/MS with a nebulizer and a moving belt was used to characterize NF and its metabolites (36). The properties and fragmentation patterns of 18 different fluorene derivatives were first examined. Without prior derivatization, all substances yielded interpretable mass spectra.

Recently, different techiques have been applied to overcome the problem with large volumes of solvents from the HPLC. These include thermospray moving belt and DLI. DLI is often used combined with micro-LC. The DLI-technique demands small volumes of solvent, especially in connection with micro-LC, which can be a problem in the final treatment of the sample in the clean-up procedure.

There are few articles in the literature on the moving belt system, especially on instruments with a nebulizer. The nebulizer, which actually dry sprays the sample on to the belt, is only heated at the site (a few mm) of spray formation and decreases the risk for decomposition of unstable metabolites.

The LC/MS system had the capacity to distinguish between seven different hydroxylated isomers of OH-AAF. In combination with the UV-analyses in the HPLC-system an identification could be performed based on the retention time from the total, or single ion current chromatograms, differences in mass spectral intensities and specific losses of fragments.

In addition, the radioactivity detector coupled on line, indicated peaks with metabolites originating from the radiolabelled NF. The UV-detector gave a signal that was linear and parallel with the signal from the ion source, demonstrating that although the two detectors measure different parameters, they do so at a constant ratio.

Thus the described system was considered to be well suited for the studies on the metabolism of NF.

## METABOLISM OF 2-NITROFLUORENE

### Metabolism after oral administration of NF

Although NF (and nitro-PAH in general) is a chemically stable molecule it is metabolized extensively in the organism. After oral administration of NF the major part of the dose is excreted within 48 h (37, 39). After 4 h approximately 1.5% of the dose has been metabolized by the intestinal microflora, absorbed, metabolized (several steps) in the liver, distributed in the circulation, filtered by the kidneys and found in urine.

The excretion of metabolites is accompanied by excretion of excreted mutagenicity. Typically, direct-acting mutagenicity (-S9) dominated over mutagenicity in the presence of S9, both in urine and feces (37,39).

The in vivo formation (37) of the potent carcinogen AAF (46) is indicated. After an oral dose of NF to conventional rats, NF is reduced to 2-aminofluorene (AF) by the intestinal microflora, acetylated and further hydroxylated in the liver resulting in OH-AAF's which can be excreted as such or in conjugated form. This metabolic route is quantitatively the most important. AAF has been a model compound

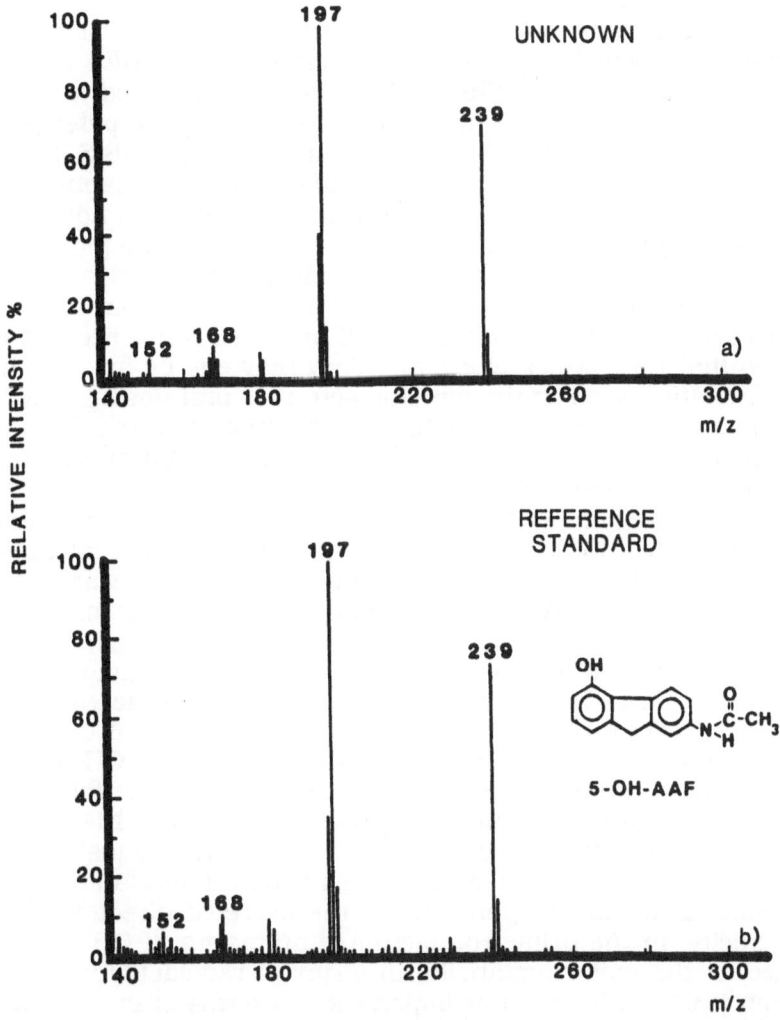

Figure 3. An example of LC-MS identification of a NF-metabolite, found in urine of rats administered NF orally

for chemical carcinogenesis since the Wilson's early finding of its carcinogenic potential in 1941 (47). AAF has been used in a number of bioassays and been characterized from many different points of view (33). AAF is not found in the environment and occupational exposure can only occur when AAF is used in research. It is thus of concern when an environmental pollutant (NF) commonly found in diesel exhaust (9,48,49,51) has the capacity to be metabolized to this potent carcinogen (AAF), in vivo. Other nitro-PAH's have also been shown to form acetylated metabolites although the biological significance of these metabolites is not known (52,53).

After oral administration of NF there exists an alternative metabolic route which results in the formation of OH-NF's in the conventional animal (37). While OH-AAF's are considered to be detoxification products (54) and have a low mutagenic potency (55). OH-NF's, on the other hand, are more mutagenic (TA 98-S9) than NF itself (39). In conventional rats (37) treated with beta-naphtoflavone (induces cytochrome P-450 c and d) prior to administration of NF, the metabolic pattern shifted towards excretion of a larger proportion of OH-NF's compared to uninduced rats, and simultanously the mutagenicity of urine increased. So far, nothing is known about the carcinogenic potential of OH-NF's but it can not be excluded that they are carcinogenic. They may e.g. be involved in the tumor formation seen in the forestomach after oral dosing of NF; no forestomach tumors are seen following administration of AF or AAF. OH-NF's may also play a role in the formation of subcutaneous tumors after skin application of NF (56,27).

The involvement of the intestinal microflora in the metabolism of NF was studied (39). An indication that the intestinal microflora were of importance in the metabolism of NF, was seen when the urinary mutagenicity from conventional and germfree animals was compared (39). The urine from germfree animals exceeded the conventional urine in direct-acting mutagenicity by a factor of approximately six. The same observation was made in feces.

The LC/MS analyses of urine and feces from germfree animals confirmed the presence of OH-NF's and the absence of OH-AAF's. NF was, to a small extent, excreted in the urine on day one following administration indicating the absorption of unreduced NF from the GI-tract. Consequently, NF circulated in the body. The major metabolic route in germfree animals was thus the formation of OH-NF's which also were responsible for the excreted direct-acting mutagenicity. In the urine from germfree animals a di-OH-NF was detected as the major metabolite in terms of radioactivity (34%), although it was only of minor importance in terms of mutagenicity (2%).

The mutagenicity of NF was increased up to 15 times after monohydroxylation (39). Further hydroxylation appeared to decrease the mutagenicity to levels below NF. The formation of OH-NF's, their potency in genotoxic assays and possible carcinogenic character indicate the need for carcinogenicity studies on this class of compounds.

Oral administration of NP to rats also resulted in the formation of reduced, acetylated and hydroxylated metabolites, but ring-hydroxylated NP's were also reported to be responsible for a higher direct-acting mutagenicity in urine of rats treated with phenobarbital (57) indicating the importance of enzyme-induction in the metabolism of nitro-PAH's. Other studies have shown that pretreatment of rats with beta-naphtoflavone increased the amount of ring-OH-AAF's in milk after intra-peritoneal administration of AAF (58). This fact raises the question whether the OH-NF's can also be excreted in milk, hereby exposing infants to a genotoxic risk, following the mothers inhalation of urban air and/or diesel exhaust.

Oxidative metabolites of NP undergo nitroreduction and subsequent DNA-binding much more readily than NP (59) leading to the conclusion of Beland et al (60) that tumorigenicity assays should be conducted not only with the parent compound and their reduced derivatives but also with their ring oxidized metabolites to assess human health risks from nitro-PAH's.

One can always argue as to whether data on animal metabolism are relevant to humans, but in the case of nitro-PAH's there are a number of reasons which indicate that the animal studies are relevant to the human situation:

1. Reduction of nitro-PAH's to amino-PAH's can be performed by anaerobic fecal bacterial suspension from humans as well as rats (61-63).
2. Human liver S9 bioactivated AF and AAF to mutagens (64).
3. Human hepatoma cell lines can perform nitroreduction as well as ring hydroxylation of NP (65).
4. Liver microsomal metabolism of AAF is similar in rats and humans (66).
5. Human lymphocytes metabolize AAF to ring- and N-OH derivatives of AAF (67).
6. AAF metabolism is similar in cultures of epithelial cells from human and rat bladder (68).
7. It has been shown that the carcinogen AAF given orally to humans results in the same urinary metabolites as in the rat (69).

Metabolism after inhalation of NF

In the isolated perfused rat lung there is a rapid metabolism to direct-acting mutagens both when NF is administered intra-tracheally and intra-vascularly. The metabolites formed are unconjugated ring-OH-NF's. Unmetabolized NF, given intratracheally, can also pass through the lung into the circulation together with metabolites (OH-NF's).

Thus, it is likely that inhalation might result in whole body exposure to circulating carcinogens (NF and OH-NF's(?)) as well as potent mutagens (NF and OH-NF's). In the study performed on the isolated perfused liver (38), the purpose was to study the type of metabolism when NF was administered intra-vascularly, as the liver can be exposed to NF after inhalation.

The liver metabolized NF to OH-NF's but excreted them - in terms of mutagenicity - in a harmless form, as glucuronides. Bile was

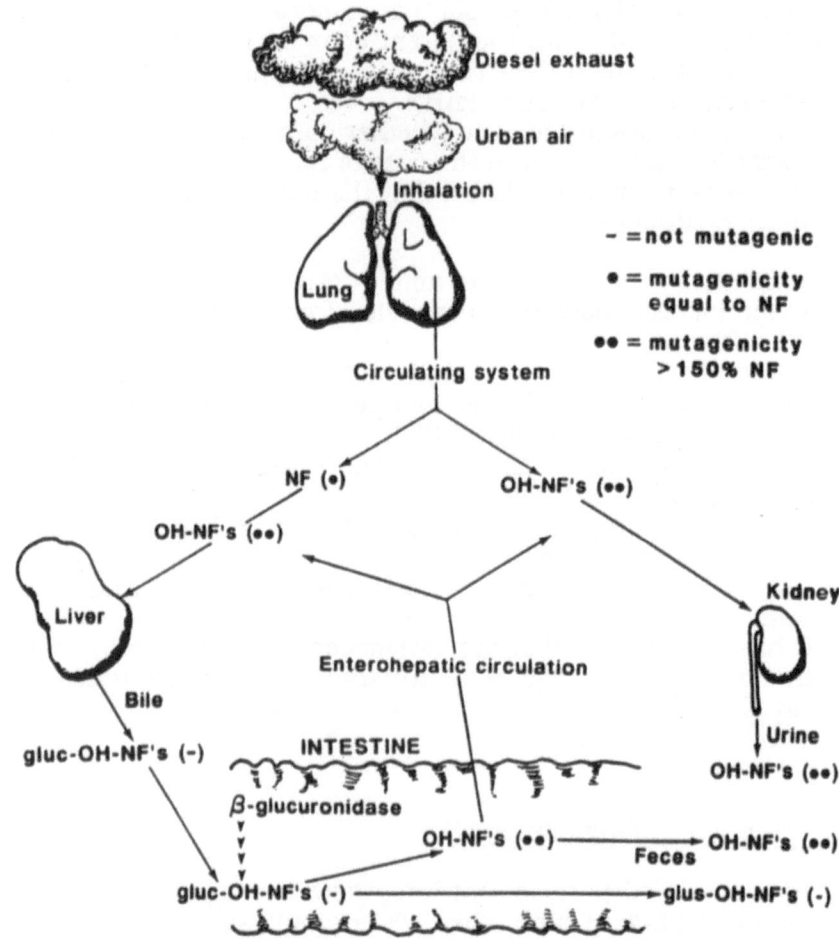

Figure 4.    Proposed routes of metabolism and target organs for inhaled NF.

not mutagenic before treatment with beta-glucuronidase which liberated the direct-acting mutagens, the OH-NF's.

Beta-glucuronidase is an intestinal enzyme, also found in man, which is induced when individuals are on an "Western diet" (high in fat and protein)(70). Thus the results indicate a chain of events. Inhaled NF is metabolized by the lung to OH-NF's or transported to the liver as NF and is there ring-hydroxylated. The liver conjugates the OH-NF's and excretes them via the bile. In the intestine the OH-NF's may be liberated, exposing the intestine to a genotoxic risk. In other words, air pollutants such as nitro-PAH's could have the colon or other organs as a target.

The results presented (38) are well in accordance with lung metabolic data on NP. Lung microsomes from rats, rabbits and hamsters metabolizes NP to mutagenic products which were ring hydroxylated (71). Interestingly, also nasal mucosa metabolizes NP to OH-NP's, a metabolic route that represented more than 90% of the metabolites (72). Isolated perfused lung metabolized NP in the same way as the nasal mucosa. The rate of metabolism of NP in the lung increased, i.e. the probability increased for production of genotoxic metabolites, following exposure to diesel exhaust (72).

INITIATING AND PROMOTING CAPACITY OF NF

NF was studied in a liver model for chemical carcinogenesis for its capacity as an initiator and a promotor (44). The liver model is based on the administration of a potent initiatior (DEN) whereafter the genotoxic lesions are further developed by partial hepatectomy and a potent chemical promotion by dietary AAF during two weeks. After, in total, six weeks initiation can be measured as preneoplastic lesions (foci) (80).

NF was found to be an initiator and gave a dose-response curve, which was approximately ten times the control at the highest dose. The difference was statistically significant.

When NF was characterized as a promotor the basic concept for the liver model was used, but the dietary AAF-promotion regimen was replaced by six intragastric administrations of NF. At the lower dose - which was similar to the doses used in the NF metabolic studies - NF and AAF were both weak promotors. At high doses AAF was a very potent promoter while NF remained at a low, but statistically significant, level of promoting activity.

In the gamma-model used for risk assessment (45) the same liver model as described above, was used, but in this case the chemical initiation was replaced by gamma-irradiation. The irradiation of the livers was performed on a surface of 40 x 30 mm with the rats under weak anaesthesia.

The reasons to only irradiate the liver were, a) the use of the liver model described above, and b) that irradiation of other organs might result in unwanted side effects.

Gamma-irradiation gave a statistically significant dose response curve regarding the formation of foci. The chemically induced foci

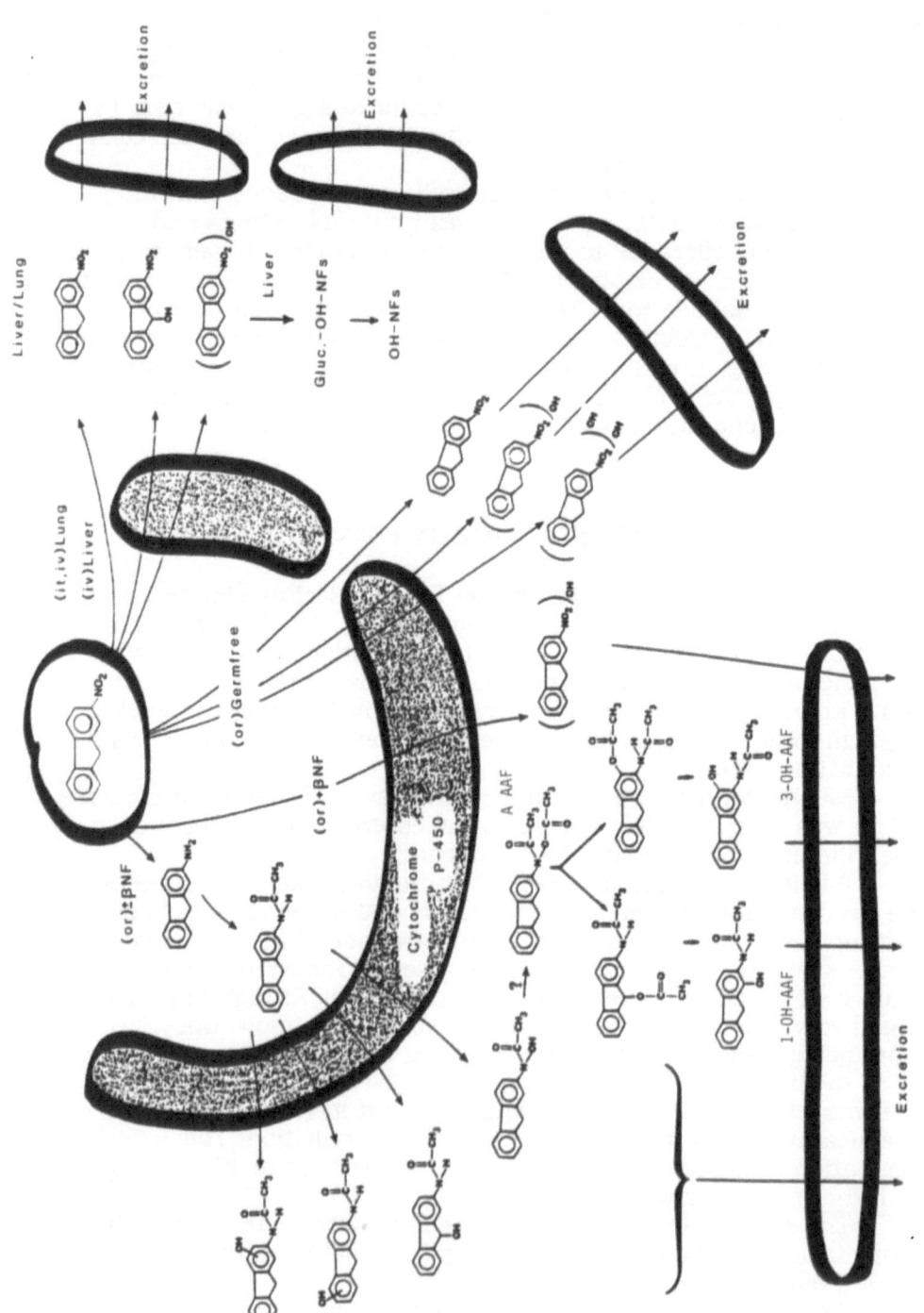

Figure 5. A general summary of the in vivo metabolism of NF.

were not possible to distinguish from foci induced by gamma-irradiation. The achieved dose-respons curve was used to convert the chemical dose which induced foci to irradiation equivalents. In that way the NF dose could be expressed in units of Gray.


RISK ASSESSMENT

## Are nitro-PAH's a human health risk?

For the group of nitro-PAH's, epidemiology or studies on specially selected groups are more or less impossible to perform since nitro-PAH's constitute only a part of the products of incomplete combustion. It is extremely hard to define a group of people which is exposed to nitro-PAH's and a control group which is exactly the same from all aspects except that it is not exposed to nitro-PAH's. Nitro-PAH's always occur together with other carcinogens (combustion products) and therefore another approach must be used.

The following discussion is an attempt to shed some light, although there will always be dark corners in such assessments, on the human risk of nitro-PAH's with the model substance NF as a basis.

The first question is:

Is there a **qualitative** health risk with NF? In table 1 the biological effects of NF are summarized which clearly indicate a variety of genotoxic risks. In table 2 it is shown that NF gave rise to tumors at several sites both after oral administration and skin application. The answer to the question must be yes. NF has genotoxic properties in many different assays, in vitro as well as in vivo.


Table 1. A summary of NF's biological effects.

| ASSAY | EFFECT | REF. |
|---|---|---|
| Mutagenicity, Nematode assay | + | 81 |
| Sister Chromatide Exchange | + | 82,3,83 |
| Initiator | + | 44 |
| Promoter | (+) | 44 |
| Carcinogenicity | + | 56,27 |
| Formation of DNA-adducts | + | 84 |
| Micronuclei assays | - | 6,86,87,88 |
| Mutation assay, mammalian cells | + | 85 |
| Bacterial mutagenicity (*Salmonella*) | + | 89,90,91 |
| Bacterial mutagenicity (*E. Coli*) | + | 92 |
| Mutagenicity, mouse lymphoma assay | + | 3,93,85 |
| LD$_{50}$ | (+) | 71 |

+   positive effect
(+)   weak effect
-   no effect

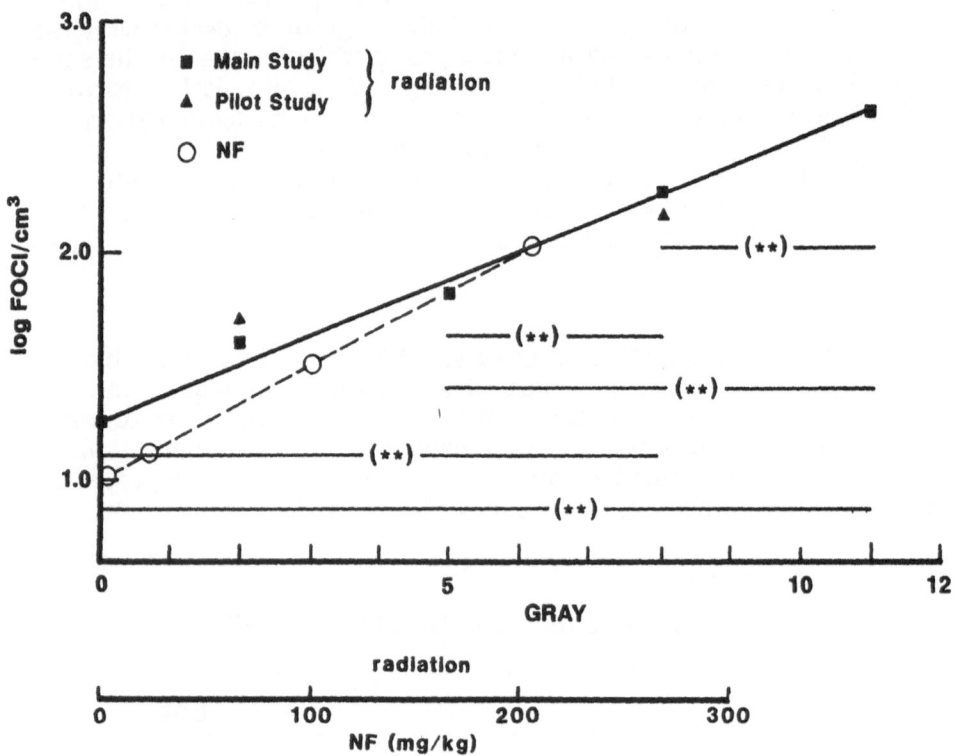

Figure 6. Dose of gamma-irradiation (filled symbols) and NF (open symbols) of foci in rat livers.

Table 2.  Sites of tumors after administration of NF to rats.

| ORGAN | ROUTE | |
| | oral[a,b] | skin[b] |
| --- | :---: | :---: |
| # Mammary gland | + | + |
| # Ear duct | + | |
| # Pituitary gland | | + |
| # Adrenal gland | | + |
| # Lung | + | + |
| # Salivary gland | | + |
| # Forestomach | + | |
| # Liver | + | |
| # Intestine | + | |
| # Subcutanous | | + |

a) Miller, 1955 (27)
b) Morris, 1950 (56)

The second question is:

If there is a qualitative risk, does it really mean anything to man in terms of **quantitative** risk? As mentioned earlier epidemiology is a weak tool to answer this question and therefore two different approaches have been used to elucidate the risk level of nitro-PAH's.

The first is an extrapolation of animal cancer data, of one metabolite (AAF) of NF, to man, and the second involves transformation of preneoplastic lesions in rat livers to radiation equivalents and further calculations on known human risk due to irradiation (45). The aims of performing calculations on the human cancer risk due to exposure to nitro-PAH's were two:

**1.** To find out the **level** of risk. Is the calculated risk at an acceptable or non-acceptable level?

**2.** What are the **weaknesses** in the calculations? Where is more scientific knowledge needed?

The dose:

Both calculations are based on the same dose data on NF and nitro-PAH's (45). As a basis for calculations, the NF-levels from Berlin (73,14) have been used, because the existing data from Japan (74) is weak, in terms of statistics, due to too few samples. The Chinese data (74) is also weak in the meaning that the data has been presented only in diagrams and Beijing is less typical for a Western city relative to Berlin.

If food is a source relevant for human exposure to nitro-PAH's in terms of particle bound nitro-PAH's in plants along roads and/or formation during certain types of cooking, the estimated human life dose has been largely underestimated. One example indicating that this might be the case is the formation of NP in a popular chicken-dish in Japan (17). One meal (approx. 200 g) corresponds to 3.5 years breathing, 24 h/day, in the streets of Tokyo with reference to NP dose (calculated from 74 and 17). For the calculated human nitro-PAH dose the synthesis of nitro-PAH's during food processing is assumed to be zero. For oral uptake of deposited particle-bound nitro-PAH's, the same distribution as the particle-bound PAH's has been used when calculating the indirect dosing of nitro-PAH's, i.e. 1-5 times the inhalated dose (75).

In total, the human lifetime dose of nitro-PAH's for an urban citizen, has been calculated to be 28 mg (45). In Table 3 the weakness and strengths regarding current knowledge on the human dose of nitro-PAH's are listed.

**The AAF-model:**

In the AAF-model combined data has been presented from the metabolic pathways, and a mega-study (96) performed on 24.000 animals to elucidate the carcinogenic potential of AAF at low life time

Table 3.  Weaknesses (-) and strengths (+) regarding current knowledge on the human exposure to nitro-PAH's.

---

(-)   **1.** Are the literature data on nitro-PAH levels representative?

(-)   **2.** Sampling temperatures are not given (might influence distribution gas/particle phase).

(-)   **3.** Limited information available on the food levels of nitro-PAH's. Could be a major source of nitro-PAH's.

(-)   **4.** Can plants (for food comsumption) along roads absorb and metabolize nitro-PAH's?

(-)   **5.** What is the relationship between oral and inhaled dose?

(-)   **6.** Are the potent direct-acting mutagens, the OH-NF's, also carcinogens?

(+)   **7.** A large number of analyses of NF in urban air.

(+)   **8.** NF seems to be a good model for nitro-PAH's in terms of dose.

(+)   **9.** An equal distribution of nitro-PAH's in the urban environment due to a very large number of sources.

---

Table 4.    Weaknesses (-) and strengths (+) of the AAF-model.

(-)    **1.** Data extrapolated from animals to man.

(-)    **2.** A part of the human dose might undergo ring-hydroxylation instead of reduction and acetylation. However, much more potent mutagens (carcinogens ?) are formed via that route.

(+)    **3.** Many animals in the cancer study (24.000).

(+)    **4.** Known metabolism for the model compound (NF).

(+)    **5.** Low lifetime doses of AAF gave rise to tumors.

(+)    **6.** The dose-response curve was linear for liver tumors on which the calculations were performed.

(+)    **7.** The metabolism in laboratory animals and man are probably similar.

(+)    **8.** The AAF-model is an vivo model.

(+)    **9.** AAF is considered to be a human carcinogen.

(+)    **10.** Enzymes necessary for the metabolic pathways are found in man.

dosing. The tumor data on the liver has been used for the extrapolation to man. The dose-response curve for tumor formation in the animal study after life time dosing was linear. In the calculations (45) the animal data has been extrapolated to man. The positive and negative comments that could be made regarding this model are found in table 4.

**The gamma-model:**

The gamma-model is totally different from the AAF-model in that known human risk data on gamma-irradiation is the basis for the risk calculation. The genotoxic effects in the rat liver (44) after exposure to NF occur as preneoplastic lesions. The same genotoxic lesions can be formed after exposure to gamma-irradiation of the rat livers (45). The dose of gamma-irradiation was exact in terms of dose maxima, exposed area, standard deviation of the dose etc., since a gamma-irradiation apparatus for human cancer treatment was used (45).

In this way the genotoxic lesions caused by NF could be converted to gamma-irradiation equivalents (Gy). The total human dose of nitro-PAH's could then be expressed in radiation units which made it possible to use human risk data on carcinogenesis after

Table 5.   Weaknesses (-) and strengths (+) of the gamma-model.

---

(-)   **1.** Are foci generated from chemicals and gamma-irradiation, respectively, the same ?

(-)   **2.** Recent human risk data on gamma-irradiation indicate that the risk is underestimated.

(+)   **3.** Relatively speaking, well known risk data on humans after exposure to gamma-irradiation exists.

(+)   **4.** Conversion of genotoxic lesions to radiation-equivalents.

(+)   **5.** The gamma-model is an in vivo model.

---

gamma-irradiation exposure to large populations (97). In Table 5, the weaknesses and strengths of this model are listed.

**The risk:**

The AAF- and gamma-models result in a risk in the range of 0.1-30 (mean values 4-16) x $10^{-6}$. In addition to that, recently published data on human cancer risk, indicate that the used data on cancer risk after exposure to gamma-irradiation might be under-estimated (79). The reasons for this are: a) a reevaluation of the relationship between gamma- and neutron-irradiation is being made, and b) an increase in human cancer risk over the whole life span, not only during the first decades after exposure, in the data that exists on the consequences of the atomic bomb on survivors in Hiroshima and Nagasaki (79). The preliminary data (the final data will be published in the near future) indicate that the risk has been underestimated by up to a factor of two to six. In the gamma(+)-model using the new human risk data, a factor of four (four selected for being in the middle of the range two to six) has been used to correct the data on gamma-irradiation for what could be the result of the reevaluation of the Japanese data.

Although no definite risk limit has been set, a risk of 1 x $10^{-6}$ is in general what is considered to be the limit for an unacceptable risk. That risk limit could be exeeded by a factor of up to 30 times in the case of nitro-PAH's.

**Possible consequences to nitro-PAH exposure besides cancer:**

In the presented risk data, additional factors which would increase the cancer risk estimate could be included in the assessment. These factors might involve:

**1.** Malformations. Nitro-PAH metabolites have been reported to cause malformations in laboratory animals (76).

**2.** Miscarriages, for the same reason as under 1.

**3.** Promotion of tumor development, NF has been shown to have promoting capacity (44).

**4.** Cocarcinogenic effects. Nitro-PAH's and PAH's always occur together. Recent data indicate that NP and B(a)P are potent cocarcinogens (94).

**5.** Humans on a "Western diet" have higher levels of intestinal beta-glucuronidase (70) which could result in an increased liberation of genotoxic metabolites in the colon (44).

**6.** Induction of cytochrome P-450 c and d (by other environmental contaminants) results in the formation of potent direct acting mutagens and possible carcinogens (37).

**7.** Certain food components can affect intestinal nitro-reductases dramatically. Since the nitroreduction is a critical metabolic step this could be of great importance (L. Möller, unpublished data).

**8.** The effects of alcohol consumption on the metabolism of nitro-PAH's. Liver microsomes from rats with prior dosing of ethanol, metabolize NP in a different manner (77) and hepatic microsomes from ethanol-fed hamsters bioactivated AF more effectively (78).

**9.** Possible effects on fertility. Nitro-compounds have been reported to cause infertility and reduced sperm count in rats (50).

So far, there does not exist enough knowledge to assess the risk for these phenomena. However, if such risks exist in humans, they must be added to the risk data on cancer presented above. The minimum and maximum levels of risk, if the three models are summarized, are in the range of $0.1 - 30 \times 10^{-6}$.

To answer to the second question, whether nitro-PAH's expose man to a quantitative risk, is not easy but the above discussed data indicate that nitro-PAH's very well could contribute to human cancer on a non-neglectable level. More research is needed where the data is weak or incomplete.

## REFERENCES

**1.** Lindvall, T. (1985). Health effects of nitrogen dioxide and oxidants. Scand. J Work Environ. Health. 11, 10-28.

**2.** Tokiwa, H. , Nakagawa, R., Morita, K. and Ohnishi, Y. (1981). Mutagenicity of nitro derivatives induced by exposure of aromatic compounds to nitrogen dioxide., Mutat. Res., 85, 195.

**3.** McCoy, E. (1984). Role of metabolism on the mutagenicity of nitroarenes. In: Biochemical Basis of Chemical Carcinogenesis, Eds., Greim, H., Jung, M., Kramer, M., Marquardt, H. and Oesch, F. Raven Press, New York, p57.

**4.** Danford, N. Wilcox, P. and Parry J.M. (1982). The clastogenic activity of dinitropyrenes in a rat liver epithelial cell line. Mutat. Res. 105, 349.

**5.** Nachtman, J.P. and Wolff, S. (1982). Activity of nitro-polynuclear aromatic hydrocarbons in the sister chromatide exchange assay with and without metabolic activation. Environ. Mutagen. 4,1.

**6.** Neal, S.B. and Probst G.S. (1983). Chemically-induced sister chromatid exchange in vivo in bone marrow of Chinese hamsters, an evaluation of 24 compounds. Mutat. Res. 113, 33.

**7.** Tucker, J.D. and Ong, T. (1984). Induction of sister chromatid exchanges and chromosome aberrations in human peripheral lymphocytes by 2,4,7-trinitro-9-fluorene. Mutat, Res. 138, 181.

**8.** Hartong, A., Kraft, J., Schulze, J., Kiess, H. and Lies, K-H. (1984). The identification of nitrated polycyclic aromatic hydrocarbons in diesel particulate extracts and their potential formation as artifacts during particulate collection. Chromatographia, 19, 269-273.

**9.** Handa, T., Yamauchi, T., Ohnishi, M., Hisematsu, Y. and Ishii, T. (1983). Detection and average content levels of carcinogenic and mutagenic compounds from the particulates on diesel and gasoline engine muffers. Environ. International, 9, 335-341.

**10.** Tokiwa, H., Nakagawa, R. and Horikowa, K. (1985). Mutagenic/carcinogenic agents in indoor pollutants: The dinitropyrenes generated by kerosene heaters and fuel gas and liquified petroleum gas burners. Mutat. Res., 157, 39-47.

**11.** Ramdahl, T., Becher, G. and Bjōrseth, A. (1982). Nitrated polycyclic aromatic hydrocarbons in urban air particles. Environ. Sci. Technol. 16, 861-865.

**12.** Gorse, R.A., Riley, T.L., Ferris, F.C., Pero, A.M., and Skewes, L.M. (1983). 1-Nitropyrene concentration and bacterial mutagenicity in on-road vehicle emissions. Environ. Sci. Technol. 17, 198-202.

**13.** Tokiwa, H., Kitamori, S., Nakagawa, R. and Ohnishi, Y. (1983). Mutagens in airborne particulate pollutants and nitro derivatives produced by exposure of aromatic compounds to gaseous pollutants. Environ. Sci. Res. 27, 555-567.

**14.** Moriske, H.-J. (1986). Polare verbindungen im Stadtaerosol. VDI Fortschrittberichte, Reike 15, no. 42, VDI-Verlag, Düsseldorf.

**15.** Sato, T., Kato, K., Ose, Y., Nagase, H. and Ishikawa, T. (1985). Nitroarenes in Suimon river sediment. Mutat. Res. 157, 135-143.

**16.** Ohnishi, Y., Kinouchi, T., Manabe Y., Tsushi, H., Otsuka, H., Tokiwa, H. and Otofujii, T. (1985). Nitro compounds in environmental mixtures and foods, in short-term bioassays in the analysis of complex environmental mixtures IV, Waters, M.D. et al., eds. Plenum Press, New York, 195.

**17.** Kinouchi, T., Hideshi, T. and Ohnishi, Y. (1986). Detection of 1-nitropyrene in Yakatori (grilled chicken). Mutat. Res. 171, 105-113.

**18.** Tokiwa, H., Nakagawa, R., Morita, K. and Ohnishi, Y. (1981). Mutagenicity of nitro derivatives induced by exposure of aromatic compounds to nitrogen dioxide. Mutat. Res. 85, 195-205.

**19.** Okinaka, R.T., Nichols, J.W., Whaley, T.W. and Strniste, G.F. (1984). Phototransformation of 2-aminofluorene into N-oxidized mutagens. Carcinogenesis 5, 1741-1743.

**20.** Ohe, T. (1984). Mutagenicity of photochemical reaction products of polycyclic aromatic hydrocarbons with nitrite. The Science of the Total Environ. 39, 161-175.

**21.** Rosenkranz, H.S., and Mermelstein, R.M. (1985). The genotoxicity, metabolism and carcinogenicity of nitrated polycyclic hydrocarbons. J. Environ. Sci. Health, 2, 221-272.

**22.** Tokiwa, H., Nakagawa, R. and Ohnishi Y. (1981). Mutagenic assay of aromatic nitro compounds with Salmonella typhimurium. Mutat. Res. 91, 321.

**23.** Pitts, J.N Jr, Harger, W., Lokensgaard, D.M., Fitz, D.R., Scorziell, G.M. and Mejia, V. (1982). Diurnal variations in the mutagenicity of airborne particulate organic matter in Californias south coast air basin. Mutat. Res. 104, 35.

**24.** Wang, Y.I.Y., Rappaport S. M., Sawyer, R.F., Talcott, R.E. and Wei, E.T. (1978). Direct-acting mutagens in automobile exhaust. Cancer Lett. 5, 39.

**25.** Pederson, T.C. and Siak, J-S. (1981). The role of nitroaromatic compounds in the direct acting mutagenicity of diesel particle extracts. J. Appl. Toxicol. 1, 54.

**26.** El-Bayoumy, K., Hecht, S. and Hoffman D. (1982). Comparative tumor initiating activity on mouse skin of 6-nitrobenzo(a)pyrene, 6-nitrochrysene, 3-nitroperylene, 1-nitropyrene and their parent hydrocarbons. Cancer Lett. 16, 333.

**27.** Miller, J.A., Sandin, R.B., Miller E.C. and Rush, H.P. (1955). The carcinogenicity of compounds related to 2-acetylaminofluorene. Cancer Res. 15, 188.

**28.** Schuetzle, D., Riley, T.L., Prater, T.J., Harvey, T.M. and Hunt, D.F. (1982). Analysis of nitrated polycyclic aromatic hydrocarbons in diesel particulate. Anal. Chem. 54, 265-271.

**29.** Henderson, T.R., Royer, R.E., Clark, C.R., Harvey, T.M. and Hunt, D.F. (1982). MS/MS analysis of diesel emissions and fuels treated with $NO_2$. J. Appl. Toxicol. 2, 231-237.

**30.** Schuetzle, D. (1983). Sampling of vehicle emissions for chemical analysis and biological testing. Environ. Health Perspec. 47, 65-80.

**31.** Xu, X., Nachtman, J., Rappaport, S. and Wei, E. (1981). Identification of 2-nitrofluorene in diesel exhaust particulates. J. Appl. Toxicol. 1, 196-198.

**32.** Nishioka, M.G., Petersen, B. and Lewtas, J. (1983). Comparison of nitro-aromatic content and direct-acting mutagenicity of passenger car engine emissions. In: Mobile Source Emissions Including Polycyclic Organic Species, Ed., Rondial, D., D. Riedel Publishing Company, p197.

**33.** Thorgeirsson, S.S., Weisburger, E.K., King, C. M. and Scribner, J.D. (1979). National Cancer Institute Monograph, no. 58.

**34.** Beije, B. and Möller, L. (1988). 2-Nitrofluorene and related compounds. Prevalence and biological effects., Mutat. Res. 196,177-209.

**35.** Schuetzle, D. and Frazier, J.A. (1986). Factors influencing the emission of vapor and particulate phase components from diesel engines. In: Ishinishi, N., Koizumi, A., McClellan, R.O. and Stöber, W. (eds), Carcinogenic and mutagenic effects of diesel engine exhaust. Elsevier, Amsterdam, pp 41-64.

**36.** Möller, L. and Gustafsson, J-Å. (1986). Liquid chromatographic mass spectrometric analysis of 2-nitrofluorene and its derivatives. Biomed. Mass. Spectrom. 13, 681-688.

**37.** Möller, L., Rafter, J. and Gustafsson, J-Å. (1987). Metabolism of the carcinogenic air pollutant 2-nitrofluorene in the rat. Carcinogenesis, 8, 637-645.

**38.** Möller, L., Törnquist, S., Beije, B., Rafter, J., Toftgård, R. and Gustafsson, J-Å. (1987). Metabolism of the carcinogenic air pollutant 2-nitrofluorene in the isolated rat lung and liver. Carcinogenesis, 8, 1847-1852.

**39.** Möller, L., Corrie, M., Midtvedt, T., Rafter, J. and Gustafsson, J-Å. (1988). The role of the intestinal microflora in the formation of mutagenic metabolites from the carcinogenic air pollutant 2-nitrofluorene. Carcinogenesis, 9, 823-830.

**40.** Törnquist, S., Möller, L., Gabrielsson, J., Gustafsson, J-Å. and Toftgård, R. (1989). 2-Nitrofluorene metabolism in the rat lung. Pharmacokinetic and metabolic effects of beta-napthoflavone treatment. Carcinogenesis, submitted.

**41.** Törnquist, S., Sundin, M., Möller, L., Gustafsson, J-Å. and Toftgård, R. (1988). Age dependent expression of cytochrome P-450 b and metabolism of the potent carcinogen 2-nitrofluorene in the rat lung. Carcinogenesis, 9, 2209-2214.

**42.** Beije, B. and Möller, L. (1986). Unscheduled DNA-synthesis in the liver and mutagenic activity in the urine of rats exposed to 2-nitrofluorene or 2-acetylaminofluorene. Environ. Mut. 8,10.

**43.** Beije, B. and Möller, L. (1988). Correlation between induction of unscheduled DNA synthesis in the liver and excretion of mutagenic metabolites in the urine of rats exposed to the carcinogenic air pollutant 2-nitrofluorene. Carcinogenesis, 9, 8, 1465-1470, 1988.

**44.** Möller, L., Torndal, U-B., Gustafsson, J-Å. and Eriksson,L.C. (1989). The air pollutant 2-nitrofluorene as initiator and promotor in a liver model for studies on chemical carcinogenesis. Carcinogenesis,10,13,435-440.

**45.** Möller, L., Lax, I., and Eriksson, L.C. (1989). Risk assessment of nitrated polycyclic aromatic hydrocarbons via the carcinogenic air pollutant and model substance, 2-nitrofluorene, by two different methods. Carcinogenesis, submitted.

**46.** Rosenkranz, H.S. and Mermelstein, R. (1983). Mutagenicity and genotoxicity of nitroarenes. All nitro-containing chemicals were not created equal. Mutat. Res. 114, 217.

**47.** Wilson, R.H., DeEds, F. and Cox, A.J. (1941). The toxicity and carcinogenic activity of 2-acetylaminofluorene. Cancer Res., 1, 595.

**48.** Campbell, R.M. and Lee, M.L. (1984). Capillary column gas chromatographic determination of nitro polycyclic aromatic compounds in particulate extracts. Anal. Chem., 56, 1026.

**49.** Bertilsson, T. and Egebäck, K-E. (1984). Swedish EPA-report, SNV PM 1739, p 19.

**50.** Lindner, R.E., Hess, R.A. and Strader, L.F. (1986). Testicular toxicity and infertility in male rats treated with 1,3-dinitrobenzene. J. Toxicol. Environ. Health, 19, 477.

**51.** Schuetzle,. and Perez, J.M. (1983). Factors influencing the emissions of nitrated-polynuclear aromatic hydrocarbons (nitro-PAH) from diesel engines. J.Air Poll. Control Ass., 33, 751.

**52.** Bond, J.A., Medinsky, M.A. and Dutcher, J.S. (1984). Metabolism of 1-[$^{14}$C] nitropyrene in isolated perfused rat livers.Toxicol. Appl. Pharmalcol., 75, 531.

**53.** Kinouchi, T., Morotomi, M., Mutai, M., Fifer, E.K., Beland, F.A. and Ohnishi, Y. (1986). Metabolism of 1-nitropyrene in germ-free and conventional rats. Japan J. Cancer Res. (Gann), 77, 356.

**54.** Weisburger, E.K. (1979). N-substituted aryl compounds in carcinogenesis and mutagenesis. National Cancer Institute Monograph, No. 58, p 1.

**55.** McCann, J., Choi, E., Yamasaki, E. and Ames, B. (1975). Detection of carcinogens as mutagens in the salmonells/microsome test: Assay of 300 chemicals. Proc. Natl. Acad. Sci., 72, 5135.

**56.** Morris, H.P., Dubnik,C.S. and Johnson,J.M. (1950). Studies of the carcinogenic action in the rat okf 2-nitro, 2-amino, 2-acetylamino, and 2-diacetylaminofluorene after ingestion and after painting. J.Natl.Cancer inst.,10,1201.

**57.** Belisario, M.A., Carrano, L., DeGiulio, A., Pecce, R. and Buonocore, V. (1987). Effect of liver enzyme inducers on metabolite excretion in rats treated with 1-nitropyrene. Toxicol. Lett., 36, 233.

**58.** Malejka-Giganti, D., Magat, W.J., Adelmann, A.M. and Decker, R.W. (1987). Metabolite profile in milk of lactating rats after treatment with a carcinogen, N-2-fluorenylacetamide. Drug Metab. Dispos., 15, 760.

59. Djuric, Z., Fifer, E.K., Howard, P.C. and Beland, F.A. (1986). Oxidative microsomal metabolism of 1-nitropyrene and DNA-binding of oxidized metabolites following nitroreduction. Carcinogenesis, 7, 1073.

60. Beland, F.A., Heflich, R.H., Howard, P.C. and Fu, P.P. (1985). In Polycyclic Hydrocarbons and Carcinogenesis, Harvey, R.G., ed. American Chemical Society, Washington D.C., 371.

61. Cerniglia, C.E., Howard, P.C., Fu, P.P. and Franklin, W. (1984). Metabolism of nitropolycyclic aromatic hydrocarbons by human intestinal microflora. Biochem. Biophys. Res. Commun., 123, 262.

62. Howard, P.C., Beland, F.A. and Cerniglia, C.E. (1983). Carcinogenesis, 4, 985.

63. El Bayoumy, K., Sharma, C., Louis, Y.M., Reddy, B. and Hecht S.S. (1983). The role of intestinal microflora in the metabolic reduction of 1-nitropyrene to 1-aminopyrene in conventional and germfree rats and in humans. Cancer Lett., 19, 311.

64. Harries, G.C., Boobis, A.R., Sesardic, D., Edwards, R.J. and Davies D.S. (1986). Food Chem. Toxicol., 24, 757.

65. Eddy, E.P., Howard, P.C., McCoy, G.D. and Rosenkranz, H.S. (1987). Mutagenicity, unscheduled DNA synthesis, and metabolism of 1-nitropyrene in the human hepatoma cell line HepG2. Cancer Res., 47, 3163.

66. Boobis, A.R., Brodie, M.J., McManus, M.E., Staiano, N., Thorgeirsson, S.S. and Davies, D.S. (1981). Metabolism and mutagenic activation of 2-acetylaminofluorene by human liver and lung. Adv. Exp. Med. Biol., 136, 1193.

67. McManus, M.E., Trainor, K.J., Morley, A.A., Burgess, W., Stupans, I. and Birkett, D.J. (1987). Metabolism of 2-acetylaminofluorene in cultured human lymphocytes. Chem. Pathol. Pharmacol., 55, 409.

68. Moore, B.P., Hicks, R.M., Knowles, M.A. and Redgraves, S. (1982). Metabolism and binding of benzo(a)pyrene and 2-acetylaminofluorene by short-term organ cultures of human and rat bladder. Cancer Res., 42, 642.

69. Weisburger, J.H., Grantham, P.H., van Horn, E., Steigbigel, N.H., Rall, D.P. and Weisburger, E.K. (1964). Activation and detoxification of N-2-fluoreneylacetamide in man. Cancer Res., 24, 475.

70. Gorbach, S.L. (1982). The intestinal microflora and its colon cancer connection. Infection, 10, 379.

71. Dybing, E., Dahl, J.E., Beland, F.A. and Thorgeirsson, S.S. (1986). Cell Biol. Toxicol., 2, 341.

72. Bond, J.A., Mauderly, J.L. and McClellan, R.O. (1984). In: Polynuclear Aromatic Hydrocarbons, Proceedings of the Ninth International on Polynuclear Aromatic Hydrocarbons Symposium, Cooke, M., Dennis, A.J., eds., Batelle Press, Columbus, p 79.

73. Moriske, H-J., Block, I. and Rüden. (1984). Über polare organische Verbindungen im Stadtaerosol und deren mutagene Wirksamkeit im Salmonella typhimurium-Test nach Ames.Forum Städte Hygiene, 35, 113.

74. Iida, Y., Daishima, S., Furuya, K., Kikushi, T., Matsushita, H., Tanebe, K., Wu, J., Wan, A-P. and Huang Y-C. (1985). Present state of air pollution in Beijing. Sekei J. Asian Pacific Studies, 111.

75. Ehrenberg, L., personal communication Dep. of Radiation Biology, Univ. of Stockholm, Sweden.

76. Harris, C., Namkung, M.J. and Juchau, M.R. (1987). Regulation of intracellular glutathione in rat embryos and visceral yolk sacs and its effect on 2-nitrofluorene-induced malformations in the whole embryo culture system. Toxicol. Appl. Pharmacol., 88, 141.

77. Howard, P.C., Demarco, G.J., Consolo, M.C. and McCoy, G.D. (1987). Differing effects of chronic ethanol consumption by mice on liver microsomal metabolism of xenobiotics: 1-nitropyrene, nicotine, aniline, and n-nitrosopyrrolidine. Mol. Toxicol., 1, 177.

78. Ioannides, C. and Steele, C.M. (1986). Hepatic microsomal mixed-function oxidases by aromatic amines and its relationship to their bioactivation to mutagens. Chem.-Biol. Interact., 59, 129.

79. Roberts, L. (1987). Atomic bomb doses reassessed. Science, 238, 1649.

80. Solt, D. and Farber, E. (1976). New Principle for the Analysis of Chemical Carcinogenesis. Nature, 262, 701.

81. Lew, K.K., Nichols, D.G. and Kolbert, A.W. (1983). Nato Conf. Series, 1, 139.

82. Nachtman, J.P. and Wolff, S. (1982). Activity of nitro-polynuclear aromatic hydrocarbons in the sister chromatid exchange assay with and without metabolic activation. Environ. Mut., 4, 1.

83. McCoy, E. (1984). Role of metabolism on the mutagenicity of nitroareas. In: Biochemical Basis of Chemical Carcinogenesis, Eds., Greim, H., Jung, R., Kramer, M., Marquardt, H. and Oesch, F., Raven Press, New York.

84. Massaro, M., McCartney, M., Rosenkranz, E.J., Anders, M., McCoy, E., Mermelstein, R. and Rosenkranz, H.S. (1983). Evidence that nitroarene metabolites form mutagenic adducts with DNA-adenine as well as with DNA-guanine. Mutat. Res., 122, 243.

**85.** Amacher, D.E., Paillet, S.C., Turner, G.N. (1979). Utility of the mouse lymphoma L5178Y/TK assay for the detection of chemical mutagens. In: Mammalian Cell Mutagenesis, Banbury Report No. 2, 277.

**86.** Suzuki, Y. (1985). Studies on development of the sensitive micronucleus test. Part 2 The in vitro method using cultured bone marrow cells.Tokyo Jikeikai Med. J., 100, 707.

**87.** Sakitani, T. and Suzuki, Y. (1986). Part 2 Mutagenic activities of air pollutants observed by micronucleus test.Tokyo Jikeikai Med. J., 101, 259.

**88.** Ohe, T. (1985). Studies on comparative decomposition rate by rat liver homogenate and on micromucleus test of nitrated polycyclic aromatic hydrocarbons. Bull. Environ. Contam. Toxicol., 34, 715.

**89.** Rosenkranz, H.S., McCoy, E.C., Frierson, M. and Klopman, G. (1985). The role of DNA sequence and structure of the electrophile on the mutagenicity of nitroarenes and arylamine derivatives. Environ. Mut. 7, 645.

**90.** Andrews, L.S., Pohl, L.R., Hinson, J.A. and Gilette, J.R. (1979). Mutagenesis of 2-nitrofluorene (NF), 2-nitrosofluorene (NOF) and 2-hydroxylaminofluorene (NHOHF) for salmonella TA 100 and TA 200 FR.Toxicol. Appl. Pharmacol. 48, A48.

**91.** Banerjee, T.S., Bhaumik, G., Yu, C-L,, Swaminathan, B., Giri, A.K., Srivastava, S. and Bhattacharjee, S.B. (1984). Evaluation of the genetoxicity of lac dye. Fd. Chem. Toxicol. 22, 677.

**92.** Doudney, C.O., Franke, M.A., and Rinaldi, C.N. (1981). The DNA damage activity (DDA) assay and its application to river waters and diesel exhausts. Environ. Internat., 5, 293.

**93.** Wangenheim, J. and Bolcsfoldi, G. (1986). Mouse lymphoma TK+/- assay of 30 compounds. Environ. Mut. 8, 90.

**94.** Moon, R.C., Rao, K.V.N. and Detrisac, C.J. (1988). Potential carcinogenicity of 1-nitro-pyrene. In:The fifth Health Effects Institute Annual Conference, Colorado Springs, Health Effects Institute, 15.

**95.** Simmon, V.F., Rosenkranz, H.S., Zeiger, E., Poirier, L.A. (1979). Mutagenic activity of chemical carcinogens and related compounds in the intraperitoneal host-mediated assay. J.Natl.Cancer inst.,62,911.

**96.** Littlefield, N.A., Farmer, J.H., Gaylor, D.W. and Sheldon, W.G. (1979). Effects of dose and time in a long-term, low-dose carcinogenic study. J. Environ. Pathol. Toxicol., 3, 17-34.

**97.** ICRP, no. 26 (1977). Recommendations of the international commission on radiological protection. Pergamon Press, Oxford, vol. 1.

NITROARENES: THEIR DETECTION, MUTAGENICITY AND OCCURRENCE IN THE
ENVIRONMENT

Joellen Lewtas and  Marcia G. Nishioka

U.S. Environmental Protection Agency
Research Triangle Park, North Carolina 27711
Battelle, Columbus, Ohio 43201

ABSTRACT

Nitroarenes and hydroxylated nitroarenes have been identified in diesel
and other combustion source emissions and ambient air using
bioassay-directed chemical analysis. Studies of the extractable organic
matter associated with airborne particles show that substantial mutagenicity
is associated with fractions which are more polar than the polycyclic
aromatic hydrocarbons (PAH). These mutagenic polar neutral and acidic
fractions contain nitroarenes. Hydroxylated nitroarenes, where the aromatic
portion consists of one- and two-rings (e.g., hydroxylated nitrobenzene,
-toluene, -naphthalene, and -biphenyl), have been found in relatively high
concentrations in the organic acids in ambient air. The distribution of
nitroarenes and hydroxylated nitroarenes and the specific isomers present
in the air appear to be related to their mechanism of formation and source
whether they are emitted directly from combustion sources or formed through
atmospheric transformation.

INTRODUCTION

Nitroarenes are defined and discussed here as those aromatic compounds
with at least one nitro ($NO_2$) functional group including nitrated 1 or 2
ring ($NO_2$-AR) compounds (e.g., nitrobenzene); nitrated polycyclic aromatic
hydrocarbons ($NO_2$-PAH)(e.g.,1-$NO_2$-pyrene); nitrated heterocyclic compounds
(e.g. $NO_2$-azaarenes) and other nitrated aromatic compounds which may have
additional functional groups such as hydroxylations (e.g., OH-$NO_2$-AR and
OH-$NO_2$-PAH). Although the simpler nitroarenes have been used for decades
as industrial chemicals (e.g., nitrobenzene and nitrotoluene) and
pharmaceutical chemicals (e.g., nitrofurans), the discoveries in 1980 of
potent mutagenic $NO_2$-PAH and di-$NO_2$-PAH in diesel exhaust particle emissions
and later in ambient air have led to a dramatic increase in interest in
these compounds, their metabolism, mutagenicity, carcinogenicity, formation
and occurrence.

IDENTIFICATION OF NITROARENES: BIOASSAY-DIRECTED FRACTIONATION AND CHEMICAL
ANALYSIS

Nitroarenes in general and $NO_2$-PAH in particular, are probably the most
important class of mutagens identified as air pollutants in the last
decade. Bioassay-directed fractionation and chemical analysis was used to

identify nitroarenes in photocopy toners (1,2), diesel exhaust particles (3,4), kerosene heater emissions (5,6), and ambient air (4). Of the approaches that have been taken to identify potential carcinogens in complex mixtures (7), bioassay-directed fractionation and chemical analysis has been the most successful (4,8). In this approach the complex mixture is fractionated and each fraction is bioassayed. Mutagenically active fractions are further fractionated, bioassayed, and characterized until the major class or specific compounds potentially responsible for the mutagenicity are identified. The Salmonella typhimurium plate incorporation assay has been extensively used in this approach to identify mutagens in air, water, food and specific emission sources (4). This approach combined with the use of bacterial tester strains selectively sensitive to nitroarenes (9) led to the identification of $NO_2$-PAH as potent mutagens in complex mixtures of certain combustion sources (8). There are several methodological advances which have recently had a significant impact on research in this area. First is the development of micromutagenesis bioassays which can be coupled directly to analytical fractionation procedures to identify the mutagens in complex mixtures (4). Second is the development of new chemical-analysis techniques which can facilitate the identification of trace species (e.g., dinitropyrenes), polar organic species (e.g., hydroxylated and nitrated aromatic hydrocarbons) and labile compounds (e.g., organic peroxides) (8). Negative chemical ionization (NCI) mass spectrometry (MS) is especially useful given the electrophilic nature of mutagens (10). NCI/MS not only provides structural and isomer-specific identification, but more importantly, for analysis of a mixture, appears to provide highly sensitive and selective detection of those compounds which are electronegative and may therefore be more likely to be mutagenic. The power of these biological and chemical methods has recently been demonstrated in the identification of $OH-NO_2$-AR and $OH-NO_2$-PAH compounds in urban air (10). Although highly directed in focus, bioassay-directed fractionation is an iterative procedure of synthesis and biological evaluation. Many, if not all, of the of the polar compounds being newly identified as potential mutagens either are not available as reference material and/or have not been evaluated for mutagenic activity. Synthesized reference compounds are used to confirm identification, quantify concentration and to assess mutagenicity. Sufficiently large quantities of reference material must be synthesized to permit evaluation of genotoxicity and carcinogenicity in mammalian systems.

As a research tool, bioassay-directed fractionation facilitates the identification of biologically reactive components in complex mixtures. When bacterial mutagenesis assays are used to direct the fractionation and analysis, then potent mutagens such as dinitropyrenes (9) present at low concentrations (e.g. <ppm) or moderately potent mutagens present at higher concentrations (e.g. >ppm) can be detected and tentatively identified. While biological studies may demonstrate that these mutagens are not carcinogens due to their mechanism of action or detoxification in vivo; this approach has identified potent mutagens which have been shown to be potent carcinogens including the dinitropyrenes (11).

OCCURRENCE AND MUTAGENICITY OF NITROARENES

Nitroarenes are widely distributed in the environment (13) and have been identified in air, water, and food. The origin of these compounds is thought to be the nitration of the parent aromatic compound, as a result of either combustion or atmospheric reactions. Nitro-polycyclic aromatic hydrocarbons ($NO_2$-PAH) are well-documented constituents of both primary combustion emissions and ambient air. Examples of reported levels of $NO_2$-PAH are given in Table 1.

Table 1. Occurrence and Mutagenicity of Nitroarenes

| Chemical (Mutagenicity) Source | Concentration μg/g extract | ng/m³ air | Reference |
|---|---|---|---|
| **3-Nitrotoluene** | | | |
| Ambient Air | | | |
|   Boise, ID  U.S. | 0.1-0.6 | | 14 |
| **1-Nitronaphthalene** | | | |
| Diesel particles | 0.3-0.7 | | 11 |
|   Ambient Air | | | |
|     Los Angeles, CA | | 2-3 | 12 |
|     Boise, ID | | 0.03-0.4 | 14 |
| **3-Nitrobiphenyl** | | | |
| Ambient Air | | | |
|     Los Angeles, CA | | 0.03-0.1 | 12 |
|     Boise, ID | | 0.6-6.0 | 14 |
| **2-Nitrofluorene** | | | |
| Diesel emissions | 71-186 | | 15 |
|   Air particles | | | |
|     Tokyo, Japan | ND-22 | ND-0.03 | 16 |
|     China | 0.03-0.7 | | 11 |
|     Germany | 0.2-5 | | 11 |
| **9-Nitroanthracene** | | | |
| Diesel | 5-94 | | 17 |
|   Ambient Air | | | |
|     Los Angeles, CA | | 0.05-0.1 | 12 |
|     Boise, ID | | 0.04-1.5 | 14 |
|     Columbus, OH | | | |
|       Outdoors | | 0.01-0.1 | 18 |
|       Indoors | | 0.04.1.3 | 18 |
| **1-Nitropyrene** | | | |
| Diesel particles | 100-200 | | 15, 19 |
| Gasoline particles | 2.5 | | 19 |
|   Air particles | | | |
|     Detroit, MI | 0.2-0.6 | 0.02-0.03 | 20 |
|     Tokyo, Japan | 0.2-1.6 | 0.02-0.1 | 16 |
|     Boise, ID | 0.06-0.1 | | 14 |
|     Los Angeles, CA | 0.03-0.04 | | 12 |
|     Columbus, OH | | | |
|       Outdoors | | 0.01 -0.05 | 18 |
|       Indoors | | 0.005-0.1 | 18 |
| **2-Nitrofluoranthene** | | | |
| Air particles | | | |
|     Boise, ID | | 0.07-0.2 | 14 |
|     Los Angeles, LA | | 0.3-0.4 | 12 |
|     Columbus, OH | | | |
|       Outdoors | | 0.03-0.2 | 18 |
|       Indoors | | 0.01-0.2 | 18 |
| **3-Nitrofluoranthene** | | | |
| Diesel particles | 0.9-7.0 | | 4 |

Table 1. (continued).

| Chemical (Mutagenicity) Source | Concentration | | Reference |
|---|---|---|---|
| | μg/g extract | ng/m$^3$ air | |
| 6-Nitrobenzo[a]pyrene | | | |
| Diesel particles | ND-50 | | 11 |
| Gasoline particles | 0.2-33 | | 11 |
| Air particles | | | |
| Michigan, US | 0.9-2.5 | 0.04-0.3 | 20 |
| 1,3-Dinitropyrene | | | |
| Diesel particles | ND-1.6 | | 11 |
| Kerosene heater | 0.5 | | |
| Air particles | | | 5 |
| Tokyo, Japan | 0.005 | | 16 |
| 1,6-Dinitropyrene | | | |
| Diesel particles | ND-1.2 | | 11 |
| Air particles | | | |
| Michigan, US | 0.004-0.05 | 0.1-4.4 | 21 |
| Tokyo, Japan | 0.005-0.1 | 0.3-8.7 | 16 |
| 1,8-Dinitropyrene | | | |
| Diesel particles | ND-3.4 | | 11 |
| Air particles | | | |
| Michigan, US | 0.002-0.5 | 0.04-3.8 | 11 |

The more volatile nitro-aromatic hydrocarbons ($NO_2$-AR; AR here for 1 and 2 ring aromatic compounds) have also been detected in air and are distributed between vapor phase and particle phase in ambient samples (12). The vapor phase constituents can be collected on solid sorbants such as XAD-2 and polyurethane foam (PUF). The $NO_2$-AR/PAH found in combustion emissions, e.g. 3-$NO_2$-fluoranthene, appear to be formed via direct electrophilic nitration of AR/PAH (22,12). In contrast, many $NO_2$-AR/PAH found in ambient air, e.g. 2-$NO_2$-fluoranthene, are thought to be formed by the nitration mechanisms elucidated in atmospheric chamber studies (23). The atmospheric reactions involve multistep reactions: daytime reactions of AR/PAH with OH radicals (OH) in the presence of $NO_x$ and nighttime reactions of AR/PAH with $N_2O_5$ (23-27). Direct nitration through combustion and atmospheric nitration reactions are proposed to produce distinctly different $NO_2$-AR/PAH isomers or differing distributions of the $NO_2$-AR/PAH isomers. Because many atmospherically produced $NO_2$-AR/PAH are found largely as vapor phase material, while many direct nitration $NO_2$-AR/PAH of similar volatility are found largely in particle extracts, distribution of specific $NO_2$-AR/PAH between gas and particle phases may be used to distinguish primary combustion emissions from products of atmospheric nitration reactions (12,14).

These differences in phase distribution and isomer substitution between the direct nitration in combustion emissions and atmospheric nitration are supported by measurements of $NO_2$-AR/PAH in ambient air under photochemically reactive and non-reactive conditions. Figure 1 shows the isomer distribution and phase distribution of $NO_2$-AR between particle (filter) phase and vapor phase (XAD-2) sample for two significantly differently ambient samples collected during the daytime in Boise, Idaho, during December and January with ambient temperatures of -5 to 7° C (14). Because of the simplicity of the winter Boise airshed, $NO_2$- and OH-$NO_2$-AR/PAH will

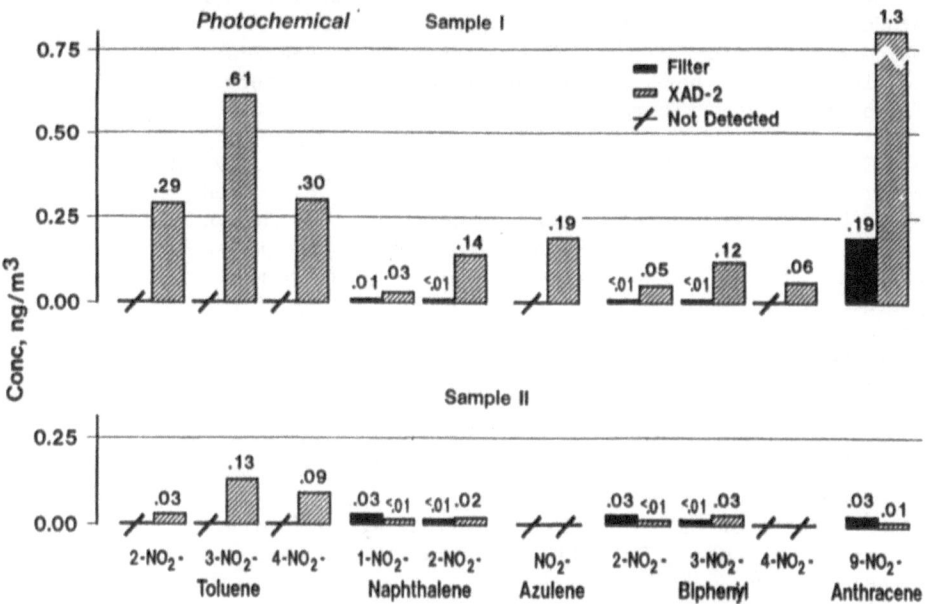

Figure 1. Comparison of winter daytime ambient air $NO_2$-Arenes in two samples collected in Boise, ID under different atmospheric conditions (14). Sample I is labeled photochemical due to the higher HONO concentrations during the sampling period as compared to Sample II.

be either direct automotive emissions or photochemical products from auto- and/or woodstove-emitted AR/PAH. Because of the low ambient temperatures, the phase distribution of these nitrated compounds was not expected to be distorted significantly by volatilization from the particles during sampling. Concurrent measurements of several source emission indicators (K for wood combustion, Pb and $NO_x$ for automotive emissions), percentage of possible sunlight (sunlight being necessary for photolysis of HONO), and HONO itself (precursor of OH radicals that initiate daytime nitration reactions), were compared with the concentrations of the nitrated species. The two selected air samples were significantly different. The nitrated species of Sample I appear to be heavily influenced by atmospheric nitration chemistry and is labeled as the photochemical sample in Figure 1 & 2. The other air sample (II) appeared to be influenced more by automotive emissions. Despite low ambient temperatures and limited sunlight, detection of specific $NO_2$-AR/PAH isomers indicated that the atmospheric nitration reactions which had occurred produced higher ambient levels of $NO_2$-AR/PAH than levels produced from the primary emission sources. These data also suggest empirically that $NO_2$-AR may serve as indicators of direct emissions and photochemical products. Several direct nitration products, including 1-$NO_2$-naphthalene, 2-$NO_2$-biphenyl and 9-$NO_2$-anthracene, were more abundant in the filter extract than in the XAD-2 extract of Sample II. Additionally, for 1-$NO_2$-naphthalene and 2-$NO_2$-biphenyl, the Sample II filter concentrations were greater than the photochemical Sample I filter concentrations by a factor of 3. In contrast, photochemical products, including 3-$NO_2$-toluene, 2-$NO_2$-naphthalene and 3-$NO_2$-biphenyl (23), were more abundant here in XAD-2 extracts than in filter extracts, with the levels in the Sample I XAD-2 extract approximately 5 times greater than the levels in the Sample II XAD-2 extract (14). The semi-volatile direct (electrophilic) nitration products are presumed to be formed after combustion, but before extensive dilution; these products condense onto particles during dilution/cooling (12). In contrast, the semi-volatile photochemical products are primarily gas-phase species and are collected on XAD-2. The phase distribution of $NO_2$-AR/PAH discussed for Samples I and II here are consistent with these mechanisms proposed from observations in chambers following atmospheric reaction.

The $NO_2$-toluene isomers (melting points of -4 to 52° C) appear to be sufficiently volatile as to be collected exclusively on XAD-2. Although the photochemically-produced isomer (3-$NO_2$-toluene) was the most abundant isomer in both XAD-2 extracts, the favored direct nitration product, 4-$NO_2$-toluene, was more abundant relative to 3-$NO_2$-toluene in Sample II compared with photochemical Sample I. The structural similarity of toluene and biphenyl is reflected in the concentration profile for XAD-2 collected isomers; the ratios between 2-, 3, and 4-$NO_2$-toluene and biphenyl isomers were 48:100:49 and 42:100:50, respectively, for the XAD-2 extract of the photochemical Sample I.

Several unexpected results from these analyses include detection of a $NO_2$-azulene isomer and detection of high levels of 9-$NO_2$-anthracene in the XAD-2 extract of Sample I (14). As indicated in Figure 1, a structural isomer of 1- and 2-$NO_2$-naphthalene was detected at substantial levels. Since 1- and 2-$NO_2$-naphthalene are the only possible mono-nitrated naphthalene isomers, we conclude that this component must be a $NO_2$-azulene isomer. Its presence in only the XAD-2 extract (when $NO_2$-naphthalenes were found on both filter and XAD-2) strongly suggests atmospheric formation. This isomer may have gone undetected in other ambient air studies because analyses have focused on the particle extracts. However, this isomer may be unique to these samples and/or these atmospheric conditions, as it does not appear in the chromatograms from analyses of polyurethane foam (PUF) plugs used for collection of ambient vapor phase organics during a high $NO_x$ episode in the Los Angeles area (12).

Substantial levels of 9-$NO_2$-anthracene in both XAD-2 and filter extracts of Sample I represents significantly different findings from the relatively low levels, found only on the filter, for a daytime sample collected in the Los Angeles area (12). While chamber studies have shown that 9-$NO_2$-anthracene can be created from anthracene as a sampling artifact at high (10 ppm) $NO_2$ levels (23), data here do not suggest artifact formation. This conclusion is drawn because the level of 9-$NO_2$-anthracene is much lower in Sample II where $NO_x$ and HONO were present at higher levels. Rather, the data here suggest that additional atmospheric mechanisms are responsible for formation of 9-$NO_2$-anthracene, and that this product may be distributed between gas and particle phases based on vapor pressure.

$NO_2$-fluoranthene/pyrene isomers are detected almost exclusively in the filter extracts, and concentrations of these species are shown in Table 1. As found in other ambient air particulate extracts (12,28,29), 2-$NO_2$-fluoranthene was the most abundant $NO_2$-fluoranthene/pyrene isomer in the Boise, ID air samples (14). The 3-$NO_2$-fluoranthene isomer was detected at very low levels in the filter extract of Sample II. Other $NO_2$-fluoranthene/pyrene isomers, including 4-$NO_2$-pyrene, x-$NO_2$-acephenanthrylene (30), and 2-$NO_2$-pyrene were tentatively identified in these air samples.


HYDROXYLATED NITROARENES

Recent studies of the gas phase reactivities of naphthalene and biphenyl with OH in the presence of $NO_x$ have shown that formation of hydroxy-AR (OH-AR) and $NO_2$-AR accounts for relatively little of the total products formed (26). Earlier smog chamber studies showed that while $NO_2$-toluene was relatively non-reactive, the OH-toluenes (cresols) reacted readily with $NO_2$, following either H atom abstraction from the OH or OH addition to the ring, to produce ortho and para substituted OH-NO$_2$-toluenes (NO$_2$-cresols or CH$_3$-NO$_2$-phenols) (31,32). These results would suggest that in both smog chambers and ambient air, OH-AR/PAH that are formed initially from AR may react further, and readily, to form OH-NO$_2$-AR. Due to the extreme complexity of ambient air particulate matter and the general difficulty in analyzing polar compounds by gas chromatography/mass spectrometry (GC/MS) methods, OH-NO$_2$-AR/PAH have only recently been detected and quantified in an ambient air sample (10). As with NO$_2$-AR/PAH, the specific OH-NO$_2$-AR/PAH isomers detected and their distribution between gas and particle phases may be used empirically to suggest differences due to primary emissions and atmospheric reactions.

Based on our understanding of the differences in these two Boise, ID air samples from NO$_x$, HONO and NO$_2$-AR/PAH data, the differences in OH-NO$_2$-AR isomers detected and their phase distribution suggest intriguing, yet empirical, observations on the role of photochemistry as a source for these compounds in the ambient air. The filter and XAD-2 concentrations of the OH-NO$_2$-AR detected in these two samples are shown graphically in Figure 2. As seen clearly there, Sample II is dominated by para-substituted OH-NO$_2$-AR that are found almost exclusively in the filter extract. In contrast, Sample I (high HONO photochemicals sample) contains para-substituted OH-NO$_2$-AR which are collected not only on the filter but on the XAD-2 as well. The ambient concentration of each filter-collected para-substituted OH-NO$_2$-AR of Sample I is approximately half the concentration of that isomer in Sample II, which also suggests that Sample I has more atmospheric reaction products and that Sample II has more direct combustion emission products.

In addition to filter- and XAD-2-collected para OH-NO$_2$-AR, and in contrast to Sample II, Sample I contains relatively high levels of ortho substituted OH-NO$_2$-AR in the XAD-2 extract. In addition to the species shown in Figure 2, the level of 4-OH-3-NO$_2$-biphenyl (or other isomer which may co-elute chromatographically) was quantified in these extracts. Because only one OH-NO$_2$-biphenyl isomer was available for confirmation of identification, specific observations based on ortho/para differences cannot be made at this time. However, concentrations of ortho substituted 4-OH-3-NO$_2$-biphenyl were similar to other ortho OH-NO$_2$-AR in that the XAD-2 level of Sample I (0.07 ng/m$^3$) was significantly greater than either the filter extract level of that sample (0.006 ng/m$^3$) or the XAD-2 extract level of Sample II (0.002 ng/m$^3$).

The filter-collected para-substituted OH-NO$_2$-AR in these samples are proposed to be due to automotive emissions and the XAD-collected ortho and para-substituted OH-NO$_2$-AR due to atmospheric photochemical reactions. A major assumption made here is that the distribution of a component between filter and XAD-2 is not significantly altered by volatilization from the filter during sampling. Given that greater percentages of XAD-2 collected OH-NO$_2$-AR were collected on the colder days (Sample I), we assume that for these two samples redistribution was not a significant factor. Studies are needed to establish the vapor pressure of each compound and to compare phase distribution as a function of temperature as has been done for PAH (33).

The levels of the OH-NO$_2$-AR measured here were, in general, higher than the levels of the NO$_2$-AR/PAH by at least a factor of 4. If similar differences are found for other ambient air samples, these OH-NO$_2$-AR may be used in addition to relatively low level NO$_2$-AR/PAH to assess relative contribution of photochemical reactions to the composition of nitrated species in ambient air. Clearly, additional field measurements and atmospheric exposure chamber studies are needed to elucidate the formation mechanisms for these OH-NO$_2$-AR.

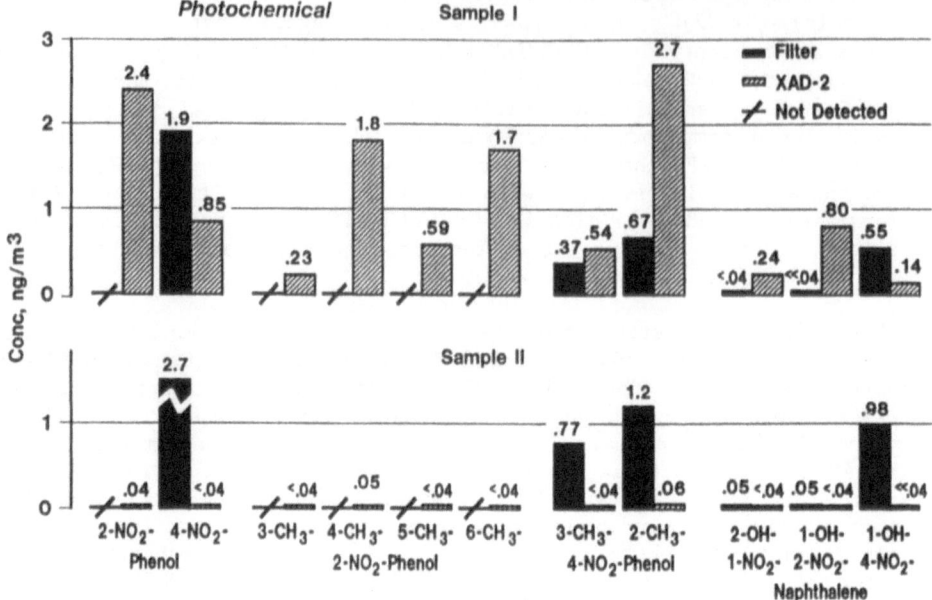

Figure 2. Comparison of winter daytime ambient air OH–NO$_2$-Arenes in two samples collected in Boise, ID under different atmospheric conditions (14). Sample I is labeled photochemical due to the higher HONO concentrations during the sampling period as compared to Sample II.

Our recent bioassay-directed fractionation studies have shown that ambient air particulate extract fractions containing OH-NO$_2$-PAH have a greater percentage of mutagenicity than fractions containing NO$_2$-PAH (10,34,35). These studies also suggest that tentatively identified OH-NO$_2$-PAH and OH-NO$_2$-azaarenes may contribute to the mutagenicity of ambient air particulate matter. The mutagenic activities of NO$_2$-PAH are now well-documented (9,11), as are the activities of the OH-NO$_2$-pyrene isomers (36). Our most recent findings indicate that many diverse OH-NO$_2$-substituted compounds may be mutagenic. Mutagenic OH-NO$_2$-AR include 3-NO$_2$-phenol, 6-CH$_3$-2-NO$_2$-phenol, 6-CH$_3$-3-NO$_2$-phenol, 2-CH$_3$-4-NO$_2$-phenol, and 2-OH-1-NO$_2$-naphthalene. These compounds, with the exception of 6-CH$_3$-3-NO$_2$-phenol, were detected in the ambient air samples analyzed here and were quantified at levels greater than corresponding NO$_2$-AR.

Conclusions

Mutagenicity assays used in bioassay-directed chemical characterization studies of environmental samples led to the identification of nitroarenes as an important class of environmental mutagens in the 1980's. Attempts to characterize the polar neutral and acidic mutagens in air pollution using new advances in bioassay-directed analysis led to the identification of hydroxylated nitroarenes in ambient air. Current research to identify the sources of airborne mutagens and carcinogens in the U.S. EPA's Integrated Air Cancer Project (37) are focusing on characterizing the influence of atmospheric reactions and specific emission sources on mutagenic and carcinogenic compounds in the air. These studies include simultaneous sampling of particle and vapor components of air at the emission source, in the ambient air indoors and outdoors under different atmospheric conditions to provide samples such as those described here for Boise, ID. The results of these studies suggest that the specific nitrated and hydroxylated isomers present in the air reflect the different mechanisms of formation and thereby the source of the nitroarene. Direct combustion emissions appear to emit nitroarene isomers formed by direct electrophilic nitration (e.g. 1-nitropyrene, 3-nitrofluoranthene) whereas atmospheric photochemical reactions involve multistep reactions with OH radicals in the presence of $NO_x$ (e.g. as indicated by HONO) resulting in the formation of different nitroarene isomers (e.g. 2-nitrofluoranthene, 2-nitropyrene). Formation mechanisms of hydroxylated nitroarenes may also differ based on evidence for different distributions of ortho and para substituted $OH-NO_2-AR$ under different atmospheric conditions. Further research is needed to understand the formation, occurence, mutagenicity and cancer risk of the hydroxylated nitroarenes.

## Acknowledgements

This article has been cleared for publication through the Health Effects Research Laboratory, U.S. EPA technical peer review and does not necessarily reflect policies of the Agency. Mention of trade names or commercial products does not constitute endorsement or recommendation for use.

## REFERENCES

1)  G. Lofroth, E. Hefner, I. Alfheim, and M. Moller, Mutagenic activity in photocopies, Science 209:1037-1039 (1980).

2)  H.S. Rosenkranz, E.C. McCoy, D.R. Sanders, M. Butler, D.K. Kiriazides, and R. Mermelstein, Nitropyrenes: Isolation, identification, and reduction of mutagenic impurities in carbon black and toners, Science, 209:1039-1043 (1980).

3)  D. Schuetzle, F.S.C. Lee, T.J. Prater, and S.B. Tejada, The identification of polynuclear aromatic hydrocarbon (PAH) derivatives in mutagenic fractions of diesel particulate extracts, Int. J. Environ. Anal. Chem. 9:93-144 (1981).

4)  J. Lewtas, Genotoxicity of complex mixtures: Strategies for the identification and comparative assessment of airborne mutagens and carcinogens from combustion sources, Fundamentals and Applied Tox. 10:571-589 (1988).

5)  T. Tokiwa, R. Nakagawa, and K. Horikawa, Mutagenic/carcinogenic agents in indoor pollutants: The dinitropyrenes generated by kerosene heaters and fuel gas and liquid petroleum gas burners. Mutat. Res. 157:39-47 (1985).

6)  G.W. Traynor, M.G. Apte, H.A. Sokol, J.C. Chuang, and J.L. Mumford, Selected organic pollutant emissions from unvented kerosene heaters in: "Proceedings of the 79th Air Pollution Control Association Annual Meeting," Minneapolis (1986).

7)  L.D. Claxton, Review of fractionation and bioassay characterization techniques for the evaluation of organics associated with ambient air particles. in: "Symposium on Genotoxic Effects of Airborne Agents," R.R. Tice, D.L. Costa, and K.M. Schaich, Eds., Plenum, New York (1982).

8)  D. Schuetzle and J. Lewtas, Bioassay-directed chemical analysis in environmental research. Anal. Chem. 58:1060A-1075A (1986).

9)  H.S. Rosenkranz and R. Mermelstein, Mutagenicity and genotoxicity of

nitroarenes: All nitro-containing chemicals were not created equal. Mutat. Res. 114:217-267 (1983).

10) M.G. Nishioka, C.C. Howard, D.A. Contos, L.M. Ball, J. Lewtas, Detection of hydroxylated nitro aromatic and hydroxylated nitro polycyclic aromatic compounds in an ambient air particulate extract using bioassay-directed fractionation, Environ. Sci. Technol. 22:908-915 (1988) and 23:248 (1989).

11) IARC, "Diesel and gasoline engine exhausts and some nitroarenes" in: "IARC Monograph on the Evaluation of Carcinogenic Risks to Humans", 46:189-387, IARC, Lyon, France (1989).

12) J. Arey, B. Zielinska, R. Atkinson, A.M. Winer, Polycyclic aromatic hydrocarbon and nitroarene concentrations in ambient air during a wintertime high-no$_x$ episode in the Los Angeles basin, Atmos. Environ. 21:1437-1444 (1987).

13) H. Tokiwa and Y. Ohnishi, Mutagenicity and carcinogenicity of nitroarenes and their sources in the environment, CRC Crit. Rev. Toxicology, 17:23-60 (1986).

14) M.G. Nishioka and J. Lewtas, Vapor and particle phase distribution of nitro- and hydroxylated nitro-aromatic/polycyclic aromatic hydrocarbons in ambient air during daytime winter samples, Atmospheric Environ. (submitted).

15) D. Schuetzle, Sampling of vehicle emissions for chemical analysis and biological testing, Environ. Health Perspect. 47:65-80 (1983).

16) K. Tanabe, H. Matsushita, C.-T. Kuo, S. Imamiya, Determination of carcinogenic nitroarenes in airborne particulates by high performance liquid chromatography (Jpn.). Taiki Osen Gakkaishi (J. Jpn. Soc. Air Pollut.), 21:535-544 (1986).

17) D. Schuetzle, J.M. Perez, Factors influencing the emissions of nitrated-polynuclear aromatic hydrocarbons (nitro-PAH) from diesel engines. J. Air Pollut. Control Assoc., 33:751-755, (1983).

18) J.C. Chuang, G.A. Mack, M.R. Kuhlman, N.K. Wilson, Polycyclic aromatic hydrocarbons and their derivatives in indoor and outdoor air in an eight-home study, Atmospheric Environ. (submitted).

19) M.G. Nishioka, B.A. Petersen, J. Lewtas, Comparison of nitro-aromatic content and direct-acting mutagenicity of diesel emissions, in: "Polynuclear Aromatic Hydrocarbons: Physical and Biological Chemistry" M. Cooke, A.J. Dennis, G.L. Fisher, eds. Battelle Press, Columbus, Ohio (1982).

20) T. L. Gibson, Nitro derivatives of polynuclear aromatic hydrocarbons in airborne and source particulate matter, Atmos. Environ. 16:2037-2040 (1982).

21) T. L. Gibson, Sources of direct-acting nitroarene mutagens in airborne particulate matter, Mutat. Res. 122:115-121 (1983).

22) M.C. Paputa-Peck, R.S. Marano, D. Scheutzle, T.L. Riley, C.V. Hampton, T.J. Prater, L.M. Skewes, T.E. Jensen, P.H. Ruehle, L.C. Bosch, W.P. Duncan, Determination of nitrated polynuclear aromatic hydrocarbons in particulate extracts by capillary column gas chromatography with nitrogen selective detection, Anal. Chem. 55:1946-1954 (1983).

23) J. Arey, B. Zielinska, R. Atkinson, S.A. Aschmann, Nitroarene products from the gas-phase reactions of volatile polycyclic aromatic hydrocarbons with the OH radical and $N_2O_5$, International J Chem Kinetics 21:775-799 (1989).

24) J.A. Sweetman, B. Zielinska, R. Atkinson, T. Ramdahl, A.M. Winer, J.N. Pitts, Jr., A possible formation pathway for the 2-nitrofluoranthene observed in ambient particulate organic matter, Atmos. Environ. 20:235-238 (1986).

25) J. Arey, B. Zielinska, R. Atkinson, A.M. Winer, T. Ramdahl, J.N. Pitts, Jr., The formation of nitro-PAH from the gas-phase reactions of fluoranthene and pyrene with the OH radical in the presence of $NO_x$, Atmos. Environ. 20:2339-2345 (1986).

26) R. Atkinson, J. Arey, B. Zielinska, S.A. Aschmann, Kinetics and products of the gas-phase reactions of OH radicals and $N_2O_5$ with naphthalene and biphenyl, Environ. Sci. Technol. 21:1014-1022 (1987).

27) B. Zielinska, J. Arey, R. Atkinson, T. Ramdahl, A.M. Winer, J.N. Pitts, Jr., reaction of dinitrogen pentoxide with fluoranthene, J. Am. Chem. Soc. 108:4126-4132 (1986).

28) J.N. Pitts, Jr., J.A. Sweetman, B. Zielinska, A.M. Winer, R. Atkinson, Determination of 2-nitrofluoranthene and 2-nitropyrene in ambient particulate organic matter: Evidence for atmospheric reactions, Atmos. Environ. 19:1601-1608 (1985).

29) T. Ramdahl, B. Zielinska, J. Arey, R. Atkinson, A.M. Winer, J.N. Pitts, Jr., Ubiquitous occurrence of 2-nitrofluoranthene and 2-nitropyrene in air, Nature 321:425-428 (1986).

30) B. Zielinska, J. Arey, R. Atkinson, P.A. McElroy, Nitration of acephenanthrylene under simulated atmospheric conditions and in solution and the presence of nitroacephenanthrylene(s) in ambient particles, Environ. Sci. Technol. 22:1044-1048 (1988).

31) R. Atkinson, W.P.L. Carter, K.R. Darnal, A.M. Winer, J.N. Pitts, Jr., Smog chamber and modeling study of the gas phase $NO_x$-Air photooxidation of toluene and the cresols, International J Chem Kinetics 12:779-836 (1980).

32) D. Grosjean, Atmospheric reactions of ortho cresol: Gas phase and aerosol products, Atmos. Environ. 18:1641-1652 (1984).

33) R.W. Coutant, L. Brown, J.C. Chuang, R.M. Riggin, R.G. Lewis, Phase distribution and artifact formation in ambient air sampling for polynuclear aromatic hydrocarbons, Atmos. Environ. 22:403-409 (1988).

34) J. Lewtas, C.C. Chuang, M.G. Nishioka, B.A. Petersen, Fractionation of gram quantities of the organic extract of NIST SRM 1649 urban air particulate matter for bioassay-directed fractionation, in press, Int. J. Environ. Anal. Chem. (1990).

35) M.G. Nishioka, C.C. Chuang, B.A. Petersen, A. Austin, J. Lewtas, Development and quantitative evaluation of a compound class fractionation scheme for bioassay-directed characterization of ambient air particulate matter, Environ. Int. 11:137-146 (1985).

36)  L.M. Ball, M.J. Kohan, L.D. Claxton, J. Lewtas, Mutagenicity of
     derivatives and metabolites of 1-nitropyrene: Activation by rat liver
     S9 and bacterial enzymes.  Mutat. Res. 138:113-125 (1984).

37)  J. Lewtas, Emerging methodologies for assessment of complex mixtures:
     Application of bioassays in the Integrated Air Cancer Project,  J.
     Toxicol. Industrial Health, 5:839-850 (1989).

# THE ATMOSPHERIC FORMATION OF NITROARENES

# AND THEIR OCCURRENCE IN AMBIENT AIR

Barbara Zielinska,[*] Janet Arey and Roger Atkinson

Statewide Air Pollution Research Center
University of California
Riverside, CA 92521

## INTRODUCTION

Over the last six years research has been carried out at the Statewide Air Pollution Research Center concerning the gas-phase atmospheric reactions of polycyclic aromatic hydrocarbons (PAH) and their nitro-derivatives, as well as on the ambient concentrations of the PAH and nitroarenes. This research has involved kinetic and product studies of their atmospherically important gas-phase reactions,[1-17] analytical methods development and field studies to determine their ambient concentrations,[18-29] and studies of the synthesis, mutagenic activities and metabolism of nitroarenes.[30-34] We present here a brief overview of this research as it relates to the atmospheric formation of nitroarenes and their occurrence in ambient air.

PAH are emitted into the atmosphere from combustion sources.[35] They are distributed between the gas- and particle-phases in the atmosphere, with the more volatile two to four-ring PAH being at least partially present in the gas phase.[36,37] In the gas phase, the potential atmospheric loss processes of the PAH and nitroarenes are photolysis and reactions with OH and $NO_3$ radicals, $N_2O_5$ and ozone.[38] Under laboratory conditions the gas-phase OH radical (in the presence of $NO_x$) and $N_2O_5$ reactions of PAH lead to the formation of nitroarenes.[3,4,6-8,11-13,17]

Hydroxyl radicals are present worldwide during daylight hours, and their annually and diurnally averaged tropospheric concentration for a 12-hr daylight period has been estimated[39] to be $1.5 \times 10^6$ molecule $cm^{-3}$.

---

[*]Present affiliation:  Energy and Environmental Engineering Center, Desert Research Institute, P. O. Box 60220, Reno, Nevada  89506.  Please address correspondence regarding this work to J. Arey or R. Atkinson.

Dinitrogen pentoxide will be present at significant concentrations only in certain polluted nighttime atmospheres, and concentrations as high as ~3 x $10^{11}$ molecule $cm^{-3}$ have been calculated[40] from observed $NO_2$ and $NO_3$ radical concentrations and the equilibrium constant for the $NO_2$ + $NO_3$ $\rightleftarrows$ $N_2O_5$ reactions. The atmospheric nitroarene formation rate will depend on the concentration of the individual reactive species (i.e., the PAH and OH radicals or $N_2O_5$), the rate constants for reaction with the PAH, and the nitroarene yields from these reactions. We summarize our laboratory data on nitroarene formation from eleven PAH, and compare the expected nitroarene products with those observed in ambient air samples collected throughout California under various ambient pollutant conditions.


EXPERIMENTAL

Nitroarene product studies of the reactions of 2 to 4-ring PAH with OH and $N_2O_5$ have been carried out in our laboratories.[3,4,6-8,11-13,17] Generally, the experiments were carried out in a 6400 L collapsible all-Teflon chamber. Products were collected from the chamber onto polyurethane foam (PUF) plugs or Tenax-GC cartridges and analyzed by gas chromatography/ mass spectrometry (GC/MS) after appropriate extraction and high performance liquid chromatography (HPLC) steps.

Ambient air samples at a number of sites in California were collected onto Teflon-impregnated glass fiber filters using standard high-volume samplers and onto PUF plugs located downstream of the filters. The filter and PUF plug extracts were fractionated by HPLC and analyzed by GC/MS as detailed elsewhere.[11,12,22,26,28,29,34] Ambient air particulate from Washington, D.C., St. Louis, MO, and Norway were also analyzed for nitrofluoranthenes and nitropyrenes.[20]


RESULTS AND DISCUSSION

The atmospheric lifetimes for eleven PAH with respect to gas-phase reaction with OH and $NO_3$ radicals, $O_3$ and $N_2O_5$ have been calculated from the rate constants we measured (or in some cases estimated) and the estimated tropospheric concentrations of the reactant species (see Table 1; structures are shown in Table 2). It is clear from this table that the most significant loss process for most of the PAH listed is by reaction with the OH radical. Therefore, the nitroarenes produced from the OH radical reactions of these PAH are expected to be formed under ambient conditions. The low concentration of $NO_2$ in ambient air relative to our laboratory experiments should not affect the nitroarene formation, as discussed in detail elsewhere.[13,14]

Table 1. Calculated Atmospheric Lifetimes of the PAH Due to Gas-Phase Reaction with OH and NO$_3$ Radicals, N$_2$O$_5$ and O$_3$ (Taken from Arey et al.[13])

| PAH | Lifetime Due to Reaction with | | | |
|-----|------|--------|---------|------|
| | OH[a] | NO$_3$[b] | N$_2$O$_5$[c] | O$_3$[d] |
| Naphthalene | 8.6 hr | e | 83 days | >80 days |
| 1-Methylnaphthalene | 3.5 hr | e | 35 days | >125 days |
| 2-Methylnaphthalene | 3.6 hr | e | 28 days | >40 days |
| Acenaphthylene | 1.7 hr | 13 min | e | ~43 min |
| Acenaphthene | 1.8 hr | 2.5 hr | 21 days | >30 days |
| Biphenyl | 2.1 days | e | >16 yr | >80 days |
| Phenanthrene | 6.0 hr | e | f | h |
| Anthracene | 1.4 hr | e | f | h |
| Fluoranthene | 3.7 hr[g] | e | 64 days[g] | h |
| Pyrene | 3.7 hr[g] | e | 21 days[g] | h |
| Acephenanthrylene | i | i | i | i |

[a]For 12-hr daytime OH radical concentration of 1.5 x 10$^6$ molecule cm$^{-3}$ (from Prinn et al.[39]).
[b]For 12-hr nighttime NO$_3$ radical concentration of 2.4 x 10$^8$ molecule cm$^{-3}$ (from Atkinson et al.[40]).
[c]For 12-hr nighttime N$_2$O$_5$ concentration of 2 x 10$^{10}$ molecule cm$^{-3}$ (from Atkinson et al.[40]).
[d]For 24-hr average O$_3$ concentration of 7 x 10$^{11}$ molecule cm$^{-3}$ (from Logan[41]).
[e]No reaction observed.
[f]No value, naphthalene rate constant used as a lower limit for product yield estimates as discussed in Arey et al.[13]
[g]Estimated as described in Atkinson et al.[17]
[h]No reaction expected.
[i]Expected to react with rate constants similar to those of acenaphthylene.

Exceptions to this domination of the OH radical reactions for atmospheric removal of the PAH occur for the three PAH which contain an external cyclopenta-fused ring, i.e., acenaphthene, acenaphthylene and acephenanthrylene. For these PAH, reaction with the NO$_3$ radical may be an important loss process, and for the latter two PAH reaction with O$_3$ may also be important. During a field study in Glendora, CA, we observed diurnal variations in the ambient concentrations of the more volatile PAH (molecular weights 128-178) which were consistent with their lifetimes due to OH radical reaction being as shown in Table 1.[29] Continuing to compare predictions from laboratory data against ambient measurements, we next address the nitroarenes formed in these OH radical and N$_2$O$_5$ reactions.

The specific nitroarenes formed and their product yields from the OH radical-initiated and N$_2$O$_5$ reactions of the PAH we have studied are given in Table 2. For comparison, the nitroarenes formed from electrophilic nitration reactions are also given. 1-Nitropyrene, the electrophilic nitration

Table 2. Summary of the Nitroarenes Produced from the Gas-Phase OH Radical-Initiated and $N_2O_5$ Reactions[13] and Electrophilic Nitration of PAH

| PAH | Structure | Position of Nitration (Yield) in Reaction with | | Position of Electrophilic Nitration [Ref.] |
| --- | --- | --- | --- | --- |
| | | OH | $N_2O_5$ | |
| Naphthalene | | 1-(0.3%); 2-(0.3%) | 1-(17%); 2-(7%) | 1->>2- [42] |
| 1-Methylnaphthalene | | 5->4->6->3->7-->2->>8- Total yield (~0.4%) | 3->5->4->8--6->7-->2- Total yield (~30%) | 4->2->5->8->7->3->6- [43] |
| 2-Methylnaphthalene | | 5->6-7--4--8->>3->1- Total yield (~0.2%) | 4->1--5->8--3->7--6- Total yield (~30%) | 1->>8->4->6->5->3->7- [43] |
| Acenaphthylene | | 4- (2%) | None observed | 1- [44] |
| Acenaphthene | | 5->3->4- Total yield (~0.2%) | 4-(40%); 3-(~2%); 5-(~2%)ᵃ | 3-; 5- [42] |
| Biphenyl | | 3- (5%) | No reaction observed | 2-; 4- [45] |

| PAH | Structure | Position of Nitration (Yield) in Reaction with | | Position of Electrophilic Nitration [Ref.] |
| --- | --- | --- | --- | --- |
| | | OH | $N_2O_5$ | |
| Phenanthrene | | Two isomers (not 9-nitrophenanthrene) Total yield ($\lesssim$0.1%) | Four isomers (including 9-nitrophenanthrene) Total yield (<1%) | 9->3-;2-;1- [42] |
| Anthracene | | 1-; 2- Total yield (~0.2%) | 1-; 2- Total yield (<2%) | 9- [42] |
| Fluoranthene | | 2-(3%);7-(~0.15%);8-(~0.15%) | 2-(~25-30%) | 3->8->7->1- [42] |
| Pyrene | | 2-(~0.35%); 4-(~0.045%) | 4-; 2- Total yield (<1%) | 1- [42] |
| Acephenanthrylene | | Two isomers (not 4- or 5-nitro-) Total yield (~0.1%) | None observed | 4-; 5- [11] |

[a]Concurrent $NO_3$ radical reaction will dominate over $N_2O_5$ reaction in ambient air.

product of pyrene, has been observed in several combustion emissions, including diesel[46] and gasoline-fueled[47] vehicle and coal-fired power plant[48] emissions, and, in general, nitroarenes which are the electrophilic nitration products of the PAH are expected to be those isomers directly emitted into the atmosphere from combustion sources. As discussed below, differences among the isomers observed in ambient air and those formed from electrophilic nitration of the PAH are compelling evidence for the atmospheric formation of certain nitroarenes. Detailed discussions of the nitroarenes predicted from our laboratory data to be formed from atmospheric transformation reactions of the PAH and the nitroarene isomers observed in ambient air, for each of the PAH listed in Tables 1 and 2, have been presented elsewhere.[13,28]

Figure 1 shows the GC/MS multiple ion detection (MID) traces for the molecular ions of the nitronaphthalenes (m/z 173), methylnitronaphthalenes (m/z 187), and nitrobiphenyls and nitroacenaphthenes (m/z 199) in an ambient air sample collected on a PUF plug during daylight hours in Glendora, CA. No significant $N_2O_5$ concentration occurs during daytime[40] and, therefore, only the OH radical-initiated pathway for atmospheric nitroarene formation was considered. As shown in Table 2, 3-nitrobiphenyl is formed with a 5% yield from the OH radical-initiated reaction of biphenyl, and is not the expected electrophilic nitration product. (It should be noted that the PAH listed in Tables 1 and 2 were simultaneously measured and all were present.) Therefore, the presence of 3-nitrobiphenyl was attributed to its atmospheric formation from the OH radical-initiated reaction of biphenyl. The lack of 2- and 4-nitrobiphenyl and nitroacenaphthenes suggests that any contribution from combustion emissions was small in comparison with atmospheric formation. Further, the lower observed ambient concentration of acenaphthene relative to biphenyl,[29] combined with the significantly lower nitroacenaphthene formation yield (~0.2%) from the OH radical-initiated reaction, is consistent with the absence of any detectable nitroacenaphthene isomers.

In a similar manner, the presence of comparable amounts of 1- and 2-nitronaphthalene[26] is consistent with their atmospheric formation. Indeed, the observed ambient concentrations of 1- and 2-nitronaphthalene and 3-nitrobiphenyl agree remarkably well with those predicted from the concentrations of the parent PAH, the OH radical reaction rate constants, the nitroarene yields and the nitroarene loss processes.[16]

The pattern of methylnitronaphthalenes observed also is consistent with their formation by OH radical-initiated reaction of 1- and 2-methylnaphthalene.[12] For example, note in Figure 1 the presence of 1-methyl-6-nitronaphthalene and 2-methyl-7-nitronaphthalene, the isomers least likely to be

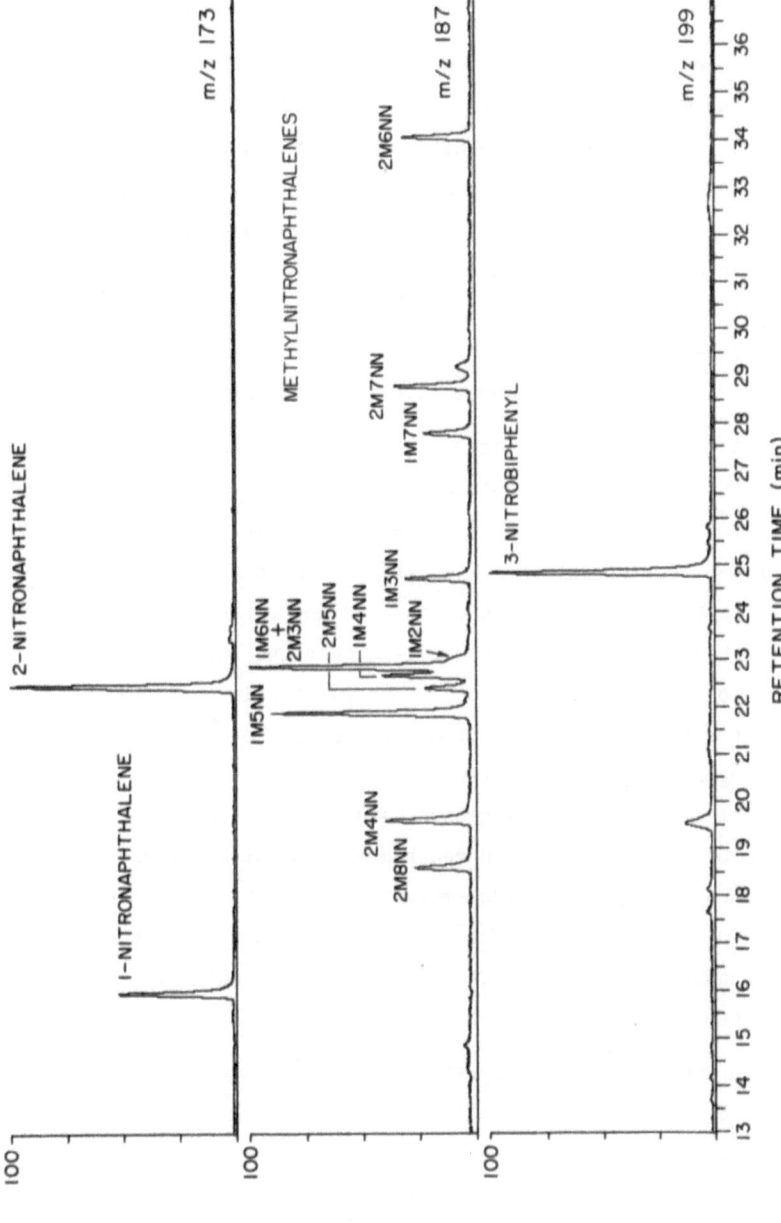

Fig. 1. GC/MS MID traces showing the molecular ions of the nitronaphthalenes (m/z 173), the methylnitronaphthalenes (m/z 187) and the nitrobiphenyls and nitroacenaphthenes (m/z 199) present in an ambient air sample collected on PUF plugs in Glendora, CA, 0800-2000 hr; August 20, 1986.

produced in electrophilic nitration reactions of 1- and 2-methylnaphthalene, respectively.

A final illustration of the importance of the atmospheric formation of nitroarenes is provided by the nitro-isomers of the four-ring PAH observed in ambient air, i.e., fluoranthene, pyrene and acephenanthrylene. Figure 2 shows the GC/MS MID traces of the molecular ions of the molecular weight 247 nitroarene species. The ambient samples shown in Fig. 2A and 2B were collected in Torrance, CA during a high $NO_x$ episode in which $N_2O_5$ would not be present.[22] The major electrophilic nitration products of pyrene and fluoranthene are 1-nitropyrene and 3-nitrofluoranthene, respectively, and we attribute their presence in the ambient samples to direct emission from combustion sources. The remaining nitro-isomers observed and their relative abundances are in agreement with their formation from the OH radical-initiated reaction of the parent PAH. Specifically, we observe 2-nitro-fluoranthene and 2-nitropyrene together with lesser amounts of 7- and 8-nitrofluoranthene, 4-nitropyrene and a nitroacephenanthrylene.[6,8,11,23,28]

The ambient sample shown in Fig. 2C was collected in Claremont, CA during a photochemical pollution episode in which the $N_2O_5$ concentration was estimated to reach ~5 ppb.[28] As may be seen from Table 2, the formation yield ratio of 2-nitrofluoranthene/nitropyrene is much higher for the $N_2O_5$ reactions with fluoranthene and pyrene than for the corresponding OH radical reactions. The high 2-nitrofluoranthene/2-nitropyrene ratio observed in this ambient sample suggests, therefore, that 2-nitrofluoranthene was formed to a significant extent from the reaction of fluoranthene with $N_2O_5$. Although PUF plug samples were not taken in Claremont, the high yields of the nitronaphthalenes and methylnitronaphthalenes from the $N_2O_5$ reactions relative to the OH radical-initiated reactions of naphthalene and the methylnaphthalenes suggest that high ambient concentrations of these nitro-arenes due to $N_2O_5$ reaction should have occurred. Under ambient conditions, reaction of acenaphthene with the $NO_3$ radical will dominate over reaction with $N_2O_5$, and hence the formation of 4-nitroacenaphthene is not expected.[13]

In summary, the atmospheric formation of nitroarenes from the OH radical-initiated reactions of gas-phase PAH has been shown. In certain polluted atmospheres the formation of nitroarenes from the $N_2O_5$ reactions with gas-phase PAH may also occur. Clearly, the formation of nitroarenes during atmospheric transport of the PAH from source to receptor must be taken into account in risk assessments of combustion-emitted PAH.

ACKNOWLEDGEMENTS

This work was supported by the U.S. Environmental Protection Agency (Cooperative Agreement CR-809247-01 and Assistance Agreement No. R-812973-

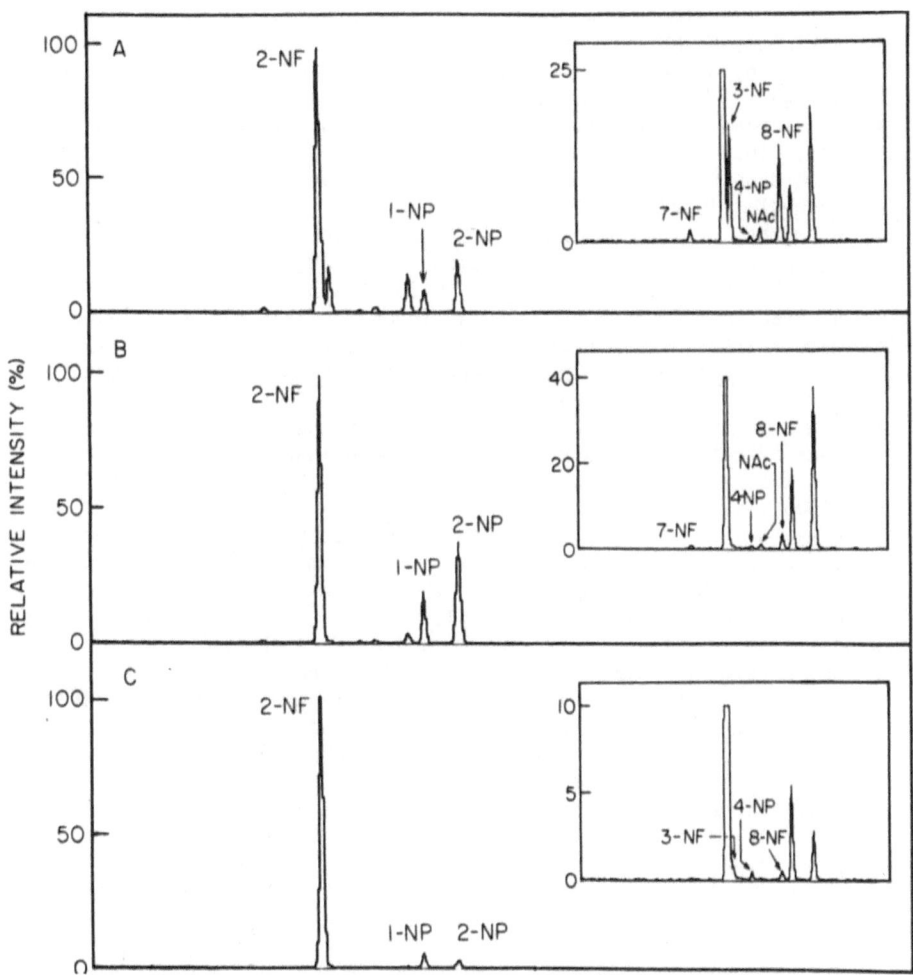

Fig. 2.  GC/MS MID traces of the m/z 247 molecular ions of the nitrofluoranthenes (NF), nitropyrenes (NP) and nitroacephenanthrylenes (NAc)
present in ambient particle samples collected at:  (A) Torrance,
CA, January 27-28, 1986, 1700-0500 hr, (B) Torrance, CA, January
28, 1986, 0500-1700 hr, and (C) Claremont, CA, September 14, 1985,
1800-2400 hr.  GC column and conditions:  60 m DB-5 capillary
column; injection in $CH_2Cl_2$ at 50 °C then programmed at 20 °C $min^{-1}$
to 200 °C, followed by programming at 4 °C $min^{-1}$.  Note scale
expansions on inserts.

01), the California Air Resources Board (Contract Nos. A3-049-32; A4-081-32;
A5-150-32; A5-185-32), the Ford Motor Co. (Contract No. 47-2-IO4475) and the
U.S. Department of Energy (Contract No. DE-ATO3-79EV10048).  The excellent
technical assistance of Ms. Sara M. Aschmann, Mr. Travis M. Dinoff, Ms.
Patricia A. McElroy, Ms. Li Li Parker and Ms. Kathleen Stritzke is
gratefully acknowledged.

# REFERENCES

1. R. Atkinson, S. M. Aschmann, and J. N. Pitts, Jr., Kinetics of the reactions of naphthalene and biphenyl with OH radicals and with $O_3$ at $294 \pm 1$ K, Environ. Sci. Technol., 18:110 (1984).

2. H. W. Biermann, H. Mac Leod, R. Atkinson, A. M. Winer, and J. N. Pitts, Jr., Kinetics of the gas-phase reactions of the hydroxyl radical with naphthalene, phenanthrene, and anthracene, Environ. Sci. Technol., 19:244 (1985).

3. J. N. Pitts, Jr., R. Atkinson, J. A. Sweetman, and B. Zielinska, The gas-phase reactions of naphthalene with $N_2O_5$ to form nitro-naphthalenes, Atmos. Environ., 19:701 (1985).

4. J. A. Sweetman, B. Zielinska, R. Atkinson, T. Ramdahl, A. M. Winer, and J. N. Pitts, Jr., A possible formation pathway for the 2-nitrofluor-anthene observed in ambient particulate organic matter, Atmos. Environ., 20:235 (1986).

5. R. Atkinson and S. M. Aschmann, Kinetics of the reactions of naphtha-lene, 2-methylnaphthalene and 2,3-dimethylnaphthalene with OH radicals and with $O_3$ at $295 \pm 1$ K, Int. J. Chem. Kinet., 18:569 (1986).

6. J. Arey, B. Zielinska, R. Atkinson, A. M. Winer, T. Ramdahl, and J. N. Pitts, Jr., The formation of nitro-PAH from the gas-phase reaction of fluoranthene and pyrene with the OH radical in the presence of $NO_x$, Atmos. Environ., 20:2339 (1986).

7. R. Atkinson, J. Arey, B. Zielinska, and S. M. Aschmann, Kinetics and products of the gas-phase reactions of OH radicals and $N_2O_5$ with naphthalene and biphenyl, Environ. Sci. Technol., 21:1014 (1987).

8. R. Atkinson, J. Arey, B. Zielinska, A. M. Winer, and J. N. Pitts, Jr., The formation of nitropolycyclic hydrocarbons and their contribution to the mutagenicity of ambient air, in: "Short-term Bioassays in the Analysis of Complex Environmental Mixtures V," S. S. Sandhu, D. M. DeMarini, M. J. Mass, M. M. Moore and J. L. Mumford, eds., Plenum Publishing Corporation (1987).

9. R. Atkinson and S. M. Aschmann, Kinetics of the gas-phase reactions of alkylnaphthalenes with $O_3$, $N_2O_5$ and OH radicals at $298 \pm 2$ K, Atmos. Environ., 21:2323 (1987).

10. R. Atkinson and S. M. Aschmann, Kinetics of the reactions of acenaph-thene and acenaphthylene and structurally-related aromatic compounds with OH and $NO_3$ radicals, $N_2O_5$ and $O_3$ at $296 \pm 2$ K, Int. J. Chem. Kinet., 20:513 (1988).

11. B. Zielinska, J. Arey, R. Atkinson, and P. A. McElroy, Nitration of acephenanthrylene under simulated atmospheric conditions and in solution, and the presence of nitroacephenanthrylene(s) in ambient particles, Environ. Sci. Technol., 22:1044 (1988).

12. B. Zielinska, J. Arey, R. Atkinson, and P. A. McElroy, Formation of methylnitronaphthalenes from the gas-phase reactions of 1- and 2-methylnaphthalene with OH radicals and $N_2O_5$ and their occurrence in ambient air, Environ. Sci. Technol., 23:723 (1989).

13. J. Arey, B. Zielinska, R. Atkinson, and S. M. Aschmann, Nitroarene products from the gas-phase reactions of volatile polycyclic aromatic hydrocarbons with the OH radical and $N_2O_5$, Int. J. Chem. Kinet., in press (1989).

14. R. Atkinson, S. M. Aschmann, J. Arey, and W. P. L. Carter, Formation of ring-retaining products from the OH radical-initiated reactions of benzene and toluene, Int. J. Chem. Kinet., in press (1989).

15. R. Atkinson, S. M. Aschmann, J. Arey, B. Zielinska, and D. Schuetzle, Gas-phase atmospheric chemistry of 1- and 2-nitronaphthalene and 1,4-naphthoquinone, Atmos. Environ., in press (1989).

16. J. Arey, R. Atkinson, S. M. Aschmann, and D. Schuetzle, Experimental investigation of the atmospheric chemistry of 2-methyl-1-nitronaph-thalene and a comparison of predicted nitroarene concentrations with ambient air data, Polycycl. Arom. Compounds, in press (1989).

17. R. Atkinson, J. Arey, B. Zielinska, and S. M. Aschmann, Kinetics and nitro-products of the gas-phase reactions of naphthalene-d$_8$, fluoranthene and pyrene with N$_2$O$_5$ and the OH radical, Int. J. Chem. Kinet., to be submitted (1989).

18. J. N. Pitts, Jr., J. A. Sweetman, B. Zielinska, A. M. Winer, and R. Atkinson, Determination of 2-nitrofluoranthene and 2-nitropyrene in ambient particulate organic matter: evidence for atmospheric reactions, Atmos. Environ., 19:1601 (1985).

19. T. Ramdahl, J. A. Sweetman, B. Zielinska, R. Atkinson, A. M. Winer, and J. N. Pitts, Jr., Analysis of mononitro-isomers of fluoranthene and pyrene by high resolution capillary gas chromatography/mass spectrometry, J. High Res. Chromatogr. Chromatogr. Commun., 8:849 (1985).

20. T. Ramdahl, B. Zielinska, J. Arey, R. Atkinson, A. M. Winer, and J. N. Pitts, Jr., Ubiquitous occurrence of 2-nitrofluoranthene and 2-nitropyrene in air, Nature, 321:425 (1986).

21. T. Ramdahl, J. Arey, B. Zielinska, R. Atkinson, and A. M. Winer, Analysis of dinitrofluoranthenes by high resolution capillary gas chromatography/mass spectrometry, J. High Res. Chromatogr. Chromatogr. Commun., 9:515 (1986).

22. J. Arey, B. Zielinska, R. Atkinson, and A. M. Winer, Polycyclic aromatic hydrocarbon and nitroarene concentrations in ambient air during a winter-time high-NO$_x$ episode in the Los Angeles basin. Atmos. Environ., 21:1437 (1987).

23. W. A. Korfmacher, L. G. Rushing, J. Arey, B. Zielinska, and J. N. Pitts, Jr., Identification of mononitropyrenes and mononitrofluoranthenes in air particulate matter via fused silica gas chromatography combined with negative ion atmospheric pressure ionization mass spectrometry, J. High Res. Chromatogr. Chromatogr. Commun., 10:641 (1987).

24. J. Arey and B. Zielinska, High resolution gas chromatography/mass spectrometry analysis of the environmental pollutants methylnitronaphthalenes, J. High Res. Chromatogr. Chromatogr. Commun., 12:101 (1989).

25. J. Arey, B. Zielinska, R. Atkinson, and A. M. Winer, Formation of nitroarenes during ambient high-volume sampling, Environ. Sci. Technol., 22:457 (1988).

26. R. Atkinson, J. Arey, A. M. Winer, and B. Zielinska, A survey of ambient concentrations of selected polycyclic aromatic hydrocarbons (PAH) at various locations in California, Final Report to California Air Resources Board Contract A5-185-32, Sacramento, CA, May (1988).

27. T. Ramdahl, B. Zielinska, J. Arey, and R. W. Kondrat, The electron impact mass spectra of di- and trinitrofluoranthenes, Biomed. Environ. Mass Spectrom., 17:55 (1988).

28. B. Zielinska, J. Arey, R. Atkinson, and A. M. Winer, The nitroarenes of molecular weight 247 in ambient particulate samples collected in southern California, Atmos. Environ., 23:223 (1989).

29. J. Arey, R. Atkinson, B. Zielinska, and P. A. McElroy, Diurnal concentrations of volatile polycyclic aromatic hydrocarbons and nitroarenes during a photochemical air pollution episode in Glendora, California, Environ. Sci. Technol., 23:321 (1989).

30. J. N. Pitts, Jr., B. Zielinska, and W. P. Harger, Isomeric mononitrobenzo[a]pyrenes: synthesis, identification and mutagenic activities, Mutat. Res., 140:81 (1984).

31. B. Zielinska, J. Arey, R. Atkinson, T. Ramdahl, A. M. Winer, and J. N. Pitts, Jr., Reaction of dinitrogen pentoxide with fluoranthene, J. Am. Chem. Soc., 108:4126 (1986).

32. B. Zielinska, W. P. Harger, J. Arey, A. M. Winer, R. A. Haas, and C. V. Hanson, The mutagenicity of 2-nitrofluoranthene and its in vitro hepatic metabolites, Mutat. Res., 190:259 (1987).

33. B. Zielinska, J. Arey, W. P. Harger, and R. W. K. Lee, Mutagenic activities of selected nitrofluoranthene derivatives in Salmonella typhimurium strains TA98, TA98NR and TA98/1,8-DNP$_6$, Mutat. Res., 206:131 (1988).

34. J. Arey, B. Zielinska, W. P. Harger, R. Atkinson, and A. M. Winer, The contribution of nitrofluoranthenes and nitropyrenes to the mutagenic activity of ambient particulate organic matter collected in southern California, Mutat. Res., 207:45 (1988).

35. K. Nikolaou, P. Masclet, and G. Mouvier, Sources and chemical reactivity of polynuclear aromatic hydrocarbons in the atmosphere - a critical review, Sci. Total Environ., 32:103 (1984).

36. R. W. Coutant, L. Brown, J. C. Chuang, R. M. Riggin, and R. G. Lewis, Phase distribution and artifact formation in ambient air sampling for polynuclear aromatic hydrocarbons, Atmos. Environ., 22:403 (1988).

37. M. P. Ligocki and J. F. Pankow, Measurements of the gas/particle distributions of atmospheric organic compounds, Environ. Sci. Technol., 23:75 (1989).

38. R. Atkinson, Atmospheric transformations of automotive emissions, in: "Air Pollution, the Automobile, and Public Health," A. Y. Watson, R. R. Bates, and D. Kennedy, eds., National Academy Press, Washington, D. C. (1988).

39. R. Prinn, D. Cunnold, R. Rasmussen, P. Simmonds, F. Alyea, A. Crawford, P. Fraser, and R. Rosen, Atmospheric trends in methylchloroform and the global average for the hydroxyl radical, Science, 238:945 (1987).

40. R. Atkinson, A. M. Winer, and J. N. Pitts, Jr., Estimation of night-time $N_2O_5$ concentrations from ambient $NO_2$ and $NO_3$ radical concentrations and the role of $N_2O_5$ in night-time chemistry, Atmos. Environ., 20:331 (1986).

41. J. A. Logan, Tropospheric ozone: seasonal behavior, trends, and anthropogenic influence, J. Geophys. Res., 90:10463 (1985).

42. P. H. Ruehle, L. C. Bosch, and W. P. Duncan, Synthesis of nitrated polycyclic aromatic hydrocarbons, in: "Nitrated Polycyclic Aromatic Hydrocarbons," C. M. White, ed., Dr. Alfred Heuthig Verlag, Heidelberg (1985).

43. C. Eaborn, P. Golborn, R. E. Spillett, and R. Taylor, Aromatic reactivity. Part XXXVII. Detritiation of substituted 1- and 2-tritionaphthalenes, J. Chem. Soc. B, 1112 (1968).

44. J. M. Goldring, L. M. Ball, R. Sangaiah, and A. Gold, Synthesis and biological activity of nitro-substituted cyclopenta-fused PAH, in: "Polynuclear Aromatic Hydrocarbons: A Decade of Progress," M. Cooke and A. J. Dennis, eds., Battelle Press, Columbus (1988).

45. J. March, "Advanced Organic Chemistry: Reactions, Mechanisms, and Structure," John Wiley & Sons, New York (1985).

46. M. C. Paputa-Peck, R. S. Marano, D. Schuetzle, T. L. Riley, C. V. Hampton, T. J. Prater, L. M. Skewes, T. E. Jensen, P. H. Ruehle, L. C. Bosch, and W. P. Duncan, Determination of nitrated polynuclear aromatic hydrocarbons in particulate extracts by capillary column gas chromatography with nitrogen selective detection, Anal. Chem., 55:1946 (1983).

47. T. L. Gibson, Sources of direct-acting nitroarene mutagens in airborne particulate matter, Mutat. Res., 122:115 (1983).

48. W. R. Harris, E. K. Chess, D. Okamoto, J. F. Remsen, and D. W. Later, Contribution of nitropyrene to the mutagenic activity of coal fly ash, Environ. Mutag., 6:131 (1984).

METABOLISM OF 1-NITROPYRENE OXIDES AND EFFECT OF NITROGEN DIOXIDE ON ARENE

ACTIVATION

Yoshinari Ohnishi,[1] Takemi Kinouchi,[1] Keiko Nishifuji,[1]
Koichi Miyanishi,[1] Takako Kanoh,[2] and Masao Fukuda[2]

[1]Department of Bacteriology, School of Medicine,
The University of Tokushima, Tokushima 770, Japan
[2]Department of Environmental Health,
Tokyo Metropolitan Research Laboratory of Public Health
Tokyo 160, Japan

INTRODUCTION

Since 1-nitropyrene (1-NP) is one of the most common nitrated poly-cyclic aromatic hydrocarbons (nitro-PAHs) found in the environment it has been studied as a model substance for mutagenic and carcinogenic nitro-PAHs affecting human health (Tokiwa and Ohnishi, 1986). 1-NP is metabo-lized by reductive and oxidative pathways (Djurić et al., 1986). Although a DNA adduct indicative of reductive metabolism, $N$-(deoxyguanosin-8-yl)-1-aminopyrene, has been identified in *Salmonella typhimurium* (Howard et al., 1983), in Chinese hamster ovary cells (Heflich et al., 1985) and in rats and mice (Stanton et al., 1985; Hashimoto and Shudo, 1985; Mitchell, 1988), it has been reported recently to be a minor adduct in rabbits and rats (Gallagher et al., 1988; Roy et al., 1989). Therefore, oxidative or oxidative-reductive metabolic pathways may be quantitatively more im-portant *in vivo*.

In this paper we present data on the formation of nitro-PAHs in mice injected intraperitoneally with PAHs during exposure to $NO_2$ gas, and then discuss our recent findings on metabolism of 1-NP oxides, which are the first active metabolites of 1-NP in the oxidative pathway.

RESULTS AND DISCUSSION

Production of Mutagenic 1-NP Derivatives from In Vivo Interaction of Pyrene with $NO_2$

The β-glucuronidase-treated urine of mice injected intraperitoneally with 800 mg of pyrene per kg of body weight during exposure to 20 ppm of $NO_2$ was found to contain highly mutagenic compounds by means of the Ames test using *Salmonella typhimurium* strain TA98 in the absence and presence of rat liver S-9 fractions. These mutagenic chemicals were 3-hydroxy-1-NP, 6-hydroxy- and/or 8-hydroxy-1-NP and probably 9-hydroxy-1-NP in addi-tion to the weak mutagen, 1-hydroxypyrene (Kanoh et al., 1987; Ohnishi et al., 1989). We have further studied the dose dependence of pyrene and $NO_2$ (Fig. 1). In mice exposed to 5 or 10 ppm $NO_2$, the mutagenicity of the urine increased as the concentration of pyrene was increased to 200 mg.

Table 1.  Mutagenicity of β-Glucuronidase-treated Urine in Mice Injected with Nonmutagenic Arenes during Exposure to $NO_2$ Gas[a]

| Arene | Gas | Mutagenicity of one-hour urine (His+/plate) | | | |
| | | TA98 | | TA100 | |
| | | (−)S-9 | (+)S-9 | (−)S-9 | (+)S-9 |
| None | Air | 0 | 0 | 17 | 0 |
| | $NO_2$ | 0 | 0 | 9 | 0 |
| Pyrene | Air | 79 | 65 | 36 | 11 |
| Pyrene | $NO_2$ | 2,160 | 1,160 | 2,900 | 530 |
| Fluoranthene | Air | 16 | 78 | 110 | 86 |
| Fluoranthene | $NO_2$ | 1,930 | 805 | 6,290 | 1,760 |
| Fluorene | Air | 0 | 34 | 6 | 22 |
| Fluorene | $NO_2$ | 1,530 | 407 | 458 | 2,290 |
| Anthracene | Air | 55 | 68 | 7 | 3 |
| Anthracene | $NO_2$ | 567 | 83 | 1,200 | 1,370 |
| Chrysene[b] | Air | 51 | 94 | 55 | 39 |
| Chrysene | $NO_2$ | 1,880 | 625 | 458 | 145 |

[a]Each mouse was intraperitoneally injected with 400 mg of one of the arenes per kg of body weight after 24-hours-exposure of air or 20 ppm of $NO_2$ and the urine was then collected for 24 hours under exposure of air or 20 ppm of $NO_2$, respectively.
[b]Chrysene itself is mutagenic.

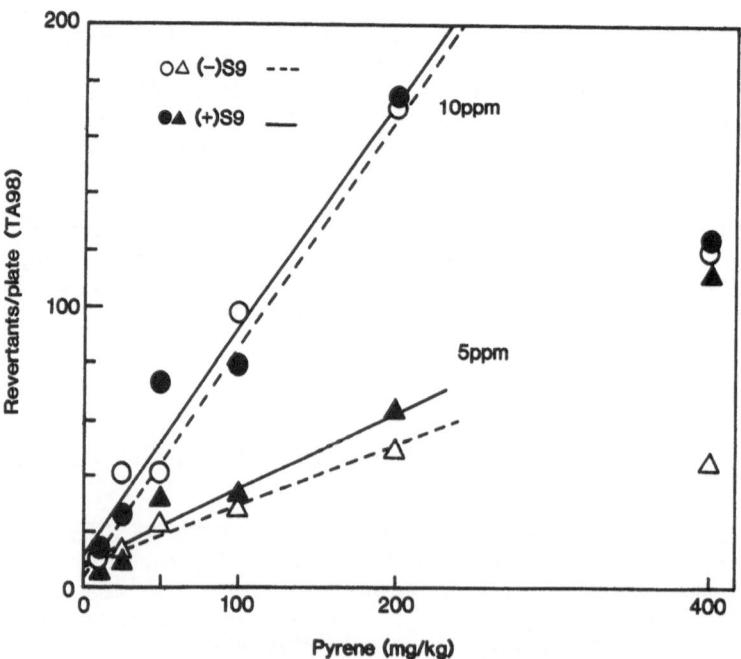

Fig. 1.  Mutagenicity of β-glucuronidase-treated urine from pyrene-injected mice under exposure to $NO_2$.

When ICR mice were intraperitoneally administered 1-hydroxypyrene and then exposed to $NO_2$ gas for 24 hours, the same kinds of hydroxylated 1-NP were detected. Treatment of mice with 400 mg of fluoranthene, fluorene, anthracene, or chrysene per kg during exposure of 20 ppm of $NO_2$ gas induced mutagenicity of their urine when assayed after treatment of β-glucuronidase (Table 1). In the case of the mutagen, chrysene, the mutagenicity was also increased after exposure of $NO_2$. These results suggest that mutagenic nitroarenes are formed from the *in vivo* interaction of nonmutagenic or weakly mutagenic PAHs with $NO_2$.

## Analysis of Oxidative Metabolites in the Bile of Rats Administered 1-NP

Although a glucuronide conjugate of 1-nitro-4,5-dihydro-4,5-dihydroxy-pyrene (1-NP 4,5-dihydrodiol) was excreted into the bile of Wistar rats administered 1-NP, glucuronide conjugates of 1-NP 9,10-dihydrodiol were not detected (Kinouchi et al., 1986). 1-NP 9,10-oxide, as well as 1-NP 4,5-oxide, however, were produced *in vitro* in nearly an equal ratio from 1-NP during the incubations of 1-NP with the rat hepatic microsomes (Ohnishi et al., 1989). Therefore, metabolites of 1-NP 9,10-oxide should be present in the bile (Fig 2). Accordingly, glutathione conjugates of 1-NP 4,5-oxide and 1-NP 9,10-oxide and their metabolites, cysteinylglycine and cysteine conjugates of 1-NP oxides, were detected in the bile of rats given [3H]1-NP by gavage (Table 2). During the 48 hours following [3H]1-NP administration, 21.4% of the biliary metabolites was excreted as glutathione, cysteinylglycine and cysteine conjugates. The amount of conjugates of 1-NP 9,10-oxide was more than that of 1-NP 4,5-oxide. *In vitro* 1-NP 4,5-dihydrodiol was produced 23-fold more than 1-NP 9,10-dihydrodiol from 1-NP oxides by the rat hepatic microsomes. This suggests that epoxide hydrolase in the rat liver has much more affinity for 1-NP 4,5-oxide than 1-NP 9,10-oxide.

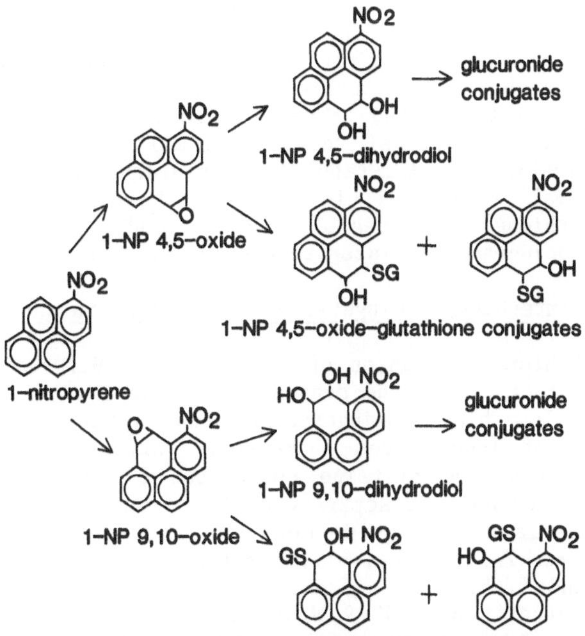

Fig. 2. Formation of glutathione conjugates in 1-nitropyrene metabolism.

Table 2.  Distribution of Radioactivity in the Bile of Rats Administered [$^3$H]1-Nitropyrene

| Peak | Conjugated chemicals | % of biliary dpm (0–48 hr) |
|------|----------------------|----------------------------|
| Solvent front | | 20.2 |
| I | | 3.9 |
| II | 1-NP 9,10-oxide-Cys-Gly | 3.2 |
| III | 1-NP 9,10-oxide-Cys | 1.5 |
| IV | 1-NP 9,10-oxide-Cys & 1-NP 4,5-oxide-Cys-Gly | 1.8 |
| V & VI | 1-NP 9,10-oxide-SG | 10.4 |
| VII | 1-NP 4,5-oxide-Cys | 1.9 |
| VIII & IX | 1-NP 4,5-oxide-SG | 2.6 |
| X | Acetylaminohydroxypyrenes[a] | 6.4 |
| XI | 1-NP 4,5-dihydrodiol[a] | 8.0 |
| XII | 1-Nitro-6/8-hydroxypyrene[a] | 9.5 |
| XIII | 1-Nitro-3-hydroxypyrene[a] | 9.4 |
| XIV | | 1.5 |

[a]Excreted as glucuronide and/or sulfate conjugates.

## Production of Cysteine Conjugates from Glutathione Conjugates of 1-NP Oxides by Treatment with Bile or Pancreatic Juice

To determine why there were cysteinylglycine and cysteine conjugates in the bile, we have measured activity of γ-glutamyltransferase (GGT) and aminopeptidase in the bile of normal rats using γ-glutamyl-p-nitroanilide and L-leucyl-p-nitroanilide, respectively, as substrates (Table 3). These activities were the same as in rats administered 1-NP. Furthermore, bile of normal rats metabolized glutathione conjugates of 1-NP oxides and produced the cysteinylglycine and cysteine conjugates. The enzyme activity of pancreatic juice of normal rats, especially GGT, was markedly high (Table 3), indicating rapid metabolism of glutathione conjugates of 1-NP oxides in the small intestine. In addition, the pancreatic juice metabolized the glutathione conjugates of 1-NP 4,5- and 9,10-oxides to the cysteinylglycine conjugates and later to the cysteine conjugates.

## Metabolism of Glutathione Conjugates of 1-NP Oxides in the Intestine

When glutathione conjugates of 1-NP 4,5- and 9,10-oxides were incubated with S-9 fractions of the contents from the small intestine of normal rats, both glutathione conjugates decreased rapidly and cysteinylglycine conjugates of 1-NP oxides increased. After the cysteinylglycine conjugates decreased and cysteine conjugates of 1-NP oxides increased gradually. The degradation activity of S-9 fractions from small intestine, cecum and large intestine contents showed 3.78, 0.85 and 0.99 nmol/hr/mg of protein, respectively, for the glutathione conjugates of 1-NP 4,5-oxide and 5.45, 1.19 and 1.20 nmol/hr/mg, respectively, for the glutathione conjugates of 1-NP 9,10-oxide. The GGT activity of the S-9 fraction of *Pseudomonas aeruginosa* strain UTC55 was very high (14.4 nmol/hr/mg for glutathione conjugate of 1-NP 4,5-oxide and 1.31 nmol/hr/mg for that of 1-NP 9,10-oxide). Other aerobic bacteria (e.g., *Proteus*

Table 3. Enzyme Activity of Bile and Pancreatic Juice in Normal Rats[a]

| Sample (hour) | Volume (ml) | Flow rate (μl/kg/min) | Protein (mg/ml) | γ-Glutamyltransferase (mU/ml) | Aminopeptidase (mU/ml) |
|---|---|---|---|---|---|
| Bile | | | | | |
| 0-6 | 3.5 | 46.7 | 2.4 | 15.5 | 8.8 |
| 6-12 | 4.4 | 58.2 | 2.8 | 51.1 | 25.3 |
| 12-24 | 4.5 | 29.6 | 2.9 | 72.0 | 42.9 |
| 24-48 | 6.8 | 22.5 | 3.5 | 68.8 | 46.1 |
| Panceatic juice | | | | | |
| 0-6 | 0.7 | 9.2 | 25.9 | 455 | 85.9 |
| 6-12 | 0.8 | 10.6 | 26.5 | 1,190 | 48.0 |
| 12-24 | 2.0 | 13.5 | 19.3 | 1,380 | 69.4 |
| 24-48 | 7.5 | 24.7 | 15.0 | 864 | 60.6 |

[a]One unit of γ-glutamyltransferase and aminopeptidase activity was difined as the formation of 1 μmol of p-nitroaniline per min from γ-glutamyl-p-nitroanilide and L-leucyl-p-nitroanilide, respectively.

mirabilis strain KUB22 and Klebsiella pneumoniae strain KUB13) also showed high GGT activity. However, anaerobic bacteria which are normal flora in the intestine showed low activity, except for Fusobacterium nucleatum strain F-1 (1.25 and 0.46 nmol/hr/mg, respectively, for both glutathione conjugates). The GGT activity of S-9 fractions of mucosa in the small intestine, cecum and large intestine was also low (<0.9 nmol/hr/mg for both glutathione conjugates of 1-NP oxides). In contrast, pancreatic juice showed high GGT activity as described above (6.33 for 1-NP 4,5-oxide-SG and 16.4 nmol/hr/mg for 1-NP 9,10-oxide-SG), indicating that pancreatic juice plays an important role in the small intestine for the metabolism of glutathione conjugates of 1-NP oxides to cysteine conjugates.

Although degradation activity of the S-9 fraction of small intestine contents for cysteine conjugates of 1-NP oxides was very low, degradation activity of the cecum and large intestine was 51.8 and 72.9 nmol/hr/mg of protein, respectively, for cysteine conjugate of 1-NP 4,5-oxide, and 105 and 258 nmol/hr/mg, respectively, for that of 1-NP 9,10-oxide. The cysteine-conjugate β-lyase activity was high in aerobic bacteria, Escherichia coli strain W3110 and Streptococcus faecalis strain UTB74, and in anaerobic bacteria, Peptostreptococcus magnus strain GAI0663 and Eubacterium limosum strian ATCC8480. Generally, the β-lyase activity of anaerobic bacteria was higher than that of the intestinal mucosa, bile and pancreatic juice.

Mutagenicity of Cysteine Conjugates of 1-NP Oxides and their Metabolic Products

Unless the cysteinyl conjugates of 1-NP oxides are absorbed in the small intestine, this must move to the cecum and large intestine where they can be metabolized to the SH compounds by anaerobic bacteria. The metabolites may act on the mucous membrane of the colon or be absorbed and transported to the liver. Otherwise, they must be excreted with intesti-

Table 4. Mutagenicity of Oxidative Metabolites of 1-Nitropyrene

| Chemicals | Dose nmol/plate | Addition | His[+] revertants/plate | |
|---|---|---|---|---|
| | | | TA98 | TA100 |
| 1-NP 4,5-oxide | 1 | | .571 | 1,000 |
| 1-NP 4,5-oxide-SG | 10 | | 34 | 35 |
| 1-NP 4,5-oxide-Cys | 10 | | 398 | 1,380 |
| 1-NP 4,5-oxide-Cys | 10 | E. limosum S-9 | 711 | 1,420 |
| 1-NP 4,5-oxide-Cys | 10 | E. limosum S-9 + AOAA | 110 | 842 |
| 1-NP 4,5-oxide-Cys | 10 | Rat liver S-9 | 456 | 1,060 |
| 1-NP 9,10-oxide | 1 | | 865 | 874 |
| 1-NP 9,10-oxide-SG | 10 | | 46 | 39 |
| 1-NP 9,10-oxide-Cys | 5 | | 41 | 26 |
| 1-NP 9,10-oxide-Cys | 5 | E. limosum S-9 | 123 | 117 |
| 1-NP 9,10-oxide-Cys | 5 | E. limosum S-9 + AOAA | 14 | 45 |
| 1-NP 9,10-oxide-Cys | 5 | Rat liver S-9 | 75 | 27 |

nal bacteria or the other components of feces. Therefore, we have studied mutagenicity of these metabolites of 1-NP (Table 4). 1-NP oxides, which are active oxidative metabolites of 1-NP, showed high mutagenicity for both *Salmonella* strains TA98 and TA100 in the absence of the rat liver S-9 fraction as reported previously (Fifer et al., 1986; Ohnishi et al., 1986). Mutagenicity of glutathione conjugates of these 1-NP oxides was more than 100-fold low when compared with that of nonconjugated original oxides. On the contrary, mutagenicity of cysteine conjugate of 1-NP 4,5-oxide, which was the metabolite from the glutathione conjugate, was more than 10-fold higher than that of the glutathione conjugate, and that of 1-NP 9,10-oxide, 2-fold higher. The presence of the S-9 fraction of *Eubacterium limosum* strain ATCC8480 increased the mutagenicity of the cysteine conjugates of 1-NP oxides, especially 1-NP 9,10-oxide. This increase must be due to cysteine-conjugate β-lyase of *E. limosum* because the presence of aminooxyacetic acid (AOAA), which is a β-lyase inhibitor, inhibited the mutagenicity. Mutagenicity of cysteine conjugate of 1-NP 4,5-oxide in the presence of the S-9 fraction of *E. limosum* was not significantly different from that in the absence of S-9 fraction, partly because *Salmonella typhimurium* itself might have β-lyase activity. These results suggest that mutagenic metabolite(s) can be formed during the process of detoxication and may act on the local intestinal mucosa, or other organs after absorption. Therefore, it is important to know whether metabolites of glutathione conjugates of 1-NP oxides are absorbed in the intestine.

## Absorption of Metabolites of Glutathione Conjugates of 1-NP Oxides in the Intestine

Fig. 3 shows excretion of radioactivity into feces and urine of rats after dosing by gavage with glutathione conjugate of each [$^3$H]1-NP oxide. About 10% of the total dose was excreted in the urine 12 hours after dosing, indicating that glutathione conjugates of 1-NP oxides or their metabolites, probably cyateine conjugates, were absorbed in the intestine. This indicates the enterohepatic circulation of 1-NP metabolites occurs *in vivo*. The radioactivity in the blood increased rapidly after 3 hours lag reached a maximum 9 hours after dosing and then gradually decreased (Fig. 4). Forty-five hours after dosing 0.4 or 0.8% of the total dose remained

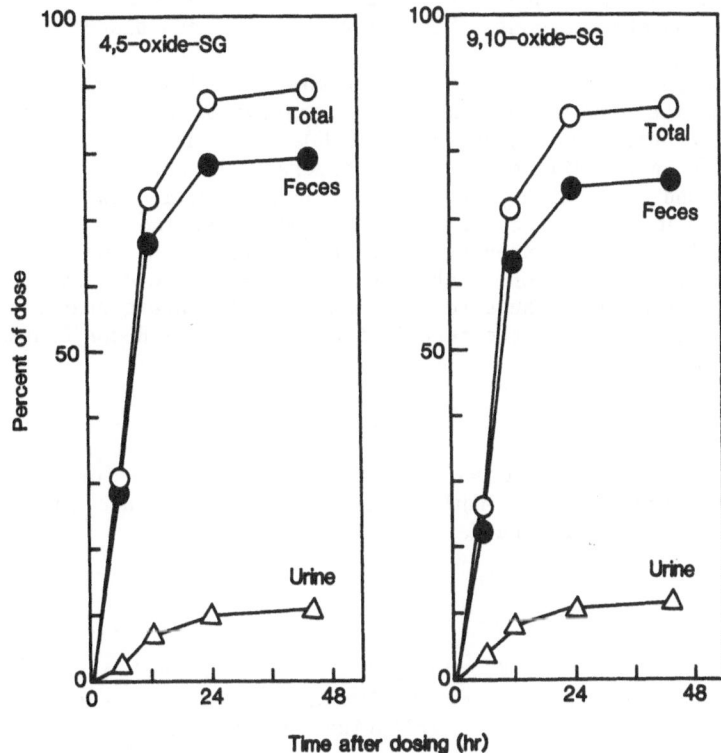

Fig. 3.    Excretion of radioactivity into feces and urine of rats after dosing by gavage with glutathione conjugate of each [³H]1-nitropyrene oxide.

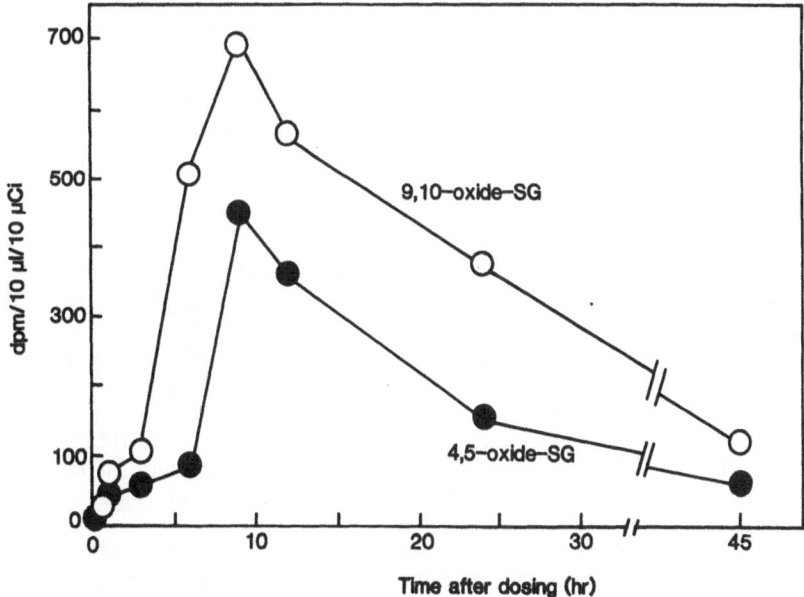

Fig. 4.    Radioactivity in the blood of rats given by gavage glutathione conjugates of [³H]1-nitropyrene oxides.

in the blood.  These results suggest that mutagenic intermediate metabolites from the glutathione conjugates of 1-NP oxides might affect the intestine and other organs, mainly liver, kidney and bladder.

In summary the formation of mutagens, probably nitroarenes, in mice was shown after intraperitoneal administration of various nonmutagenic or weakly mutagenic PAHs and exposure of $NO_2$ gas.  By analyses of oxidative metabolites in the bile of rats administered 1-NP and *in vitro* experiments we have found glutathione conjugates of 1-NP 4,5-oxide and 1-NP 9,10-oxide and their metabolites, cysteinylglycine and cysteine conjugates of 1-NP oxides, which were produced by pancreatic juice in the small intestine and finally absorbed there or metabolized to mutagenic substances in the cecum and large intestine.  These results indicate the enterohepatic circulation of 1-NP metabolites and the reactivation of detoxified conjugates.

## ACKNOWLEDGMENT

This work was supported in part by grants-in-aid for scientific and cancer research from the Ministry of Education, Science and Culture and the Ministry of Health and Welfare of Japan.

## REFERENCES

Djurić, Z., Fifer, E. K., Howard, P. C., and Beland, F. A., 1986, Oxidative microsomal metabolism of 1-nitropyrene and DNA-binding of oxidized metabolites following nitroreduction, Carcinogenesis, 7:1073.

Fifer, E. K., Howard, P. C., Heflich, R. H., and Beland, F. A., 1986, Synthesis and mutagenicity of 1-nitropyrene 4,5-oxide and 1-nitropyrene 9,10-oxide, microsomal metabolites of 1-nitropyrene, Mutagenesis, 6:433.

Gallagher, J. E., Robertson, I. G. C., Jackson, M. A., Dietrich, A. M., Ball, L. M., and Lewtas, J., 1988, [32]P-postlabeling analysis of DNA adducts of two nitrated polycyclic aromatic hydrocarbons in rabbit tracheal epithelial cells, in: "Carcinogenic and Mutagenic Responses to Aromatic Amines and Nitroarenes," C. M. King, L. J. Romano, and D. Schuetzle, eds., Elsevier, New York.

Hashimoto, Y., and Shudo, K., 1985, Modification of nucleic acids with 1-nitropyrene in the rat: identification of the modified nucleic acid base, Jpn. J. Cancer Res. (Gann), 76:253.

Heflich, R. H., Fifer, E. K., Djurić, Z., and Beland, F. A., 1985, DNA adduct formation and mutation induction by nitropyrenes in Salmonella and Chinese hamster ovary cells: relationships with nitroreduction and acetylation, Environ. Health Perspect., 62:135.

Howard, P. C., Heflich, R. H., Evans, F. E., and Beland, F. A., 1983, Formation of DNA adducts *in vitro* and in *Salmonella typhimurium* upon metabolic reduction of the environmental mutagen 1-nitropyrene, Cancer Res., 43:2052.

Kanoh, T., Fukuda, M., Mizoguchi, I., Kinouchi, T., Nishifuji, K., and Ohnishi, Y., 1987, Detection of mutagenic compounds in the urine of mice administered pyrene during exposure to $NO_2$, Jpn. J. Cancer Res. (Gann), 78:1057.

Kinouchi, T., Morotomi, M., Mutai, M., Fifer, E. K., Beland, F. A., and Ohnishi, Y., 1986, Metabolism of 1-nitropyrene in germ-free and conventional rats, Jpn. J. Cancer Res. (Gann), 77:356.

Mitchell, C. E., 1988, Formation of DNA adducts in mouse tissues after intratracheal instillation of 1-nitropyrene, Carcinogenesis, 9:857.

Ohnishi, Y., Kinouchi, T., Nishifuji, K., Fifer, E. K., and Beland, F. A., 1986, Metabolism of mutagenic 1-nitropyrene in rats, in:

"Carcinogenic and Mutagenic Effects of Diesel Engine Exhaust," N. Ishinishi, A. Koizumi, R. O. McClellan, and W. Stöber, eds., Elsevier Science Publishiers, Amsterdam.

Ohnishi, Y., Kinouchi, T., Nishifuji, K., Fifer, E. K., Beland, F. A., Kanoh, T., Fukuda, M., and Mizoguchi, I., 1989, Metabolic formation and inactivation of 1-nitropyrene oxides and the effect of pyrene administration during nitrogen dioxide exposure on the formation of 1-nitropyrene metabolites, in: Proceedings of the Eleventh International Symposium on Polynuclear Aromatic Hydrocarbons, in press.

Roy, A. K., El-Bayoumy, K., and Hecht, S. S., 1989, $^{32}$P-postlabeling analysis of 1-nitropyrene-DNA adducts in female Sprague-Dawley rats, Carcinogenesis, 10:195.

Stanton, C. A., Chow, F. L., Phillips, D. H., Grover, P. L., Garner, R. C., and Martin, C. N., 1985, Evidence for N-(deoxyguanosin-8-yl)-1-aminopyrene as a major DNA adduct in female rats treated with 1-nitropyrene, Carcinogenesis, 6:535.

Tokiwa, H., and Ohnishi, Y., 1986, Mutagenicity and carcinogenicity of nitroarenes and their sources in the environment, CRC Crit. Rev. Toxicol. 17:23.

# MUTAGENIC AND BIOCHEMICAL CONSEQUENCES OF THE REACTION

# OF ARYLAMINES WITH DNA

Charles M. King, Thomas M. Reid, Nobuya Tamura
and Pawan K. Gupta

Department of Chemical Carcinogenesis
Michigan Cancer Foundation
110 East Warren Avenue
Detroit, Michigan

## INTRODUCTION

The carcinogenic activities of the arylamines have been linked to the abilities of the target tissues to transform these agents to derivatives that are capable of reacting with nucleic acid. Thus, the O-esterification of arylhydroxylamines and arylhydroxamic acids has been related to the production of C8 substituted guanine adducts and the subsequent formation of tumors (King, 1985). An important objective of attempts to elucidate the molecular mechanisms involved in tumorigenesis is the exploration of the extent to which these adducts perturb the processing and function of the DNA. The distortion imposed on DNA structure by bulky adducts can be manifested by mutagenic responses and changes in the metabolic disposition of the DNA (Singer and Grunberger, 1983). We have approached these phenomena in two ways by use of extrachromosomal DNA in E. coli cells. C8 adducts of 2-aminofluorene or 4-aminobiphenyl, and their N-acetylated derivatives have been studied (Figure 1). Our objectives have been to develop systems that might be used for comparison of structural differences on the mutagenic effects of site-specific arylamine DNA adducts and the biochemical effects these adducts have on DNA metabolism. Randomly modified DNA has been used in experiments designed to explore the effects of randomly introduced adducts on the ability of these vectors to be replicated, as well as mutagenic consequences. Single, site specific adducts have been employed to determine their mutagenic potentials.

## RANDOM ADDUCTS

### Preparation and characterization

The random introduction of adducts into biologically active DNA is readily accomplished by incubation of the DNA with synthetic, tritiated reactive derivatives that will yield the desired adduct (Figure 2) (Tang et al., 1982; Gupta et al., 1988). By altering the reaction conditions it is possible to obtain DNA samples that differ in the number of adducts per DNA molecule, thus permitting the design of experiments that can relate adduct content to response (Tamura and King, 1990). Highly reactive reagents, such as the N-trifluoroacetyl-N-acetoxy-derivatives of 2-aminofluorene and 4-aminobiphenyl, that react rapidly with DNA to yield the desired adducts also react with the aqueous solvents used for DNA modification

*Nitroarenes*, Edited by P. C. Howard *et al.*
Plenum Press, New York, 1990

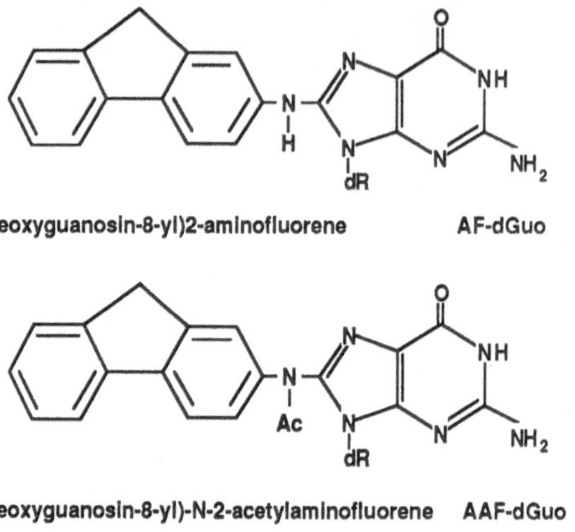

N-(2'-deoxyguanosin-8-yl)2-aminofluorene          AF-dGuo

N-(2'-deoxyguanosin-8-yl)-N-2-acetylaminofluorene    AAF-dGuo

Fig. 1  2-Aminofluorene and 2-acetylaminofluorene
adducts with guanine.

(Lee and King, 1981). The desired levels of DNA modification with these compounds can usually best be obtained by altering the quantities rather than the concentrations of the reagent. With less reactive derivatives, such as the N-acetyl-N-acetoxy- derivatives of 2-aminofluorene and 4-aminobiphenyl, satisfactory preparations can usually be obtained more readily in reactions in which the highest possible concentrations are used for extended periods. Following removal of unbound reagent by extraction with organic solvents, the adduct level is established by determinations of the isotope content and the ultraviolet absorption of the DNA. Adduct characterization has been accomplished by hydrolysis of the DNA with trifluoroacetic acid and subsequent HPLC with appropriate synthetic reference adducts derived from dGMP. In practical terms, it has usually been convenient to employ reagents with specific activities of 50 to 100 mCi/mmole with quantities of 10 to 100 $\mu$g of DNA.

Plasmid survival in wild type and repair-deficient cells

As an extension of the intial studies of Tang et al. (1982), we have compared the survival of plasmid pBR322 adducts in uvr deficient E. coli strains that carried 2-aminofluorene, 2-acetylaminofluorene, 4-aminobiphenyl or 4-acetylaminobiphenyl adducts (Tamura and King, 1989). Consistent with Tang's observations, 2-acetyl-aminofluorene adducts were more effective than the 2-aminofluorene derivatives in inactivating the plasmid in both wild type and uvr deficient strains (Table 1). The acetylated adducts exhibited severely reduced survival in all three uvr deficient strains (i.e. uvrA, uvrB and uvrC cells), as compared to wild type cells, while the survival of 2-aminofluorene- modified DNA was reduced more severely in uvrC cells than the other two uvr deficient strains. The 4-aminobiphenyl and 4-acetylaminobi-phenyl adducts were essentially equivalent to 2-aminofluorene in their abilities to inactivate the plasmid on transformation of wild type cells. However, in contrast to the 2-aminofluorene adducts, both biphenyl-derived adducts exhibited severe reductions of plasmid survival in all three uvr deficient strains.

The reduction in survival of the adducted plasmids is believed to result from the inability of the cellular enzymes to utilize the modified DNA as a template. Whether this reduction in template activity represents the direct ability of bypass by polymerase, or whether the inability of the polymerase to bypass the altered template leads to bypass through recombinational mechanisms is not known. As judged by the relative survivals of the adducts, it may be concluded that, in wild type

96

Table 1. Relative $D_{37}$ values of arylamine-pBR322 DNA adducts in wild type E. coli and uvr mutants ($D_{37}$ = adducts/DNA to reduce survival to 37%)

| ADDUCT | WILD TYPE | uvrA | uvrB | uvrC |
|---|---|---|---|---|
| ABP | 25 | 9 | 9 | 9 |
| AABP | 25 | 11 | 11 | 11 |
| AF | 23 | 17 | 20 | 8 |
| AAF | 12 | 1.3 | 1.5 | 2.1 |

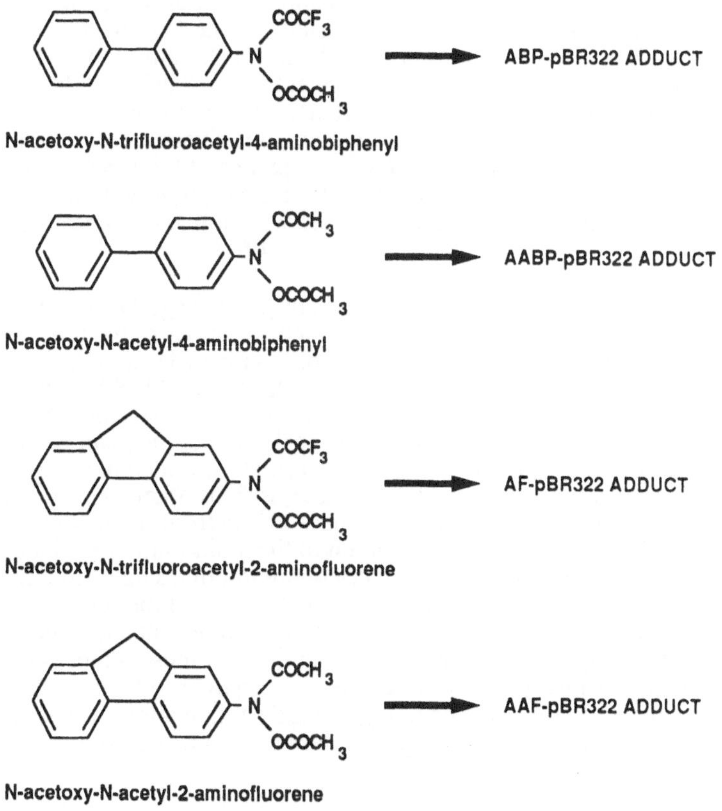
COCF3
N
OCOCH3
ABP-pBR322 ADDUCT
N-acetoxy-N-trifluoroacetyl-4-aminobiphenyl
COCH3
N
OCOCH3
AABP-pBR322 ADDUCT
N-acetoxy-N-acetyl-4-aminobiphenyl
COCF3
N
OCOCH3
AF-pBR322 ADDUCT
N-acetoxy-N-trifluoroacetyl-2-aminofluorene
COCH3
N
OCOCH3
AAF-pBR322 ADDUCT
N-acetoxy-N-acetyl-2-aminofluorene

Fig. 2 Modification of plasmid DNA with different arylamine derivatives.

cells, the 2-aminofluorene and both biphenyl adducts are bypassed with about the same ease by cellular enzymes. In cells that are deficient in the ability of the uvr endonuclease system to protect against these adducts, these three adducts exhibit approximately equivalent plasmid inactivation. These observations support the idea that all three adducts produce conformational alterations that are sufficiently alike that they reduce bypass to a similar extent. However, the dissimilarities in survival of the biphenyl and 2-aminofluorene plasmid adducts in uvr deficient cells suggests that differences in the planarity of the two aromatic structures may account for these characteristics. The greater survival of the acetylated biphenyl-adducted DNA in all of the E. coli strains, as compared to the 2-acetylaminofluorene-modified plasmid, demonstrates that the effect of the N-acetyl moiety plays a lesser role than does the aromatic ring structure. These data establish that the uvr endonuclease complex recognizes each of these four adducts. Furthermore, both the N-acetyl and the aryl moities affect the conformation of the DNA.

## Bacteriophage mutagenicity

Fuchs and his colleagues pioneered the use of plasmid DNA for the study of the mutagenicity of both 2-aminofluorene and 2-acetylaminofluorene adducts (Fuchs et al., 1981; Koffel-Schwartz et al, 1984, 1987; Bichara and Fuchs, 1985). This approach has been extended to bacteriophage DNA which enables one to explore potential differences between single and double stranded DNA that contained 2-aminofluorene adducts (Gupta et al., 1988). In both cases mutagenicity was studied by detection of changes in phenotype of the cells that contained these extrachromosomal vectors. The pBR322 plasmid renders the cell resistant to both ampicillin and tetracycline; mutagenic events were detected by loss of tetracycline resistance. The M13 bacteriophage DNA complements the defective $\beta$-galactosidase of the host cell; mutation of the viral DNA sequence encoding the required segment of the $\beta$-galactosidase gene results in colorless rather than blue plaques when plated on media that contain appropriate $\beta$-galactoside substrates. In both cases, putative mutations identified by these procedures are confirmed and the sequence of the DNA segments potentially responsible for the change in phenotype are sequenced to identify the molecular event responsible for the mutation. Under optimal circumstances both approaches can yield mutation frequencies in excess of 1%.

These experiments, with both plasmid and bacteriophage DNA, established that adduct mutagenicity was enhanced if the experiments were carried out in cells that had previously been treated with ultraviolet irradiation in order to induce error-prone DNA repair. Mutations produced with double stranded DNA were usually characterized by changes in DNA sequence at G:C pairs, i.e. they could be rationalized in terms of the mutation having occurred at the site of the guanine adduct. In contrast, mutations arising from transfection of single stranded bacteriophage DNA adducts were often due to changes in sequence at positions other than guanine. Although these "untargeted" mutations could have arisen from the inadvertent detection of spontaneous mutations not related to the aminofluorene adduct, this is believed unlikely since the mutation frequency of unmodified single stranded M13 DNA is so low (e.g. ~1 in 2,000). An alternative explanation is that the "untargeted" mutations are produced from minor adducts with bases other than guanine. While no other adducts were detected, given the very low amount of these minor adducts it is difficult to rule out this possibility by direct chemical analysis. One possible explanation was that the effects of the bulky aromatic amine adducts were being exerted at bases in addition to that complementary to the position of the adduct.


# SINGLE, SITE-SPECIFIC ADDUCTS

## Preparation and characterization

The ambiguity of the relationship between the actual site of the adduct and the eventual change in base sequence in experiments with randomly introduced adducts prompted the development of systems that provided for the incorporation of a single adduct at position in biologically active DNA (Johnson et al., 1986, 1987).

Separation of modified oligonucleotides could readily be achieved by HPLC. Hydrolysis with trifluoroacetic acid permitted an evaluation of the adduct content. Radiolabelling of the oligonucleotide at the 5' hydroxyl group with $P^{32}$ followed by electrophoresis on polyacrylamide provided further evidence of the purity of the oligonucleotide adduct. The ability to modify a single guanine of a synthetic oligonucleotide provided the potential to incorporate this segment into a gapped, heteroduplex DNA that could then be introduced into a cell. It was felt desirable to employ gapped heteroduplex molecules rather than to clone short, synthetic heteroduplex molecules with sticky ends into linear, double stranded DNA, since it is generally more difficult to obtain the larger quantities of DNA required for characterization by the latter technique.

An important consideration in the study of this type of molecule was the need to be able to determine the fate of each successful DNA molecule. Although most such studies have relied on changes in phenotype as an efficient method for the detection of mutagenic events. In contrast, we reasoned that the use of appropriate oligonucleotide probes would permit us to determine changes in sequence that might not be manifested as a change in phenotype. We have accordingly adopted this more labor intensive technique in our studies in an effort to evaluate the mutagenic disposition of a greater area of the genome. This approach has permitted us to screen a sequence of 12 to 13 nucleotides with a probe 15 bases in length by determining the sequence of those plaques or clones that lost the probe on raising the temperature to a few degrees above the calculated melting temperature (Gupta et al., 1989).

Bacteriophage mutagenicity

Our initial efforts to explore the mutagenicity of aromatic amine adducts in site-specific systems were undertaken by use of the M13 bacteriophage in collaboration with Drs. Louis J. Romano and Dana Johnson of the Department of Chemistry, Wayne State University (Johnson et al., 1986, 1987). The methodology employed for the construction of an appropriate site-specific adduct is shown in Figure 3. Extensive characterization of this molecule revealed that it exhibited the expected resistance to HincII and susceptibility to BamHI (Gupta et al., 1989). Analysis of the efficiency of ligation disclosed that greater than 90% of the molecules contained adducted oligonucleotides that were ligated at both the 3' and 5' termini. For mutagenicity studies, frameshifts were detected by changes in phenotype as described above; base substitutions were detected by hybridization.

Transfection of these molecules revealed that both 2-aminofluorene and 2-acetylaminofluorene produced base substitution and frameshift mutations (Gupta et al., 1989). Prior exposure of the host cells to ultraviolet irradiation increased both types of mutation. Mutation frequencies as high as 5% were observed in some cases. The sites of the mutations were located as far as five nucleotides from the adduct position. Interestingly, the position of the mutations were dependent, in part, on the type of host cell into which the DNA was incorporated. Control DNA constructed from unmodified oligonucleotides did not yield these mutations. Experiments with DNA that had nicks in the adduct-containing strand at the 5' terminus of the oligonucleotide were shown to be less mutagenic than the molecules that were >90% ligated at both ends of the oligonucleotide.

These studies demonstrated the feasibility of employing this type of DNA adduct for mutagenicity studies and substantiated our belief that it would be necessary to adequately evaluate the mutagenic consequences of such preparations. Despite this success, this study raised several questions. Were the mutations at positions other than the site of adduct produced by the adduct, or by the presence of blocking groups that had not been removed from the oligonucleotide? This possibility seems remote, since the DNA employed in these studies was prepared from a single source for the preparation both adducted and unmodified oligonucleotides. Another concern was that the frameshifts that were observed were in a run of guanine residues that abutted the ligation site 5' to the adduct. In order to avoid the possibility that

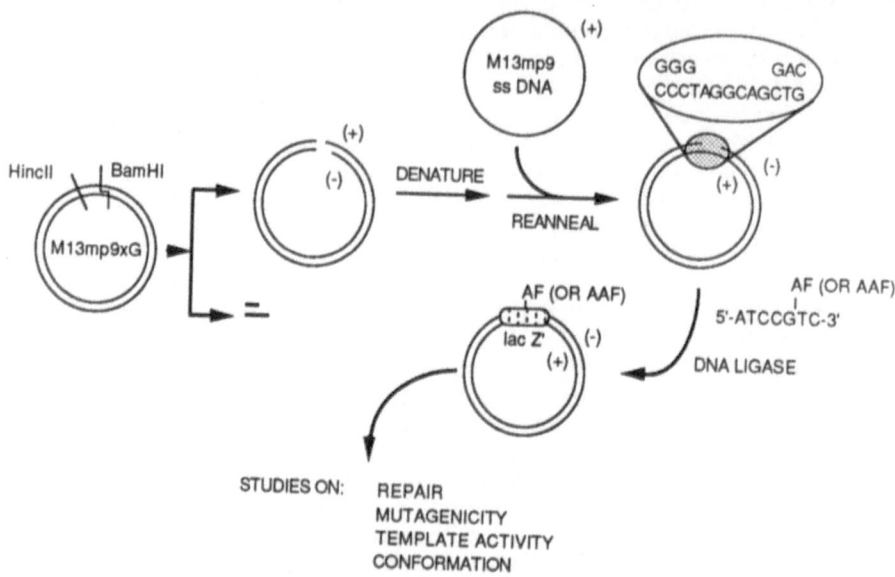

Fig. 3  Construction of single AF and AAF guanine adducts at a
specific site in biologically active M13mp9 phage DNA.

ligation sites near the adduct might some how be contributing to the mutagenicity,
we elected to develop a adduct vector based on a longer modified oligonucleotide.

Plasmid mutagenicity

The availability of small plasmids that have both the viral and pBR origins of
replication permit the preparation of single-stranded DNA which is essential for the
construction of gapped heteroduplex DNA.  A sequence was devised that, when cloned
into plasmid pIBI25 would would enable us to prepare a gapped heteroduplex that
would accommodate a 17 base long oligonucleotide (Figure 4) (Reid et al., 1989).
The single guanine in the oligonucleotide is within the unique SpeI restriction site
of this plasmid that we have designated pMC17.  Experiments with heteroduplexes and
the plasmid established that the presence of either fluorene adduct would prevent
the enzyme from cleaving the DNA.  Although hybridization is intended to be the
primary mutation screen employed with this system, the length of the sequence has
been designed so as to produce an effective +frameshift.  Thus, negative frameshift
mutants are able to complement the defective b-galactosidase of host cells that
may be employed in mutagenicity studies.  It has also been possible to construct the
plasmid with uracil in the strand opposite the adduct by preparing the single
stranded DNA in the appropriate mutant cells (Reid and King, submitted).  Transfection of double stranded DNA that contains uracil in one strand into wild type
cells leads to an extreme bias against the use of the uracil-containing strand in
the cell, as demonstrated by the contruction of base pair mismatches.  Thus, these
uracil-containing molecules would be expected to preferentially utilize the
complementary adduct-containing strand.

2-Aminofluorene adducts in non-uracil containing DNA, i.e. normal DNA, were
shown to yield SOS-inducible frameshift mutations at a frequency of approximately
0.3% (Reid et al., 1989).  The majority of the mutations were negative frameshifts
at the adduct site; the others were two to three bases from the adduct position.
Neither unmodified oligonucleotides nor the 2-acetylaminofluorene adducts exhibited
mutagenicity, with or without prior induction of SOS functions.

Transformation of the same strain of E. coli cells, NM522, with uracil-containing DNA adducts resulted in the detection of SOS-dependent mutations with both fluorene adducts (Reid and King, submitted). 2-Amino-fluorene adducts yielded nearly 3% base substitutions, all of which were at the adduct position. Mutations produced by 2-acetylaminofluorene adducts produced a mutation frequency just over 1% and were comprised of both base substitutions and frameshifts, most of which were not at the original position of the adduct. Unmodified oligonucleotides did not produce any mutations, even with SOS-induction of the host cells. Transformation of a uvrA deficient strain, AB1886, produced only a low frequency of mutations in SOS-induced cells with 2-aminofluorene adducts, i.e. ~0.3%. Neither the acetylated adduct nor the unmodified oligonucleotide produced mutants in SOS-induced cells.

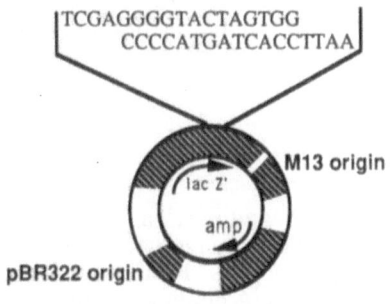

Fig. 4 Genetic map of pMC17 vector.

CONCLUSIONS

These studies clearly establish the feasibility of constructing and employing single, bulky aromatic amine adducts in mutagenicity studies by use of techniques that have the potential of identifying each of the mutagenic events produced. Given the production of mutations at sites other than the site originally occupied by the adduct suggests that the use of randomly introduced adducts will not yield unambiguous results. However, the implication of the SOS-functions in the production of these mutations suggests that the site-specific and random adducts are acting by similar mechanisms.

Similar concerns may be expressed regarding the use of biased selection systems that can reveal only limited information regarding the mutagenic potential of the adducts. Given the dependence of the mutagenic response on both the host cell and the sequence context of the adduct, the use of mutation selection techniques that are biased are likely to yield only limited information. This may impede progress in our better understanding of the complex conformational and enzymatic processes that are responsible for these mutagenicity processes.

It would seem that future studies are likely to be directed towards the clarification of the effects of adduct structure and cellular systems of mutagenic potential. The techniques described here should serve this function well. It is also likely that attempts will be made to employ site-specific adducts in the study of questions of greater biological interest in both prokaryotic and mammalian cells. For example, the sequences addressed above were chosen because of the ease of incorporation of the adduct into DNA and the easily identified structural features that aided in the characterization of these molecules. If rigorous, unbiased screening procedures were to be applied to more complex sequences that required more extensive manipulation of the DNA and yielded preparations that could not be characterized as extensively, it is likely that the acquisition of complex and unexpected results might not be readily accepted. Given the relatively unpredicatable nature of the results that have been obtained in the study of the mutagenicity of the aromatic amines thus far, it would seem advisable to attempt to establish the fundamental factors that determine these mutagenic responses (Ames et al., 1973; Mitchell and Stohrer, 1986; Lasko et al., 1988; Moriya et al., 1988; Burnoff et al., 1989).

ACKNOWLEGEMENTS

This report from the A. Alfred Taubman Facility for Environmental Carcinogenesis was supported by NIH grant CA45639 and by an institutional grant from the United Foundation of Detroit.

REFERENCES

Ames, B.N., Gurney, E.G., Miller, J.A. and Bartsch, H., 1973, Carcinogens as frameshift mutagens: metabolites and derivatives of 2-acetylamino- fluorene and other aromatic amine carcinogens. Proc. Natl. Acad. Sci., USA, 72:1254.

Bichara, M. and Fuchs, R.P.P., 1985, DNA binding and mutation spectra of the carcinogen N-2-aminofluorene in *E. coli*. A correlation between the conformation of the premutagenic lesion and the mutational specificity, J. Mol. Biol., 183:341.

Burnoff, D., Koehl, P. and Fuchs, R.P.P., 1989, Single adduct mutagenesis: strong effect of the position of a single acetylamino- fluorene adduct within a mutation hotspot, Proc. Natl. Acad. Sci., USA, 86:4147.

Fuchs, R.P.P., Koffel-Schwartz, N. and Daune, M., 1981. Hotspots of frameshift mutations induced by the ultimate carcinogen N-acetoxy-N-2-acetylamino-fluorene, Nature, 294:657.

Gupta, P.K., Johnson, D.L., Reid, T.M., Lee, M-S, Romano, L.J. and King, C.M., 1989, Mutagenesis by single site specific arylamine DNA adducts: Induction of mutations at multiple sites, J. Biol. Chem., in press.

Gupta, P.K., Lee, M-S, and King, C.M., 1988, Comparison of mutagenesis induced in single- and double-stranded M13 viral DNA by treatment with N-hydroxy-2-aminofluorene, Carcinogenesis, 9:1337.

Johnson, D.L., Reid, T.M., Lee, M-S, King, C.M., and Romano, L.J., 1986, Preparation and characterization of a viral DNA molecule containing a site specific 2-aminofluorene adduct: A new probe for mutagenesis by carcinogens. Biochemistry, 25:449.

Johnson, D.L., Reid, T.M., Lee, M-S, King, C.M., and Romano, L.J., 1987, Chemical stability of oligonucleotides containing the acetylated and deacetylated adducts of the carcinogen N-2-acetylaminofluorene, Carcinogenesis, 8:619.

King, C.M., 1985, Metabolism and the "Initiation" of tumors by chemicals, in: "Arachidonic Acid Metabolism and Tumor Initiation", L.J. Marnett ed., Martinus Nijhoff Publishing, Boston.

Koffel-Schwartz, N., Maenhaut-Michel, G. and Fuchs, R.P.P., 1987, Specific strand loss in N-2-acetylaminofluorene modified DNA, J. Mol. Biol., 193:651.

Koffel-Schwartz, N., Verdier, J.M., Bichara, M., Freund, A.M., Daune, M.P. and Fuchs, R.P.P., 1984, Carcinogen induced mutation spectrum in wild type, uvrA and umuC strains in *E.coli*: strain specificity and mutation prone sequences, J. Mol. Biol., 177:33.

Lasko, D.D., Harvey, S.C., Malaikal, S.B., Kadlubar, F.F. and Essigmann, J.M., 1988, Specificity of mutagenesis by 4-aminobiphenyl, J. Biol. Chem., 263:15429.

Lee, M.-S. and King, C.M., 1981, New synthesis of N-(guanosin-8-yl)-4-amino-biphenyl and its 5'-monophosphate, Chem. Biol. Interact., 34:239.

Mitchell, N. and Stohrer, G., 1986, Mutagenesis originating in site-specific DNA damage, J. Mol. Biol., 191:177.

Moriya, M., Takeshita, M., Johnson, F., Peden, K., Will, S. and Grollman, A.P., 1988, Targeted mutations induced by a single acetylaminofluorene DNA adduct in mammalian cells and bacteria, Proc. Natl. Acad. Sci., USA 85:1586.

Reid, T.M., Lee, M-S and King, C.M., 1990, Mutagenesis by site-specific arylamine adducts in plasmid DNA: Enhancing replication of the adducted strand alters mutation frequency, Biochemistry. In press.

Singer, B. and Grunberger, D., 1983 "Molecular Biology of Mutagens and Carcinogens" Plenum Press, New York.

Tamura, N. and King, C.M., 1990, Comparative survival of aminobiphenyl and aminofluorene substituted plasmid DNA in *Escherichia coli* Uvr endonuclease deficient strains, Carcinogenesis, 11: No.4, in press.

Tang, M-S, Lieberman, M.W. and King, C.M., 1982, Uvr genes function differently in repair of acetylaminofluorene and aminofluorene DNA adducts, Nature, 299:646.

# MUTAGENESIS INDUCED BY A SINGLE ACETYLAMINOFLUORENE

# ADDUCT WITHIN THE NARI SITE IS POSITION DEPENDENT

Patrice Koehl, Dominique Burnouf and Robert P.P. Fuchs

Groupe de Cancérogénèse et de Mutagénèse Moléculaire et Structurale
IBMC du CNRS
15 rue René Descartes
67084 Strasbourg Cedex, France

ABSTRACT

N-Acetoxy-N-2-acetylaminofluorene, a model for the potent rat liver carcinogen N-2-acetylaminofluorene, covalently binds to the C(8) position of guanine residues. It has been shown to mainly induce frameshift mutations in the bacteria E.Coli. These mutations occur within specific DNA sequences, the so-called mutation hot-spots. Among these, the NarI sequence (GGCGCC) is especially susceptible to -2 frameshift mutations (GGCGCC --> GGCC). Due to the nature of the NarI sequence, $G_1G_2CG_3CC$, three different molecular events can give rise to the observed end point GGCC (i.e. deletions of $G_2C$, $CG_3$ and $G_3C$).

In order to compare the actual role of each of the three possible guanine-AAF adducts in the NarI site in inducing the -2 frameshift mutation event, we constructed double stranded plasmid molecules containing a single AAF bound to one of the three guanine positions. Using these plasmids, we found that only AAF bound to $G_3$ induces a -2 frameshift mutation event in SOS induced bacteria. This result is discussed in terms of local DNA conformation.

## INTRODUCTION

There is now considerable interest in understanding at the molecular level interactions of chemical and physical agents with DNA and the fate of the resulting adducts, the so-called premutagenic lesions.

Most chemical mutagens bind covalently to DNA and form a variety of premutagenic lesions that are processed by cellular error-free repair mechanisms, such as excision repair or recombinationnal repair (1). Some of the adducts escape these repair processes, however, and are converted into mutations, either by direct miscoding or by cellular processes that are induced when the replication fork is blocked. This block in the replication process triggers in *E. coli* the induction of the SOS regulon. The SOS response contains several co-regulated genes, all under the control of the RecA and LexA encoded proteins, and whose gene products are involved in DNA repair or in the mutagenic processing of adducts in DNA (1).

*Nitroarenes*, Edited by P. C. Howard *et al.*
Plenum Press, New York, 1990

In our laboratory, we study the ultimate carcinogen N-acetoxy-N-2-acetyl-aminofluorene (N-Aco-AAF), a model for the strong rat liver carcinogen N-2-acetylaminofluorene (AAF). N-Aco-AAF binds covalently to DNA, mainly at position $C(8)$ of guanine (2). The mutation spectrum was determined by use of the tetracycline resistance gene inactivation assay (3), and it was found that AAF adducts mainly induce frameshift mutations (more than 90%) (3,4). These mutations occur within two types of sequences, the so-called mutation hot spots: i) runs of guanines, ii) the sequence GGCGCC, which is the recognition site of the Nar I restriction enzyme. The mutations in the run of guanines are primarily -1 frameshift mutations and can be described by a slippage-type mechanism proposed by Streisinger and coworkers (5,6) in which one strand of the DNA can slip with respect to the other strand during replication or repair synthesis. The -2 frameshift mutations occuring within the Nar I sequence (GGCGCC --> GGCC) do not seem to result from the same mechanism, in that the two pathways have different genetic requirements (4,7). Briefly, mutagenesis in the repetitive sequences is dependent on $umuDC^+$ gene function, whereas mutagenesis in the Nar I sequence is not, although this pathway also requires induction of the SOS system (7).

The spectrum of binding of AAF to guanine residues in double stranded DNA was found to be roughly random (8). The comparison between the mutation spectrum and the binding spectrum for AAF suggested to us that the high mutation frequency within the AAF hot spots may result from the processing of an unusual DNA conformation induced by the premutagenic lesion (8).

Due to the nature of the sequence of the NarI site, $G_1G_2CG_3CC$, three different molecular events, each involving the deletion of two contiguous base pairs (i.e.: $G_2C$, $CG_3$, $G_3C$), can give rise to the observed end point (GGCC). We have used a single-adduct mutagenesis strategy to investigate further the mechanism of AAF induced mutations within the NarI sequence. The strategy involved designing a plasmid, pSM 14, that provides a simple phenotypic test which specifically detects -2 frameshift mutations within the NarI sequence.

In this paper,we described the construction of a series of plasmids, all bearing a single AAF adduct located in a NarI site. The mutagenic properties after transformation of these plasmids into E. coli is also reported.


STRATEGY FOR SINGLE ADDUCT MUTAGENESIS

This method has been extensively exposed elsewhere (9, 10).

The general strategy for the construction of plasmids containing a single AAF adduct in the NarI site is described in figure 1. Briefly, two related plasmids are linearized by means of two different restriction enzymes and used to reconstitute a circular duplex molecule containing a single-stranded region, in which an oligonucleotide, modified or not, is hybridized and ligated. The covalently closed circular molecules resulting from a double ligation event are then purified on cesium chloride gradients, and used in mutagenesis experiments in E.coli.

The construction of the plasmid is such that it enable us to detect -2 frameshift mutations in the following way : the parent plasmid, pEMBL8(-) (11), encodes part of the β-galactosidase gene, and produces blue colonies when propagated in an adequate host on X-gal plates. A 14 base-pair oligonucleotide d(TCACCGGCGCCACA) containing the NarI site was cloned into the HincII site of the polylinker region of pEMBL8(-), thus producing a +2 frameshift in the reading frame of the β−galactosidase gene. Bacteria transformed with this new plasmid, pSM14, give rise to white clones when plated on X-gal plates. Any -2 frameshift event that occur within the NarI site will restore the reading frame of the LacZ gene, thus giving rise to blue colonies (figure 1).

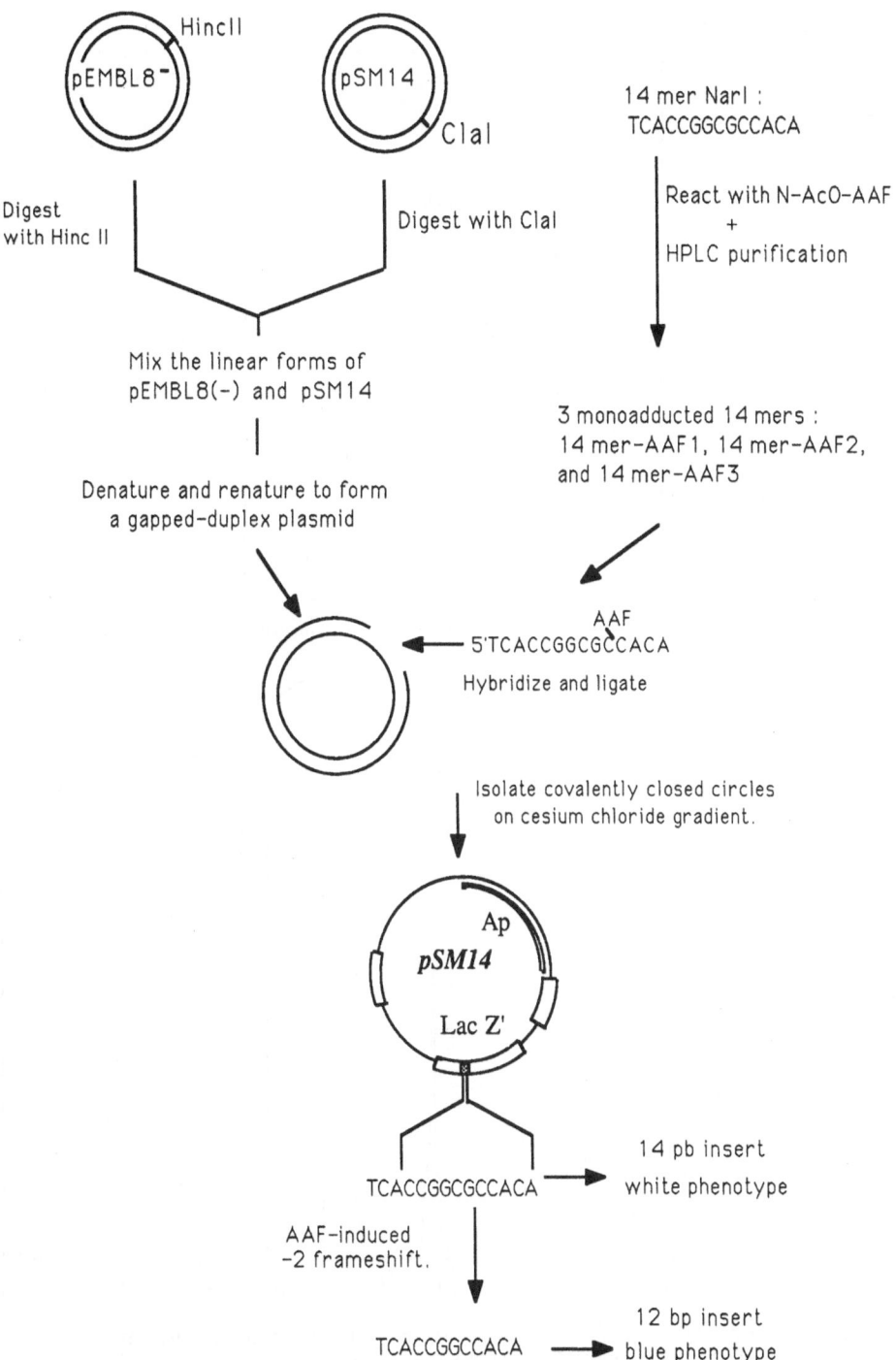

Figure 1 : Strategy for the construction of pSM 14 plasmids containing a single -AAF adduct on one of the guanine residues of the NarI sequence. pSM 14 was designed as a probe for detecting -2 frameshift mutations within the NarI sequence.

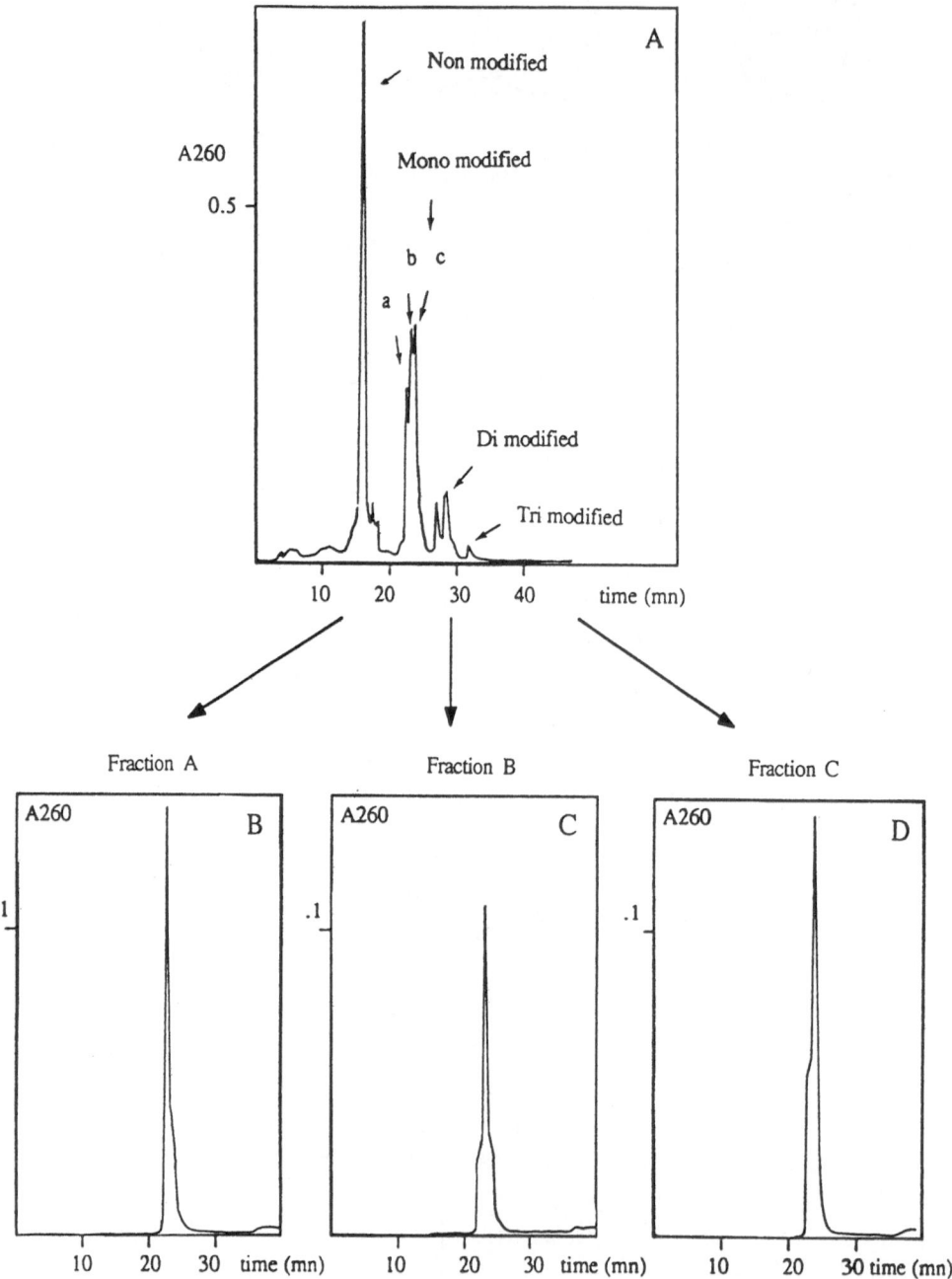

Figure 2 : Panel A present a H.P.L.C. profile of the crude mixture obtained after reaction of N-Aco-AAF with d(TCACCGGCGCCACA). The peak of mono-adducted oligonucleotides was collected in three different fractions, noted (a), (b) and (c), which were then repurified by H.P.L.C. as presented in panel B,C, and D respectively. Solvents and elution conditions required were described elsewhere (9)

# RESULTS AND DISCUSSION.

## AAF-monoadducted oligonucleotides are purified by means of HPLC chromatography

        The in vitro reaction of the 14 mer 5' TCACCGGCGCCACA 3' with the ultimate carcinogen N-AcO -AAF gives rise to several reaction products: one tri-adducted, three di-adducted and three mono-adducted oligonucleotides. Figure 2A presents the HPLC purification of the crude reaction mixture. At this step, the three mono-adducted 14 mer elutes in overlapping peaks. To separate each of the three isomers of position, the products recovered from these peaks were repurified by HPLC (figs 2B,C,D). After this final purification, the exact position of the AAF adduct on the oligonucleotide and the purity of each monomodified oligonucleotide were checked using the T4 DNA polymerase assay. The 3' -> 5' exonuclease activity of T4 DNA polymerase is stopped one nucleotide 3' to the AAF adduct (8). Figure 3 shows the result of the digestion experiment analysed on a 20% acrylamide gel. Each band represent the length of a DNA molecule, corresponding to the position where the T4 polymerase stopped. Lanes c, d and e reveal the purity of each monomodified 14 mer, which is considered to be greater than 90%.

TCACCGGCGCCACA
1 2 3

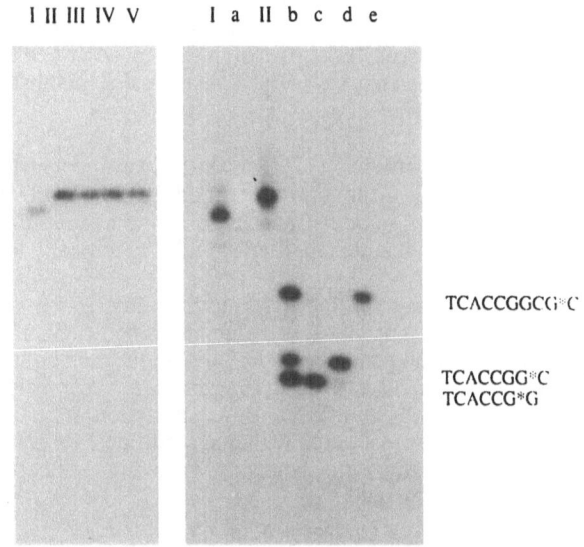

TCACCGGCG*C

TCACCGG*C
TCACCG*G

Figure 3 : T4 DNA polymerase analysis of the purified mono-adducted isomers of the 14-mer d(TCACCGGCGCCACA). All nucleotides were ($^{32}$P) 5' end-labeled.The analysis is performed on a 20% polyacrylamide sequencing gel. Left panel: Lane I shows the nonmodified oligonucleotide, and lane II shows a pool of mono-adducted oligonucleotides. The AAF adduct is responsible for a slightly slower migration for the mono-adducted 14 mers (lane II, III, IV and V). The mono-adducted isomers are not resolved on the sequencing gel (same migration in lanes II, III, IV and V). Right panel: undigested DNA is shown in lane I and II, designated as in the left panel. Lanes a shows that the unmodified oligomer is completely digested by T4 DNA polymerase. Lane b, c, d, and e show the DNA fragments produced when the 3'->5' exonuclease activity of the T4 DNA polymerase is blocked one nucleotide 3' of the AAF adduct, hence revealing the position of the modified guanine in the 14 mer (8).

## Construction and purification of AAF-monoadducted plasmids

pEMBL 8(-) and pSM 14 were linearized by the restriction enzymes HincII and ClaI respectively. The two double-stranded linear plasmids were mixed, denatured and renatured. The renatured DNA mixture contains both the linear forms of the two plasmids and two species of gapped duplex molecules. Only one of the two species will have a gap with the sequence complementary to the oligonucleotide. In a typical experiment, we estimate that in the crude gapped-duplex mixture, at least 10% of the DNA molecules are gapped duplexes with the adequate structure.

The unmodified 14 mer and the three mono-adducted position isomers were 5'-phosphorylated, and the $^{32}$P-labeled molecules were separated from the unphosphorylated molecules on a 20% acrylamide gel. They were added to the crude gapped-duplex mixture, and hybridization and ligation steps were performed, as described elsewhere (9).

The formation of fully ligated molecules is certified by the appearance, on a ethidium bromide containing agarose gel, of a new band which migrates as covalently closed plasmids. We purified the covalently closed molecules by centrifugation in CsCl gradients containing ethidium bromide in order to use them in mutagenesis experiments.

## Only AAF-G3 lesion induces a -2 frameshift mutation

The reconstructed covalently closed plasmids designated as pSM14-1, pSM14-2 and pSM14-3, each containing a single AAF adduct on the guanine 1, 2 or 3 of the NarI site respectively, were used to transform SOS-induced or uninduced JM103 bacteria. Mutants clones were selected on the basis of the phenotypic assay described in figure 1. These mutants were confirmed by restriction enzyme analysis, namely by establishing the loss of the unique NarI restriction site, or by sequencing.

In previous experiments (12), plasmids were constructed to investigate the segregation of each strand as a function of its modification level. It was concluded that the presence of blocking lesions in one strand leads to the uncoupling of the two polymerase holoenzyme complexes, and that the modified strand is specifically lost.

This observation is expected to introduce a bias in the single adduct mutagenesis experiments, because of the specific loss of the monomodified strand. In order to avoid this bias, we irradiated the pSM14 parental plasmid before the denaturation-renaturation step, at a UV dose yielding approximately 10 to 12 pyrimidine dimers per plasmid. UV irradiation of the complementary strand did not interfere with our mutation assay (table 1). In addition, the controls with the plasmids reconstructed with non modified 14 mer show that the background mutation frequency introduced by the chemistry of the oligonucleotide and the engineering of the gapped-duplex is quite low (table 1).

As expected from our previous results with randomly modified plasmids (3,4), the NarI mutant pathway is strongly SOS-dependent (table 1). As a main result, we found that only -AAF modification at the $G_3$ position induces the -2 frameshift event that characterizes the NarI mutation pathway. No mutant clones were recovered with the constructions in which the -AAF adduct is positioned on guanine $G_1$ or $G_2$. Modification at $G_2$ could potentially trigger the -2 event. Indeed, three different molecular events (i.e., deletion of $G_2C$, $CG_3$ or $G_3C$) could have given rise to the observed end point (GGCGCC -> GGCC). We suggest that AAF induced mutagenesis at NarI site hot spots is related to local conformational changes caused by the binding of an AAF adduct at $G_3$ residue. Such an unusual DNA conformation is not induced upon binding of -AAF to $G_1$ or $G_2$.

## Mutagenesis at the NarI site is related to an unusual DNA conformation

Alternating purine-pyrimidine DNA sequences have been shown to undergo a B to Z transition when modified with N-AcO-AAF (13-17). This effect can be related to the

Table 1:  Mutation frequencies of the AAF induced -2 frameshift event in the NarI site.

| | plasmids | | | |
|---|---|---|---|---|
| | pSM14 | pSM14-1 | pSM14-2 | pSM14-3 |
| No SOS induction | ND | **<19** | **<9** | **4** |
| | | (0/510) | (0/1119) | (1/2315) |
| With SOS induction (50 to 70J/m$^2$) | **<1** | **<6** | **<10** | **102** |
| | (0/10080) | (0/1650) | (0/978) | (17/1670) |

All plasmids bear UV lesions (5 to 6) in the strand complementary to the AAF-modified one.  No AAF lesion was introduced in the control plasmid pSM14.  pSM14-1, pSM14-2 and pSM14-3 have a unique AAF adduct on G1, G2 or G3, respectively. Mutation frequencies are expressed per $10^4$ transformants. Terms in brackets give the number of mutant clones over the number of total transformants. Mutagenesis experiments were done in both SOS-induced or non-inducedJM103 cells. ND: not done.

rotation of the guanine bases into a syn conformation, which is the conformation of the deoxyguanosine found in the Z helix structure. This has been observed for $(dC.dG)_n.(dC.dG)_n$ even at low salt concentration, and for $(dT.dG)_n.(dC.dA)_n$ under salt conditions that are ineffective to induce a transition for the non modified polynucleotide. Unlike these sequences, the NarI site does not hold a perfect purine pyrimidine alternance. However, modeling studies have shown that stretches as short as one dinucleotide dCpG can hold a Z conformation within a B type helix (18). This would suggest that the CpG3 dinucleotide is in Z conformation when its guanine is modified with AAF. We suggest that this conformationnal change to Z DNA is recognized by the proteins that process the lesion and that AAF mutagenesis occurs during a repair type of processing of the lesion on double stranded DNA rather than during the course of replication.

Nar I site mutagenesis is a model for a more general mutation pathway.

        The NarI mutation pathway is not restricted to NarI site: related sequences of alternating CG's show the same characteristics (19, R.Schaaper, N.Koffel-Schwartz and R.P.P. Fuchs, personnal communication), i.e. they are mutationnal hot spots for AAF induced mutagenesis, mutations are -2 base pairs deletions and occur independently of the umuDC functions. Moreover this pathway presents the same characteristics as the reversion of the hisD3052 allele in Salmonella typhimurium. In the Ames tester strains TA1538 and TA98, a large fraction of the reversion of this allele occurs by the loss of two base pairs in a $(CG)_4$ track, independently of the muc AB genes, which are functionally homologous to umuDC. Thus it seems that the NarI mutation pathway is part of a broader mutationnal pathway, which could be involved, in different organisms, in the processing of unusual structures induced within specific sequences.

REFERENCES.

1.  Walker, G. C., "Mutagenesis and Inducible Responses to Deoxyribonucleic Acid Damage in *Escherichia coli.*" , Microbiol. Rev. 48: 60-93. (1984)

2.  Kriek, E., J.A. Miller, U. Juhl, and E.C. Miller"8-(N-2-Fluorenylacetamido) guanosine, an Arylamidation Reaction Product of Guanosine and the Carcinogen N-Acetoxy-N-2-fluorenylacetamide in Neutral Solution", Biochemistry 6: 177-182. (1967)

3.  Fuchs, R.P.P., N. Schwartz and M.P. Daune,"Hot spots of frameshift mutations induced by the ultimate carcinogen N-acetoxy-N-2-acetylaminofluorene" Nature 294: 657-659 (1981)

4.  Koffel - Schwartz, N., J.M. Verdier, M. Bichara, A.M. Freund, M. Daune, R.P.P.Fuchs "Carcinogen-induced Mutation Spectrum in Wild-type, uvrA and umuC Strains of *Escherichia coli.* ", J. Mol. Biol. 177: 33-51 (1984)

5.  Streisinger, G., Okada, Y., Emrich, J., Newton, J., Tsugita, A., Terzaghi, E. & Inouye, M. "Frameshift mutations and the genetic code", Cold Spring Harbor Symp. Quant. Biol., 31, 77-89 (1966)

6.  Streisinger, G. and J. Owen (Emrich) "Mechanisms of spontaneous and induced Frameshift Mutation in Bacteriophage T4", Genetics 109: 633-659 (1985)

7.  Koffel-Schwartz, N., Fuchs, R.P.P., "Genetic control of AAF induced mutagenesis at alternating sequences: an additional role of RecA ", Molec. Gen. Genetics, 215: 306-311. (1989)

8.  Fuchs, R.P.P. DNA Binding Spectrum of the Carcinogen N-acetoxy-N-2- acetylamino-fluorene Significantly Differs from the Mutation Spectrum" , J. Mol. Biol. 177: 173-180. (1984)

9.  Koehl, P., Burnouf, D. and Fuchs, R.P. P. "Construction of plasmids containing a unique Acetylaminofluorene adduct located within a mutation hot spot; a new probe for frameshift mutagenesis ", J. Mol. Biol. 207: 355-364 (1989)

10. Burnouf, D., Koehl, P. & Fuchs, R.P.P. "Single adduct mutagenesis : strong  effect of the position of a single acetylaminofluorene adduct within a mutation hot spot", Proc. Natl. Acad. Sci. USA, 86: 4147-4151 (1989)

11. Dente, L., G. Cesarini, and R. Cortese "pEMBL: a new family of single stranded plasmids.", Nucleic Acids Res. 11: 1645-1655. (1984)

12. Koffel-Schwartz, N., G. Maenhaut - Michel, and R.P.P. Fuchs "Specific strand loss in N-2-acetylaminofluorene modified DNA", J. Mol. Biol. 193: 651-659. (1987)

13. Sage, E.,  and M. Leng "Conformation of poly(dG-dC) modified by the carcinogens N-acetoxy-N-acetyl-2-aminofluorene and N-hydroxy-N-2-aminoflurorene", Proc. Natl. Acad. Sci. USA. 77: 4597-4601. (1980)

14. Sage, E., and M. Leng "Conformation of poly(dG-dC) modified by the carcinogens N-acetoxy-N-acetyl-2-aminofluorene", Nucl.Acids Res. 9: 1241-1249. (1981)

15. Santella, R.M., D. Grunberger, S. Broyde, and B.E. Hingerty "Z-DNA conformation of N-2-acetylaminofluorene modified poly(dG-dC).poly(dG-dC) determined by reactivity with anti cytidine antibodies and minimized potential energy calculations", Nucl.Acids Res. 9: 5459-5467. (1981)

16. Santella, R.M., D. Grunberger, I.B. Weinstein,  and A. Rich "Induction of the Z conformation in poly(dG-dC).poly(dG-dC) by binding of N-2-acetylaminofluorene to guanine residues", Proc. Natl. Acad. Sci. USA. 78: 1451-1455. (1981)

17. Wells, R.D., J.J. Miglietta, J. Klysik, J.E. Larson, S.M. Stirdivant, and W. Zaccharias "Spectroscopic Studies on Acetylaminofluorene-modified $(dT-dG)_n.(dC-dA)_n$  suggest a Left-handed Conformation", J.Mol.Biol. 257: 10166-10171. (1982)

18. Arnott,S., Chandrasekaran, R., Hall, I.H., Puigjaner, L.C., Walker, J.K. and Wang, M., "DNA secondary strucutres : helices, wrinkles, and junctions", Cold Spring Harbor Symp.Quant. Biol. 47: 53-56 (1982)

19. Burnouf, D., and R.P.P. Fuchs "Construction of frameshift mutation hot spots within the tetracycline resistance gene of pBR322", Biochimie 67:  385-389. (1985)

# UNUSUAL HYDROGEN BONDING PATTERNS IN 2-AMINOFLUORENE (AF) AND

# 2-ACETYLAMINOFLUORENE (AAF) MODIFIED DNA

S. Broyde[*], B. E. Hingerty[**], R. Shapiro[+] and D. Norman[++,a]

Biology Dept.[*] and Chemistry Dept.[+] New York University, NY, NY 10003, Health and Safety Research Division[**] Oak Ridge National Laboratory, Oak Ridge, TN 37831, Biochemistry Dept.[++] Columbia University College of Physicians and Surgeons, NY, NY 10032

## INTRODUCTION

It has long been known that base pairing schemes differing from the canonical Watson-Crick type are possible both in DNA and RNA fragments and in polynucleotides. Indeed, experimental evidence for such hydrogen bonding options has been in the literature for decades. Hoogsteen, reverse Hoogsteen and reverse Watson-Crick are among the more well-known forms, but many others, not necessarily between Watson-Crick partners, have been delineated[1]. However, it is only recently that non-Watson-Crick pairing has attracted really widespread interest. One important reason for this renewed interest is the recent availability of crystal structures and high resolution NMR data; these are beginning to offer atomic resolution views of DNA duplexes which manifest novel pairing schemes. The crystal structure of the antibiotic triostin complexed with a DNA octamer[2] was of particular importance in this regard, because it called attention to the possibility of a Hoogsteen pair between a syn guanine and an N3 protonated cytosine (Figure 1a). This work, as well as solution studies by others has led to an appreciation that base protonation in duplex oligomers can occur at physiological pH, because the pK for protonation can be much higher in DNA duplexes than it is in the free base or nucleoside[3]. Consequently, pairing schemes that require protonation can no longer be deemed irrelevant to the biological situation.

Accordingly, the time appears ripe for the consideration of unusual base pairs in structures of DNAs modified by carcinogens, and we have begun to investigate such forms. In the present work we report on initial studies that reveal structures of DNAs modified by aromatic amines in which novel base pairs occur at the modification site. In one study a duplex 11-mer was investigated in which a central G was modified at carbon-8 and the opposite base was adenine. High resolution NMR data, obtained in the laboratory of Dinshaw Patel at Columbia University was available for this investigation when the modification was 2-aminofluorene (AF). These data were employed to guide the search for unconstrained energy minimized structures consistent with it[4]. Another, purely theoretical study was carried out for a modified duplex 9-mer. In this case the minimizations were carried out without reference to NMR results.

---

[a] Present address: Biochemistry Dept., Oxford University, Oxford, England

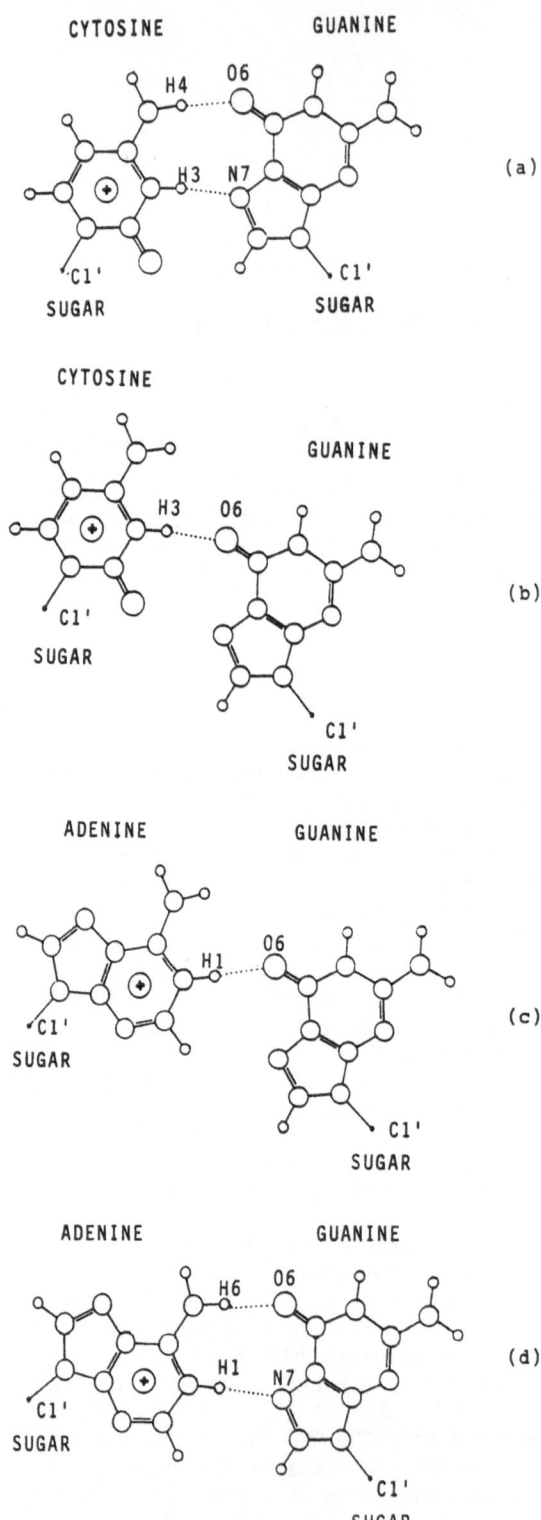

Figure 1. Hydrogen bonding schemes investigated at AF and AAF modified guanines.

METHODS

Our calculations are carried out with the program DUPLEX, described in full detail recently[5]. The potential energy set is similar to one devised by Srinivasan and Olson[6]. It includes Van der Waals, electrostatic, torsional, deoxyribose strain and anomeric gauche terms. In addition, counterion condensation and solvent are mimicked by employing reduced partial charges on the non-linked phosphate oxygens and a distance dependant dielectric constant, respectively. For duplexes, a hydrogen bond potential similar to that of Olson[7] is employed. Further, a hydrogen bond forcing function is used in all first stage minimizations to locate hydrogen bonded minima of any selected type. This function, as well as constraints imposed to locate structures that match NMR data, are released in terminal minimizations, so that final structures are unconstrained energy minima. The constraints are employed as a device for dealing with the multiple minimum problem, ie to guide the minimizer in the location of specific forms. Full details of the methods employed to locate modified duplex oligomers with selected base pairing schemes are given in references[4,5,8,9]. References 4 and 9 detail our procedures for incorporating distance constraints from NMR measurements in the first stages of the minimizations.

DUPLEX performs minimizations in torsion angle rather than cartesian space. This has enormous advantages in the search for novel forms because only the torsion angles about which major conformational rearrangements occur are flexible during the minimizations. Bond lengths, bond angles, and fixed dihedral angles, such as those in a planar, aromatic moiety are not mobile. This approximation is valid in the first instance since bond length, bond angle and dihedral angle variations are in fact very much smaller than torsion angle variations, which can run the gamut from 0 to 360 degrees. The result of this approximation is a vast diminution in the number of variables (three times the number of atoms in cartesian space, but only seven—six for the backbone and one for the sugar pucker—for a DNA nucleotide residue in torsion angle space). Consequently, the minimizer is much more powerful in effectuating large movements from the starting conformation. In addition, geometric distortions such as warping of planar aromatic moieties, which is a constant hazard when all degrees of freedom are permitted to flex during the minimization, are not possible. Of course, all degrees of freedom must be allowed in studies where pathways between conformers are under investigation, since bond length, bond angle and "fixed" dihedral angle movements are then important.

Our computational strategy involves broad searches of the conformation space of small modified DNA subunits, such as deoxydinucleoside monophosphates (the DNA conformational building blocks) and duplex trimers, followed by building to larger structures by various strategies. No computer or hand model building to match particular types of data is carried out to generate starting structures. Instead, the starting structures for the minimizations are simply combinations of preferred rotamers for each torsion angle. The strategy for locating a duplex structure with a given hydrogen bond at the modified site begins with an unmodified duplex trimer. Energy minimization produces a duplex trimer with Watson-Crick pairs at the termini and the unusual pairing scheme at the center. Then the carcinogen is added to the central guanine, and sixteen energy minimizations (at 90 degree intervals of the carcinogen-base linkage angles $\alpha'$ and $\beta'$, see Figure 2) are made to search the conformation space of the carcinogen-base linkage. The lowest energy forms from this search are then built to larger structures, either by embedding in a longer B-form duplex and minimizing the energy again, or by build-up with addition of one or more residues at a time (with energy minimization at each step), or by other strategies that have been detailed[8,10].

The hydrogen bonding patterns that have been investigated in this work are shown in Figure 1. The sequences investigated were d(CCATCG*CTACC).d(GGTAGAGATGG) (Sequence I) a G-A mismatch, and d(GCGCG*CGCG).d(CGCGCGCGC) (Sequence II). The asterisk designates the modification site. The modified guanines were syn (except where an anti guanine structure is reported) and all other torsion angles were of the B-DNA[11] type at the start of the minimizations.

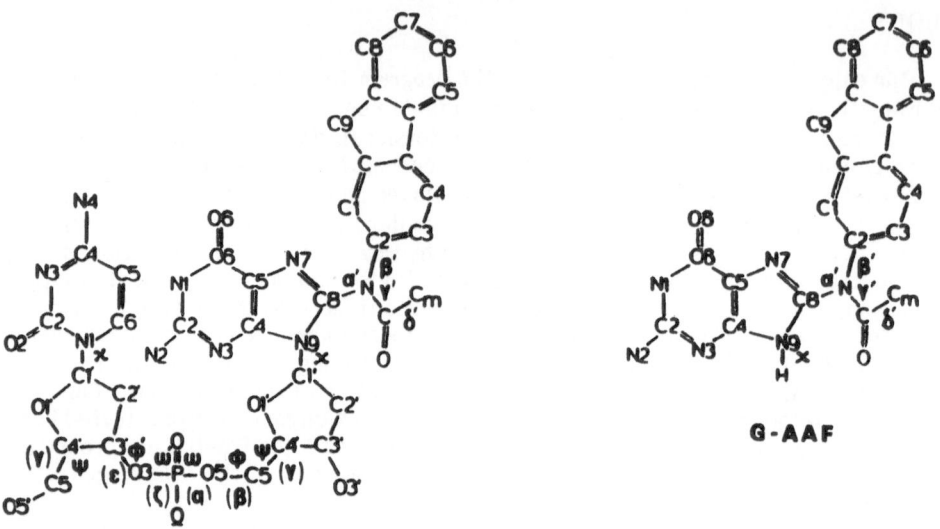

**G-AAF**

Figure 2. Structure, numbering scheme and variable conformational angle designations for the N-acetylaminofluorene (AAF) adduct to guanine C-8 in d(CpG). The dihedral angles A–B–C–D are defined as follows:

$\chi$ (pyr): O1'-C1'-N1-C6; $\chi$(pur): O1'-C1'-N9-C8; $\Psi$ : C3'-C4'-C5'-O5'; $\varphi'$ : P-O3'-C3'-C4'; $\varphi$ : C4'-C5'-O5'-P; $\omega'$ : O5'-P-O3'-C3'; $\omega$ : C5'-O5'-P-O3'; $\alpha'$ : N9-C8-N-C2; $\beta'$ : C8-N-C2-C1; $\gamma'$: C8-N-C-Cm; $\delta'$: N-C-Cm-H.

The angle A–B–C–D is measured by a clockwise rotation of D with respect to A, looking down the B–C bond. A eclipsing D is $0°$. Sugar pucker in the calculations is defined by the pseudorotation parameter $P^{31}$. IUPAC torsion angle designations are given in parentheses. In the IUPAC convention 180 degrees is added to $\chi$. In the aminofluorene (AF) adduct, the $COCH_3$ group is replaced by a hydrogen.

RESULTS

The AF modification was first investigated for Sequence I, for which detailed NMR results were available[4]. The unconstrained AF modified duplex 11-mer of lowest energy that matched all the available experimental data is shown in Figure 3. The key observation about this structure is the fact that the AF is sandwiched into the minor groove, which is clamped down on the aromatic amine in such a way as to minimize solvent exposure of the hydrophobic moiety. This places the modified G and its opposite A in a position where the single hydrogen bond between GO6 and H1$^+$N1A (Figure 1c) can form when the N1 of A is protonated. The AF is in proximity to the opposite strand, the modified guanine is syn, and the overall DNA structure is B-form, with the adenine residue opposite the modification in the less prevalent B-II conformation and other residues B-I[12]. The narrowing of the minor groove to bury the AF produces a bend in the DNA duplex. The structure shown in Figure 3 is the neutral pH structure, not protonated at adenine N1, so the hydrogen bond of Figure 1c cannot form; however, the GO6...N1 distance is such that the hydrogen bond could form if N1 were protonated, which the data indicate occurs at acid pH. Small pivotal movements of the carcinogen accompany the protonation but the overall structure is similar. Thus, important stabilization to the structure is afforded by the hydrophobic interactions that bury the carcinogen in the minor groove in an orientation that permits the hydrogen bond to form when N1 is protonated.

A second structure for AF modified Sequence I was computed with the hydrogen bonding scheme of Figure 1d. As may be seen from Figure 4, the syn modified G also

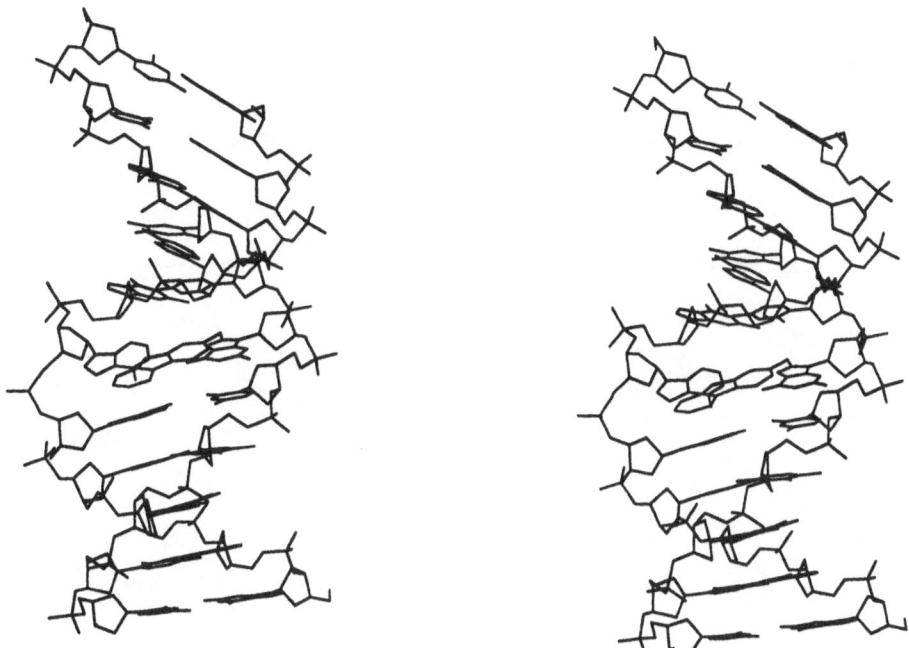

Figure 3. AF modified Sequence I, containing a G-A mismatch at the modification site, in conformation matching the NMR data. The modified G and its opposite A are positioned so that the hydrogen bond of Figure 1c can form upon protonation. Torsion angle details are given in reference 4. Coordinates and the full torsion angle set for all structures are available from the authors. Stereo views are prepared for use with a stereo viewer. An excellent stereo viewer suitable for any size image, can be obtained from NU-3D-VU Co., 71 E. 28 St., Eugene, Oregon 97405, telephone (503) 484-6176. To view with crossed eyes, left and right images must be interchanged.

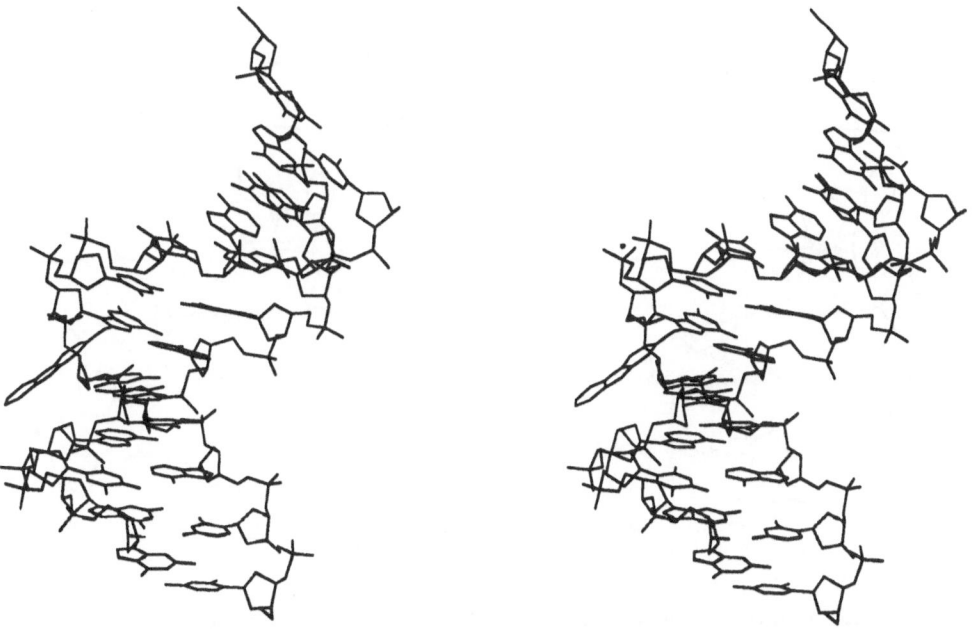

Figure 4. AF modified Sequence I, containing a G-A mismatch at the modification site, with modified G hydrogen bonding to A in pattern of Figure 1d. $\alpha' = 59^{\circ}$, $\beta' = 5^{\circ}$, $\chi(6) = 246^{\circ}$ $\mathbf{P}(5) = 36^{\circ}$, $\mathbf{P}(6) = 132^{\circ}$ (numbering begins from 5' end of modified strand and continues from 5' end of unmodified strand). All other torsion angles are within B-DNA ranges[32]. See caption to Figure 3.

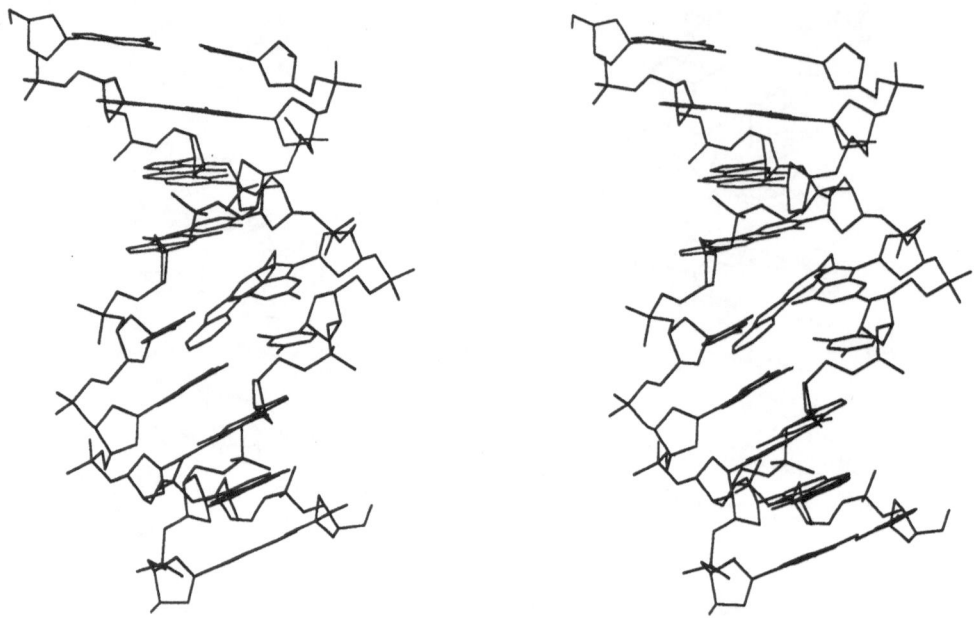

Figure 5. AF Modified Sequence II, with modified G hydrogen bonding to C in pattern shown in Figure 1b. $\alpha' = 228°$, $\beta' = 322°$, $P(5) = 137°$, $\chi(5) = 246°$; residue 14 is in B-II DNA conformation $\sigma'(14) = 257°$, $\omega'(14) = 196°$, $\varphi(14) = 124°$. All other torsion angles are within B-DNA ranges[32]. See caption to Figure 3.

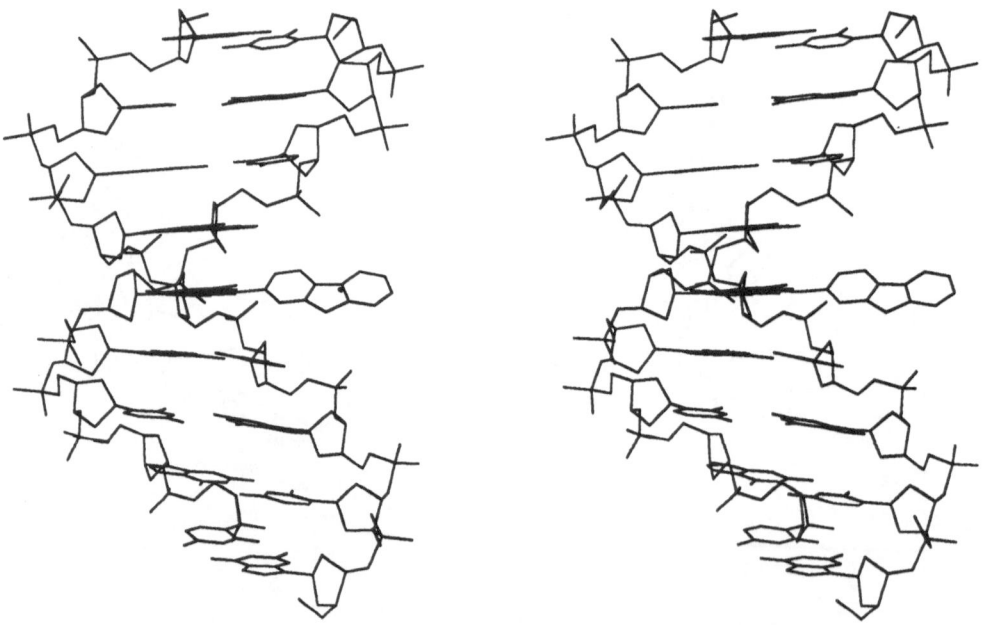

Figure 6. AF Modified Sequence II, with modified G hydrogen bonding to C in standard Watson-Crick pattern. All torsion angles are within B-DNA ranges[32] $\alpha' = 179°$, $\beta' = 43°$. See caption to Figure 3.

places the AF in the minor groove of the B helix, but not in proximity with the opposite strand, as warranted by the NMR data[4]. Interestingly, the two forms are close in energy, with the protonated version of the structure that matches the data favored by about 4 kcal/mole.

An additional important structure for AF modified DNA was obtained for Sequence II which has C opposite the modified G, employing the hydrogen bonding scheme of Figure 1b. Figure 5 reveals that this structure looks very much like the analogous one for the G-A mismatch of Figure 3, suggesting that the wedging of AF into the minor groove may be important for C opposite modified G, as well as for A. Interestingly, this syn guanine form is only 4 kcal/mole higher in energy than a guanine anti major groove model (Figure 6) which is similar to one computed previously for an AF modified alternating C-G duplex dodecamer[13,14]. (Energies of protonated and unprotonated forms are compared here by removing the proton from the protonated structure and re-minimizing the energy. The structures change very little when the proton is removed. The energies of the unprotonated variants are then used for comparison).

Finally, we computed an AF modified DNA structure with Sequence II employing the hydrogen bonding scheme of Figure 1a. Its energy is about 11 kcal/mole higher than the anti guanine major groove model of Figure 6. A strongly bent duplex accommodates the AF in the minor groove again here, but one face of the aromatic amine is exposed to solvent (Figure 7).

Energy minimized structures for the acetylaminofluorene (AAF) adduct like those of Figures 5 and 7 have also been obtained. (The AF structures for Sequence II were derived from the analogous AAF ones, detailed in reference 8, by removal of the acetyl and reminimizing). Interestingly, the minor groove models with all base pairs intact are only about 2-4 kcal/mole higher in energy than Z-DNA type models in Sequence II for AAF[8] modification. However, major groove models with guanine anti and all base pairs intact are not feasible for AAF[14,15].

DISCUSSION

A number of new structures have been computed for AF and AAF modified DNA which feature novel hydrogen bonding patterns, syn guanines and placement of the aromatic amine in the minor groove with hydrogen bonding retained between all base

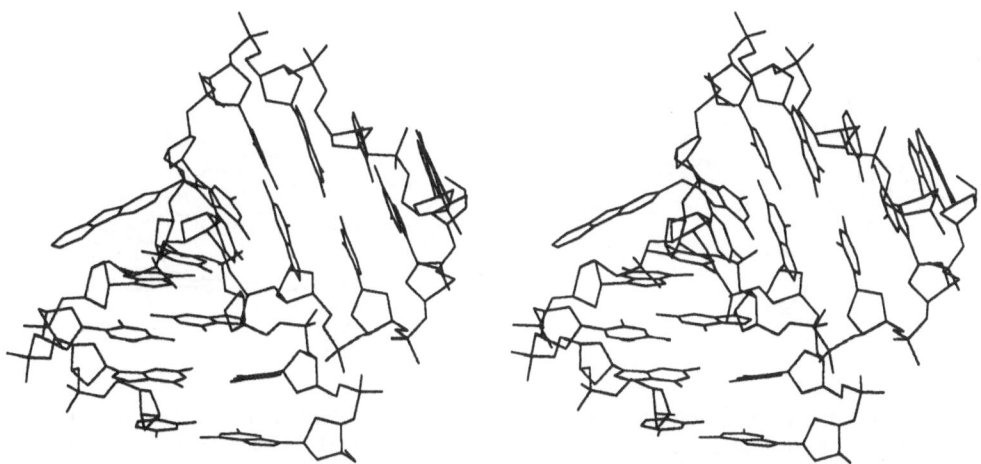

Figure 7. AF Modified Sequence II, with modified G in hydrogen bonding pattern shown in Figure 1a. $\alpha' = 92^{\circ}$, $\beta' = 309^{\circ}$, P(1) = 103$^{\circ}$, $\chi(5) = 235^{\circ}$, P(14) = 26$^{\circ}$. All other torsion angles are within B-DNA ranges[32]. See caption to Figure 3.

pairs in the protonated structures. In addition, our previous anti guanine model for AF modified DNA[13,14] is energetically important for the deacetylated aromatic amine. Intercalation-denaturation type structures are higher energy (about 20 or more kcal/mole above the lowest energy form) in duplexes with a single AF or AAF[8] lesion. However, such structures may well play an important mutagenic role at the mobile, single stranded replication fork.

Our studies have suggested some possible conformational views of aromatic amine modified DNA prior to, during and following replication. One possibility, appropriate for AF modified DNA, places the AF in the major groove of an essentially undeformed B-DNA duplex prior to replication (Figure 8a). If the carcinogen escapes repair before

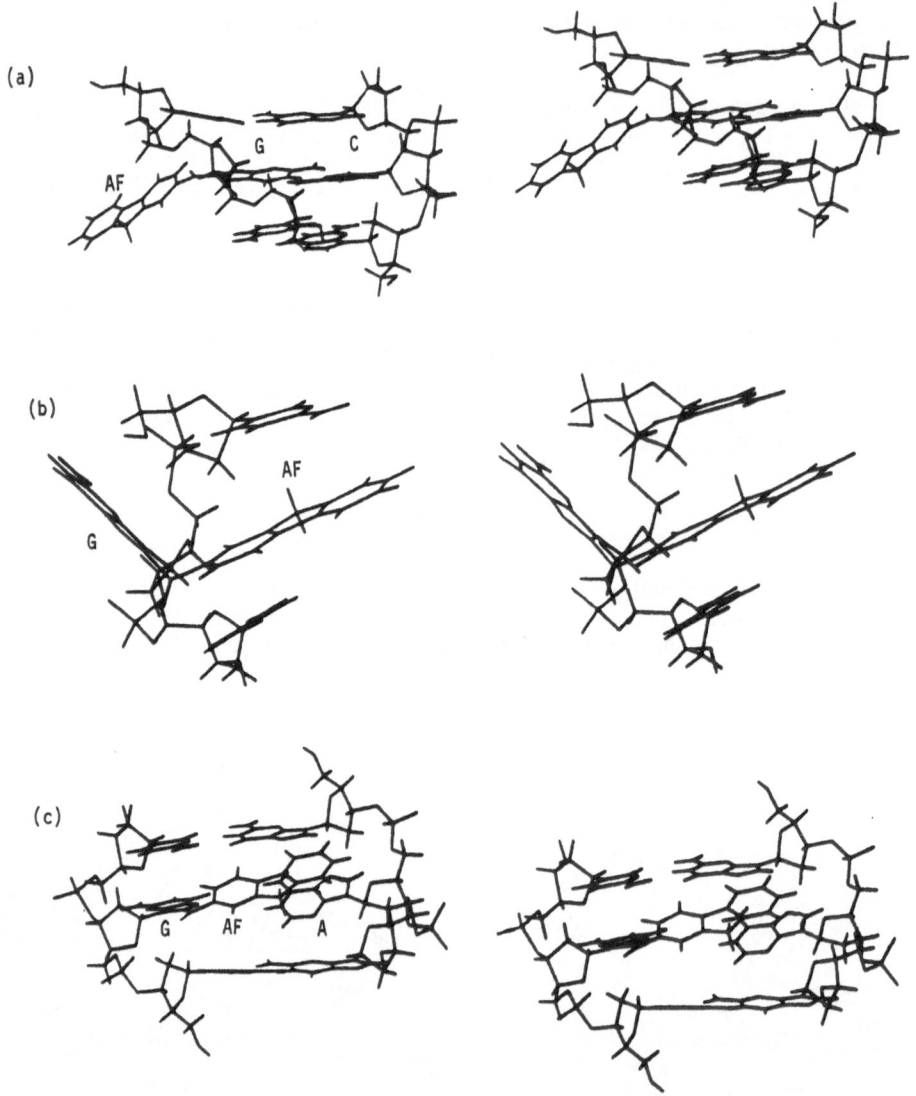

Figure 8. (a) d(CG-AF-C) d(GCG) in energy minimized B-DNA conformation similar to that of Figure 6. (b) Single stranded d(CG-AF-C) with AF stacked between C's. The modified displaced guanine is syn. Torsion angles are similar to those of reference 10, Figure 4b. (c) d(CG-AF-C) d(GAG) energy minimized trimer employed in build-up of structure shown in Figure 4. The conformation is very similar to the trimer in the 11-mer.

replication, a conformational rearrangement might take place at the mobile, single stranded replication fork, in which the modified guanine is rotated to the syn domain, either in a carcinogen-base stacked conformation (Figure 8b), or with all bases stacked, as in the modified strand of Figure 5. An error by the replication machinery could then produce an AF-modified G-A mismatch structure (Figure 8c). Alternatively, perhaps the carcinogen is placed in the minor groove of a B form duplex prior to replication (Figure 5). This is feasible for both AF and AAF[8] modified DNA. During replication the modified guanine would again be syn at the replication fork, and all bases could be stacked, or the carcinogen could be intercalated (Figure 8b). The replication machinery might then put an A opposite the modified G as in the AF-modified G-A mismatch structure, or cause a -2 deletion (Figure 9), as has been observed for AAF modification in an alternating C-G sequence[16].

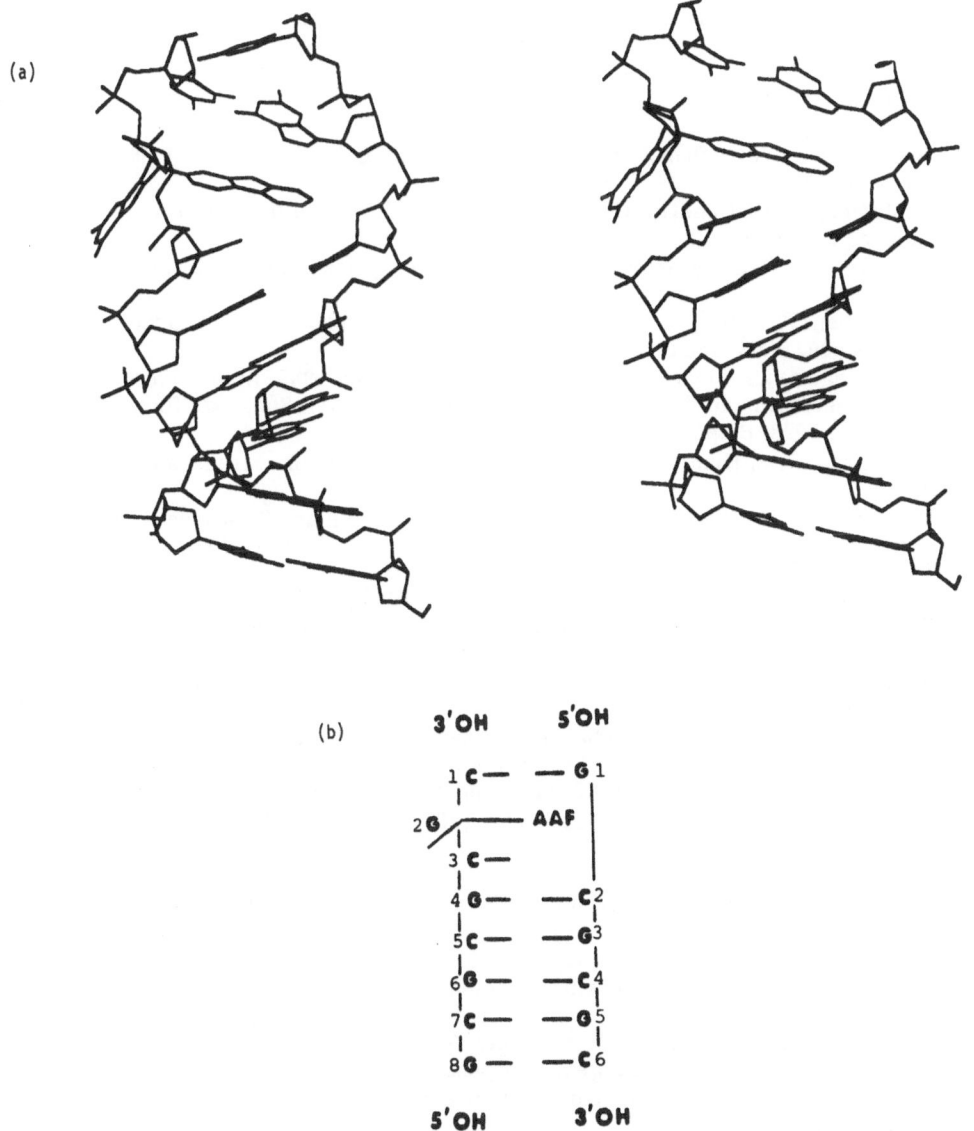

Figure 9. (a) Molecular view of a -2 deletion in Sequence II. (b) Interpretive diagram of (a). Adapted from reference 10, Figure 7.

Of course, many other mutagenic pathways are possible, including depurination[17], error-prone repair, perhaps induced by a Z form[16], and large genetic rearrangements[18], among others. Here we merely offer some possibilities suggested by our work to date. It is not inconceivable that numerous mechanisms do occur, and indeed the current literature on mutagenic outcome of AF and AAF modification[16,18-30], which indicates a diversity of mutational types in diverse systems, may also be indicative of diverse mechanisms.

ACKNOWLEDGEMENT

This research was supported by NIH CA-28038, DOE Contract DE-AC02-91ER60015 and DE-FG02-90ER60931, NSF DMB-8416009, and DOE Contract DE-AC05-84OR21400 with Martin-Marietta Energy Systems (BEH). SB gratefully acknowledges a Grand Challenge Award of computing time from DOE. SB also thanks Professor Dinshaw Patel and Professor Dezider Grunberger, Columbia University College of Physicians and Surgeons, for the opportunity to collaborate in the work described in reference 4, which first revealed the wedge type structure, in an AF-G-A mismatch.

REFERENCES

1. Saenger, W., 1984, "Principles of Nucleic Acid Structure", Springer Verlag, New York, p.119.
2. G. J. Quigley, G. Ughetto, G. A. van der Marel, J. H. van Boom and A. Rich, 1986, Science 232:1255.
3. L. Sowers, B. Ramsay-Shaw, M. Veigl and W. Sedgwick, 1987, Mutation Res. 177:201.
4. D. Norman, P. Abuaf, B. E. Hingerty, D. Live, D. Grunberger, S. Broyde and D. J. Patel, 1989, Biochemistry 28:7462.
5. B. E. Hingerty, S. Figueroa, T. Hayden and S. Broyde, 1989, Biopolymers 28:1195.
6. A. R. Srinivasan and W. Olson, 1980, Fed. Proc, Fed Am. Soc. Exp. Biol., 39:2199.
7. W. Olson, 1978, Biopolymers 17:1015.
8. R. Shapiro, B. E. Hingerty, and S. Broyde, 1989, J. Biomol. Struct. and Dynam., 7:439.
9. T. Schlick, B. E. Hingerty, C. S. Peskin, M. L. Overton and S. Broyde, 1990, "Theoretical Biochemistry and Molecular Biophysics", D. Beveridge and R. Lavery Eds., Adenine Press, Schenectady, NY.
10. S. Broyde and B. E. Hingerty, 1987, Nucleic Acids Res., 15:6539.
11. S. Arnott, P. Campbell and R. Chandrasekharan, 1976, "Handbook of Biochemistry and Molecular Biology", G. Fasman, Ed., CRC Press, Cleveland, Ohio, p. 411.
12. R. Dickerson, 1987, "Unusual DNA Structures," R. Wells and S. Harvey, Eds., Springer Verlag, NY.
13. S. Broyde and B. Hingerty, 1983, Biopolymers 22:2423.
14. B. E. Hingerty and S. Broyde, 1986, J. Biomolec. Struct. and Dynam, 4:365.
15. B. Hingerty, and S. Broyde, 1982, Biochemistry, 21:3243.
16. D. Burnouf, P. Koehl and R. P. P. Fuchs, 1989, Proc. Natl. Acad. Sci. USA, 86:4147.
17. L. A. Loeb, 1985, Cell 40:483.
18. A. Carothers, G. Urlaub, R. Steigerwalt, L. Chasin and D. Grunberger, 1986, Proc. Natl. Acad. Sci. USA 83:6519.
19. M. Bichara and R. P. P. Fuchs, 1985, J. Mol. Biol. 183:341.
20. N. Koffel-Schwartz, G. Maenhaut-Michel and R. P. P. Fuchs, 1987, J. Mol. Biol. 193:651.
21. L. J. Romano, D. L. Johnson, P. K. Gupta, T. M. Reid, M. S. Lee and C. M. King, 1988, "Carcinogenic and Mutagenic Responses to Aromatic Amines and Nitroarenes", C. M. King, L. J. Romano and D. Schuetzle, Editors, Elsevier, New York, p. 389.
22. M. Moriya, F. Takeshita, K. Peden, S. Will and A. P. Grollman, 1988, Proc. Natl. Acad. Sci. USA, 29:105.

23. P. K. Gupta, M-S. Lee and C. M. King, 1988, Carcinogenesis 9:1337.
24. D. Beranek, G. White, R. Heflich and F. Beland, 1982, Proc, Natl. Acad. Sci. USA 79:5175.
25. D. Beranek, J. Hardin, R. Heflich and F. A. Beland, (1989), Proc. Amer. Assoc. Cancer Res. 30:132.
26. V. Maher, J-L. Yang, M. C. M. Mah and J. J. McCormick, 1989, Mutation Res. 220:83.
27. W. Melchior, M. Marques and F. A. Beland, 1989, Proc. Amer. Assoc. Cancer Res. 30:130.
28 P. Gupta, D. Johnson, T. Reid, M-S. Lee, L. Romano and C. M. King, 1989, J. Biol. Chem. 264:20120.
29. S. Shibutani, M Moriya and F. Johnson, 1989, Proc. Amer. Assoc. Cancer Res. 30:128.
30. A. Carothers, A. Steigerwalt, G. Urlaub, L. Chasin and D. Grunberger, 1989, J. Mol. Biol. 208:417.
31. C. Altona and M. Sundaralingam, 1972, J. Am. Chem. Soc. 94:8205.
32. H. Drew, R. Wing, T. Takano, C. Broka, S. Tanaka, K. Itakura and R. Dickerson, 1981, Proc. Natl. Acad. Sci. USA 78:2179.

# DNA ADDUCT FORMATION DURING CHRONIC ADMINISTRATION OF AN AROMATIC AMINE

Miriam C. Poirier,[1] Nancy F. Fullerton,[2]
Henrik S. Huitfeldt,[3] Beverly A. Smith,[2] Henry C. Pitot,[4]
John M. Hunt,[5] and Frederick A. Beland[2]

[1]National Cancer Institute, Bethesda, Maryland 20892

[2]National Center for Toxicological Research,
Jefferson, Arkansas 72079

[3]The National Hospital, Oslo, Norway

[4]McArdle Laboratory for Cancer Research
University of Wisconsin, Madison, Wisconsin 53706

[5]University of Texas Medical School, Houston, Texas 77225

## INTRODUCTION

DNA adduct formation is considered to play an integral part in the initiation of tumors. Since most human exposure to chemical carcinogens is chronic, it is important to investigate DNA adduct formation in relation to preneoplasia and tumorigenesis in animal models using a similar type of exposure. In this communication, two different experimental systems are described in which rodents were continuously fed 2-acetylaminofluorene (AAF). In both experiments, DNA adducts were detected with antisera specific for the acetylated [$\underline{N}$-(deoxyguanosin-8-yl)-2-acetylaminofluorene; dG-C8-AAF] and deacetylated [$\underline{N}$-(deoxyguanosin-8-yl)-2-aminofluorene; dG-C8-AF] C8-substituted deoxyguanosine adducts of AAF. The first study compared the tumor incidence in the livers and bladders of mice chronically fed seven different doses of AAF for up to 33 months with the DNA adducts in these tissues after one month of feeding AAF at similar doses. In the second investigation, four different hepatocarcinogenesis protocols were used to induce enzyme-altered, preneoplastic foci in the livers of rats. The animals were then fed AAF for five to six days and the concentration of AAF-DNA adducts was determined in the foci and adjacent tissues. Taken together, these studies suggest that initiating events involving DNA adducts occur relatively early in the carcinogenic process, and that cells progressing to form tumors may not necessarily contain or be able to form DNA adducts. Furthermore, the number of adduct-related events required for tumor initiation appears to differ among tissues of the same animal.

## MATERIALS AND METHODS

### Mouse Experiments

In the first experiment, weanling female BALB/c mice (four to five per group) were fed either 30 or 150 mg [$^3$H]AAF (982 mCi per mmol) per kg diet for 21 days. In the second experiment, the animals were fed 30 or 150 mg AAF per kg diet for up to 28 days, while in the third study they were fed 0, 5, 10, 15, 30, 45, 60, 75, 100 or 150 mg AAF per kg diet for 28 days. Upon sacrifice, nuclei were prepared from livers as described in Basler et al. (1), and DNA was isolated from whole bladders and hepatic nuclei as previously reported (2). Hepatic DNA from mice fed radiolabeled AAF was hydrolyzed enzymatically to nucleosides, the adducts were separated by HPLC as described in Heflich et al. (3) and quantified by liquid scintillation counting. The amount of radioactivity associated with the bladder DNA was insufficient for analysis by HPLC; therefore, these samples were assayed by $^{32}$P-postlabeling using the contact transfer method of Lu et al. (4), after adduct enrichment by extraction with n-butanol as described in Gupta (5). Adduct levels in livers and bladders from mice fed nonradiolabeled AAF were determined by radioimmunoassay, as previously described (6), using a polyclonal rabbit antiserum elicited against N-(guanosin-8-yl)-2-acetylaminofluorene (G-C8-AAF) that was specific for both dG-C8-AF and dG-C8-AAF.

## Rat Experiments

Phenotypically altered liver foci were induced in male F344 rats by four different hepatocarcinogenesis protocols as described in Huitfeldt et al. (7). Briefly, protocol 1 involved the continuous feeding of 200 mg AAF per kg diet for 56 days. In protocols 2 and 3, rats were given a single initiating dose of 10 mg diethylnitrosamine per kg body weight 24 hours after a partial hepatectomy and then promoted with either phenobarbital (protocol 2) or sequential administration of ciprofibrate and phenobarbital (protocol 3). In protocol 4, donor rats were initiated with a single dose of 200 mg diethylnitrosamine per kg body weight and then given a Solt-Farber (8) selection regimen consisting of AAF and a partial hepatectomy. Donor liver cells were then transplanted into the livers of host rats that had been subjected to the AAF and partial hepatectomy selection regimen (9). Five to six days before sacrifice, rats in the last three protocols were fed 200 mg AAF kg diet. Immediately upon sacrifice, liver tissue was frozen and four serial 6-μm cryostat sections were prepared from each tissue block. The sections were stained for γ-glutamyltranspeptidase (GGT), canalicular adenosine triphosphatase (ATPase), glucose 6-phosphatase (G6Pase) and DNA adducts. The first three stains were histochemical; the latter was immunohisto-chemical, using an anti-G-C8-AF antiserum (7) with a fluorescein-labeled second antibody. Images of the tissue sections and enzyme-altered foci were traced, digitized and processed with a Hewlett Packard 9845B computer to make an image overlay of the three serial sections. These overlays were used to locate foci on the adduct-stained fluorescent slides. Fluorescein staining for adducts was examined with a Nikon Labophot microscope equipped with an epifluorescence attachment (HMX-HBO 100 watt lamphouse).

## RESULTS AND DISCUSSION

## DNA Adducts and Tumorigenesis in Mouse Liver and Bladder as a Result of Chronic AAF Feeding

The investigation of DNA adducts in mouse liver and bladder was modeled after a study that involved the chronic feeding of AAF to 20,880 female BALB/c mice at several relatively low doses (10). During that study, preneoplastic and neoplastic lesions were quantified after 18, 24 and 33 months of feeding. In the livers, a linear relationship was observed between the administered dose and the tumor incidence (Figure

1).    This was not the  case in the bladders  because at the four  lowest
doses the tumor incidence was similar to that seen in the controls, while
it rose sharply, approaching 100%, at the three highest doses (Figure 2).
In investigating DNA adducts with this animal model, it was our intention
to  obtain evidence of a biologically-effective dose  in two tissues that
following  the same carcinogen  exposure exhibited different  profiles of
tumorigenesis (11,12).

In  order to analyze critically the  relationship between DNA adduct
formation  and tumorigenesis in livers and  bladders of mice continuously
fed AAF it was necessary to  establish the nature of the  adducts formed
and  the kinetics of adduct accumulation  during continuous feeding.  The
first  question was addressed in animals fed radiolabeled AAF at doses of
30 and 150 mg AAF per kg diet for 21 days.  HPLC analysis of
enzymatically hydrolyzed liver DNA indicated the presence  of a single
adduct,  dG-C8-AF (not shown).   Due to the  limited quantity of  bladder
DNA,  similar analyses could not  be conducted with these  samples; thus,
the  bladder DNA was  analyzed by $^{32}$P-postlabeling,  which gave a single
adduct corresponding to that observed in a dG-C8-AF-modified DNA standard
(12).   Although adduct analyses were  conducted with DNA from  the whole
bladder,  immunohistochemical examinations demonstrated that the majority
of  adducts were associated with the bladder epithelial cells (Huitfeldt,
unpublished data).  In order to assess the kinetics of adduct formation,
mice  were sacrificed at various  times during one month  of AAF feeding.
With  both  tissues,  there was a rapid increase in  adduct levels  for
approximately 21 days of feeding,  followed by a much slower  increase
approached  that steady-state  conditions during  the last  week (12).

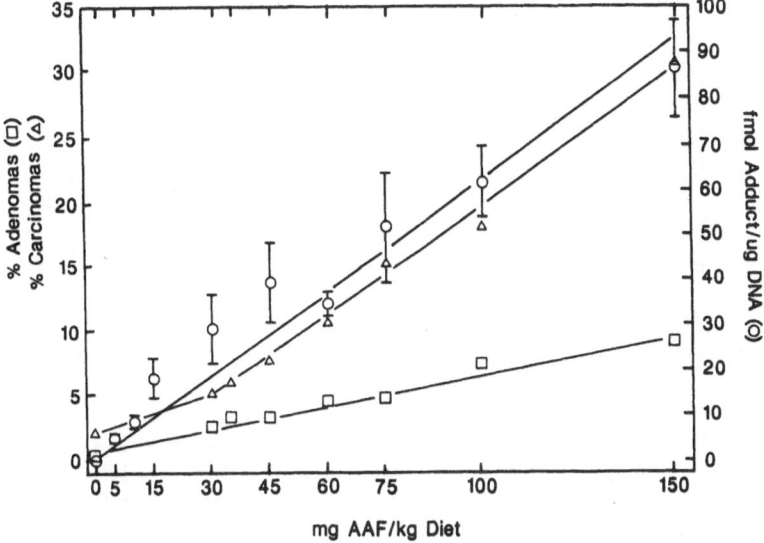

Fig.  1. Relationship between the dose  of AAF fed to female  BALB/c mice
and  the concentration of dG-C8-AF in hepatic  DNA after 28 days
of  feeding ( O ),  and  the adenoma ( □ )  and carcinoma ( Δ )
incidence  after 24 months of  feeding.  The adduct levels  were
determined  by radioimmunoassay and are expressed  as the mean +
standard  deviation for four to five mice (11).  The adenoma and
carcinoma  data are from animals sacrificed or dying between 638
and  863 days of  feeding AAF at  the doses indicated (13).

Therefore, a 28-day feeding period was selected for these studies, and radioimmunoassays were designed to measure only dG-C8-AF.

In the livers of mice chronically fed AAF for 28 days, a linear increase with dose was observed in the steady-state DNA adduct levels (11; Figure 1). Eight doses were employed in these experiments, three of which were lower than those used in the original tumor study. Figure 1 also shows the 24-month incidence of liver adenomas and carcinomas as a function of dose (13). It is evident from these curves that liver adenomas, carcinomas and adducts are all linearly related to dose. Similar data are presented in Figure 2 for the bladder, except that bladder hyperplasia is shown instead of adenomas. Again, DNA adduct concentrations were linearly correlated to the administered dose through the entire dose range, with the adduct levels generally being two to three-fold higher than the ones observed in the liver (11). The incidence of bladder hyperplasia (14) and carcinoma (15) at 24 months showed nonlinear profiles that were similar to each other, with hyperplasia being more frequent. These profiles were very different from those observed in the liver. At 30, 35 and 45 mg AAF per kg diet, the incidence of bladder hyperplasia and carcinoma was virtually the same as that in unexposed controls, while it increased sharply at higher doses

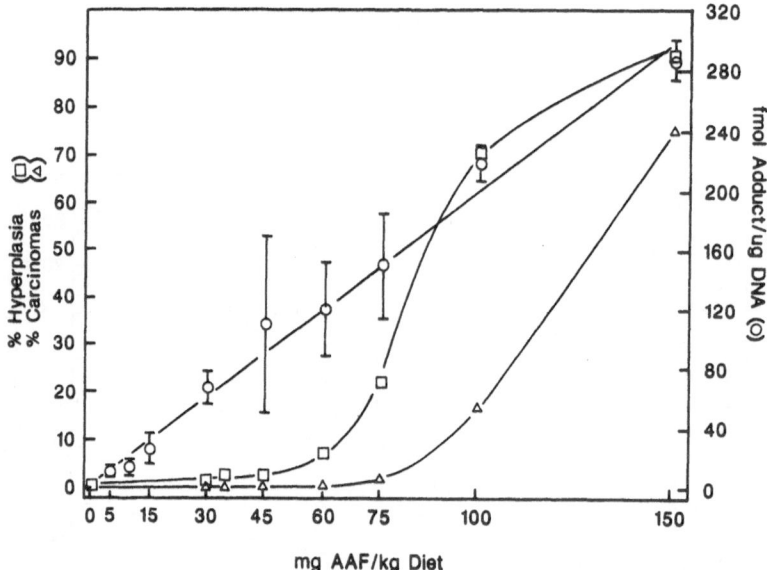

Fig. 2. Relationship between the dose of AAF fed to female BALB/c mice and the concentration of dG-C8-AF in bladder DNA after 28 days of feeding (O), and the hyperplasia (□) and carcinoma (Δ) incidence after 24 months of feeding. The adduct levels were determined by radioimmunoassay and are expressed as the mean ± standard deviation for four to five mice (11). The hyperplasia data are from animals sacrificed after 24 months of feeding AAF at the doses indicated (14). The carcinoma data are from animals sacrificed or dying between 638 and 864 days of feeding AAF (15).

with the slopes being much steeper than that observed for DNA adducts. Thus, much higher levels of adducts are associated with a particular tumor incidence in the bladder as compared to the liver, and DNA adducts can be present at doses that are essentially nontumorigenic in the bladder. Since the shapes of the dose-response curves for adenomas in the liver and hyperplasia in the bladder are similar to those for carcinomas in the liver and bladder, respectively, it appears that the adduct-related events important in initiation have occurred by the time the benign lesions are detected.

The nonlinearity of tumorigenesis in the bladder suggests that multiple adduct-related events are necessary to complete initiation. The number of DNA adduct-related events required for initiation can be derived from the following equation (16,17):

$$-\ln(1-P) = A^n$$

where P is the fraction of mice with tumors, A is the adduct concentration and n is the number of DNA adduct-related events. If the log of $-\ln(1-P)$ [expressed here as % tumor incidence] is plotted against the log of A, the slope of the resultant line will be n. When such an analysis is conducted for mouse liver adenomas and carcinomas, the slopes of the lines are 1.1 and 1.6, respectively (Figure 3). For mouse bladder, the slopes of the lines for hyperplasia and transitional cell carcinomas are 3.8 and 4.2 (Figure 3). Therefore, these calculations imply that approximately three-fold more adduct-related events are required in the

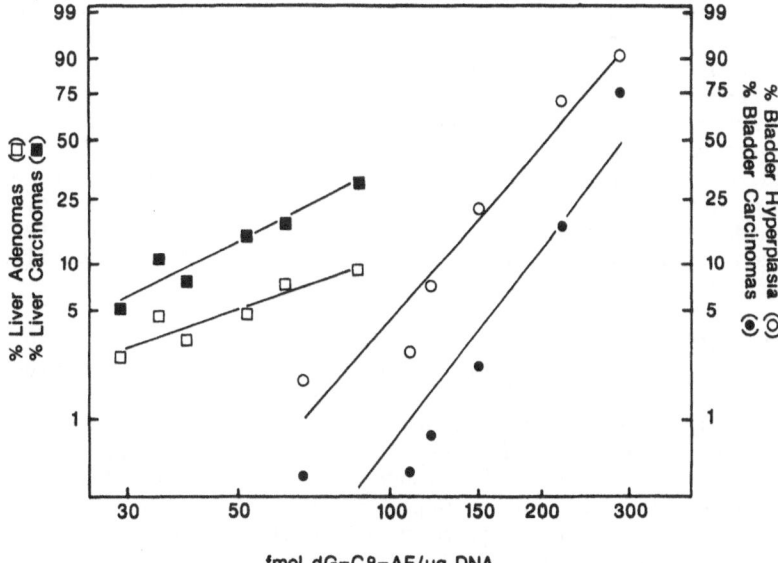

fmol dG-C8-AF/ug DNA

Fig. 3. Determination of the number of adduct-related events necessary for the induction of liver adenomas (□) and carcinomas (■), and bladder hyperplasia (O) and carcinomas (●) in female BALB/c mice fed AAF.

bladder to induce the same incidence of benign or malignant tumors as found in the liver. It is possible that other events not related to AAF-DNA adducts are contributing to tumor initiation in the liver, especially since the spontaneous tumor incidence in this tissue was quite high (17%) by 33 months (15). These calculations also suggest that the adduct-related events necessary for initiation were complete prior to the formation of hyperplasia in the bladder and adenoma in the liver.

## Lack of AAF-DNA Adducts in Rat Liver Enzyme-Altered Preneoplastic Foci

In the livers of rats chronically fed AAF or given various initiation and promotion protocols, enzyme-altered foci, which can only be detected by histochemical staining, form in morphologically normal liver. These foci are considered to be preneoplastic, are clonal in origin and have an apparent growth advantage (18). Enzyme-altered foci can contain several biochemical changes including elevated GGT, decreased canalicular ATPase and decreased G6Pase. In addition to the phenotypic changes that are detected by enzyme staining, the foci have lost some of their normal capacity for oxidative carcinogen metabolism and have increased detoxification pathways (18). Thus, we have investigated whether or not AAF-DNA adducts are formed in these foci when AAF was fed for a period of several days after the foci were established.

In the present experiments, foci were induced in rat livers by four conventional hepatocarcinogenesis protocols (see Materials and Methods). To test the capacity of foci and morphologically-normal liver to form DNA adducts, the rats were fed AAF for five to six days prior to sacrifice. Upon sacrifice, four serial sections were obtained from frozen liver blocks; three were stained for GGT, ATPase and G6Pase, while the fourth was used for DNA adduct detection by fluorescent immunohistochemistry. Patterns of enzyme-altered foci were obtained by image overlay, and each focus was scored for each of the three histochemical markers to give a total of seven possible phenotypic combinations (see Table 1). By using the image overlays to localize areas containing enzyme-altered foci on the fluorescent slides, comparisons could be made between the adduct concentrations in enzyme-altered areas and adjacent normal tissue. A representative example is shown in Figure 4. A total of 572 foci were detected and all appeared dark as compared to the fluorescence observed in the surrounding liver (Table 1). Due to technical problems, not all of the foci that were identified histochemically could be located on the fluorescent slides, but all of the foci that could be identified were negative for AAF-DNA adducts.

This study demonstrates that adduct quantities in enzyme-altered preneoplastic liver foci were below the limit of immunohistochemical detection, which is approximately 30 fmol per µg DNA (19). Since the adduct levels in whole livers under these conditions are in the range of 200-300 fmol per µg DNA, the adduct concentrations in tappear to be at least ten-fold lower than those in the surrounding liver. Similar results have been obtained by Gupta et al. (20) using [32]P-postlabeling for adduct detection. Thus, phenotypically-altered hepatic foci, induced by different carcinogenesis protocols, appear to have lost the capacity to metabolize the carcinogen AAF to intermediates that bind to DNA. The growth advantage exhibited by these putatively preneoplastic foci may be related to their loss of the capacity to metabolize this and other xenobiotics by phase I microsomal activating enzymes, accompanied by an increase in phase II detoxifying and conjugating enzymes (18). In any event, the overall impression obtained is that cells that later become tumors do so by changes in phenotypic expression resulting in a

Table 1. AAF-DNA Adduct Formation in Rat Liver Enzyme-Altered Foci of Seven Different Phenotypes Induced by Four Different Hepato-carcinogenesis Protocols[a]

| Enzyme phenotype | Number of AAF-DNA adduct-negative foci | | | |
|---|---|---|---|---|
| | Protocol 1 | Protocol 2 | Protocol 3 | Protocol 4 |
| GGT$^+$, ATPase$^-$, G6Pase$^-$ | 19 | 75 | 10 | 24 |
| GGT$^+$, ATPase$^-$ | 13 | 126 | 4 | 7 |
| GGT$^+$, G6Pase$^-$ | 5 | 0 | 7 | 6 |
| ATPase$^-$, G6Pase$^-$ | 12 | 6 | 15 | 8 |
| GGT$^+$ | 16 | 49 | 30 | 8 |
| ATPase$^-$ | 4 | 44 | 23 | 7 |
| G6Pase$^-$ | 6 | 4 | 36 | 8 |
| Total | 75 | 304 | 125 | 68 |

[a]Enzyme-altered foci were induced by the protocols described in Materials and Methods. Serial sections were prepared for staining for the enzymes indicated and for incubation with an anti-G-C8-AF antiserum and a fluorescein-labeled second antibody.

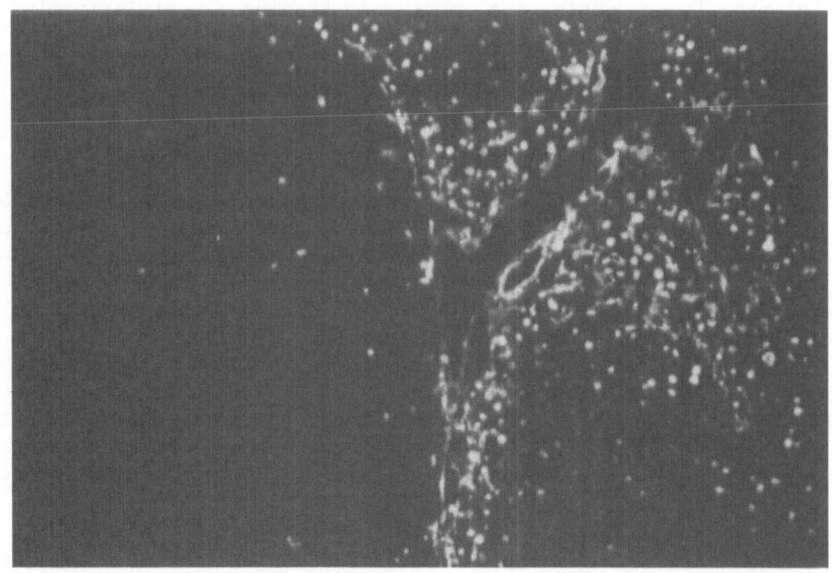

Fig. 4. Enzyme altered-focus from a rat treated by protocol 1. Nuclear AAF-DNA adducts appear as fluorescent areas. The focus is on the left side of the figure, and is nearly devoid of adducts compared to the normal surrounding liver, which is on the right.

subsequent inability to form DNA adducts. Presumably the initial events that induced the observed changes in phenotypic expression were DNA adduct-related and occurred very early during the carcinogen exposure.

## ACKNOWLEDGMENT

We thank Cindy Hartwick for helping prepare this manuscript. This work was supported in part by Grants CA07175 and CA22484 from the National Cancer Institute to H.C. Pitot.

## REFERENCES

1. Basler, J., N.D. Hastie, D. Pietras, S-I. Matsui, A.A. Sandberg, and R. Berezney (1981). Hybridization of nuclear matrix attached deoxyribonucleic acid fragments. Biochemistry 20:6921-6929.
2. Beland, F.A., N.F. Fullerton, and R.H. Heflich (1984). Rapid isolation, hydrolysis and chromatography of formaldehyde-modified DNA. J. Chromatography 308:121-131.
3. Heflich, R.H., Z. Djurić, Z. Zhuo, N.F. Fullerton, D.A. Casciano, and F.A. Beland (1988). Metabolism of 2-acetylaminofluorene in the Chinese hamster ovary cell mutation assay. Environ. Molecul. Mutagen. 11:167-181.
4. Lu, L-J.W., R.M. Disher, M.V. Reddy, and K. Randerath (1986). $^{32}$P-Postlabeling assay in mice of transplacental DNA damage induced by the environmental carcinogens safrole, 4-aminobiphenyl, and benzo(a)pyrene. Cancer Res. 46:3046-3054.
5. Gupta, R.C. (1985). Enhanced sensitivity of $^{32}$P-postlabeling analysis of aromatic carcinogen:DNA adducts. Cancer Res. 46:5656-5662.
6. Poirier, M.C., J.M. Hunt, B. True, B.A. Laishes, J.F. Young, and F.A. Beland (1984). DNA adduct formation, removal and persistence in rat liver during one month of feeding 2-acetylaminofluorene. Carcinogenesis 5:1591-1596.
7. Huitfeldt, H.S., J.M. Hunt, H.C. Pitot, and M.C. Poirier (1988). Lack of acetylaminofluorene-DNA adduct formation in enzyme-altered foci of rat liver. Carcinogenesis 9:647-652.
8. Solt, D. and E. Farber (1976). New principle for the analysis of chemical carcinogenesis. Nature 263:701-703.
9. Hunt, J.M., M.T. Buckley, B.A. Laishes, and H.A. Dunsford (1985). Immunological approaches to the purification of putative premalignant hepatocytes from genotypic mosaic rat livers. Cancer Res. 45:2226-2233.
10. Staffa, J.A. and M.A. Mehlman (1979). Innovations in Cancer Risk Assessment (ED$_{01}$ Study), J. Environ. Path. Toxicol. 3:1-246.
11. Beland, F.A., N.F. Fullerton, T. Kinouchi, and M.C. Poirier (1988). DNA adduct formation during continuous feeding of 2-acetylaminofluorene at multiple concentrations. In: H. Bartsch, K. Hemminki and I.K. O'Neill (eds.), Methods for Detecting DNA Damaging Agents in Humans: Applications in Cancer Epidemiology and Prevention, IARC Scientific Publications No. 89, Lyon, pp. 175-180.
12. Beland, F.A., N.F. Fullerton, T. Kinouchi, B.A. Smith, and M.C. Poirier (1989). DNA adduct formation in relation to tumorigenesis in mice chronically fed 2-acetylaminofluorene. In: D.E. Stevenson, R.M. McClain, J.A. Popp, J.M. Ward, P.J. Slaga, and H.C. Pitot (eds.), Mouse Liver Carcinogenesis: Mechanisms and Species Comparisons, Alan R. Liss, New York, in press.
13. Frith, C.H., R.L. Kodell, and N.A. Littlefield (1979). Biologic and morphologic characteristics of hepatocellular lesions in BALB/c female mice fed 2-acetylaminofluorene. J. Environ. Path. Toxicol. 3:121-138.

14. Littlefield, N.A., D.L. Greenman, J.H. Farmer, and W.G. Sheldon (1979). Effects of continuous and discontinued exposure to 2-AAF on urinary bladder hyperplasia and neoplasia. J. Environ. Path. Toxicol. 3:35-54.

15. Farmer, J.H., R.L. Kodell, D.L. Greenman, and G.W. Shaw (1979). Dose and time response models for the incidence of bladder and liver neoplasms in mice fed 2-acetylaminofluorene continuously. J. Environ. Path. Toxicol. 3:55-68.

16. Frei, J.V. and P.D. Lawley (1980). Thymomas induced by simple alkylating agents in C57BL/Cbi mice: kinetics of the dose response. J. Natl. Cancer Inst. 64:845-856.

17. Frei, J.V. (1980). Methylnitrosourea induction of thymomas in AKR mice requires one or two "hits" only. Carcinogenesis 1:721-723.

18. Farber, E. and D.S.R. Sarma (1986). Chemical carcinogenesis: the liver as a model. Pathol. Immunopathol. Res. 5:1-28

19. Huitfeldt, H.S., E.F. Spangler, J.M. Hunt, and M.C. Poirier (1986). Immunohistochemical localization of DNA adducts in rat liver tissue and phenotypically altered foci during oral administration of 2-acetylaminofluorene. Carcinogenesis 7:123-129.

20. Gupta, R.C., K. Earley, and F.F. Becker (1988). Analysis of DNA adducts in putative premalignant hepatic nodules and nontarget tissues of rats during 2-acetylaminofluorene carcinogenesis. Cancer Res. 48:5270-5274.

AERIAL OXIDATION OF ACETYLAMINOFLUORENE–DERIVED DNA ADDUCTS

Shinya Shibutani, Robert Gentles, and Francis Johnson*

Departments of Pharmacological Sciences and Chemistry
State University of New York at Stony Brook
Stony Brook, New York 11794

Introduction:

The first report of chemical carcinogenesis by an aromatic amine emanated from Japan (1), and this together with the separate findings in the United States fifty years ago, that both 2-naphthylamine (2) and N-acetyl-2-aminofluorene (1) (3) are carcinogenic in mammalian species, has led to extensive biological studies on this class of material (4). The chemical basis of tumor induction involves metabolic activation of the arylamines to N-hydroxy compounds followed by enzymic acetylation or sulfation, illustrated in the case of 1 by the formation of 2a-2c. Compounds of the latter type are regarded as the ultimate carcinogens which on solvolysis in (physiological) solution generate arylnitrenium ions that form adducts with DNA bases. These adducts and possibly their further transformation products are regarded as being responsible ultimately for the subsequent biological events leading to cancer.

The metabolic products in the case of 1 have been shown to react with the guanine moiety in DNA, via the nitrenium ion, at C-8 (C-aminoarylation) and at the 2-amino group via the tautomeric 3-arylcation (N-arylation) (5-8). Further studies of the C-8 adducts either as the nucleosides (3 and 4 derived from 2), or as residues in oligomeric DNA, have indicated (9,10) that although they are stable at neutral pH they undergo a rapid degradation in even mild alkali to give ring-opened products that were characterized in the case of 4 as the C-1' epimers 5a and 5b (9), and which appear to be resistant to further alkaline degradation. Other adducts derived from the interaction of 2 with DNA have not been characterized but a labile N-7 guanine adduct has been postulated by Tarpley et al (11) to account for the fact that some of the derivatives released at pH 7.4 from the treated DNA are formed in what appears to be a depurination process.

Our interest in the chemistry of these DNA adducts sprang from our attempts to deacetylate, under basic conditions, the AAF group in single-stranded oligomers containing one dG(C-8)AAF residue (postsynthetic introduction of AAF). We found that this led to substantial chain cleavage similar to that observed by Stohrer during the ammonia deprotection step, when he attempted the *de novo* synthesis of oligomers containing a dG(C-8)AF residue (12-14). However, we found in our reactions (as did Stohrer) that the inclusion of 2-mercaptoethanol in the hydrolysis mixture leads to almost

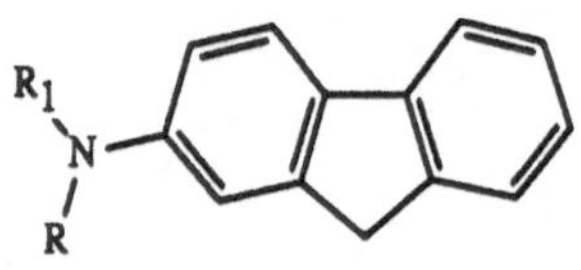

1   R = H,      R$_1$ = H,
2a  R = H,      R$_1$ = OH,
2b  R = H,      R$_1$ = OAc or OSO$_3$H,
2c  R = Ac,     R$_1$ = OAc,

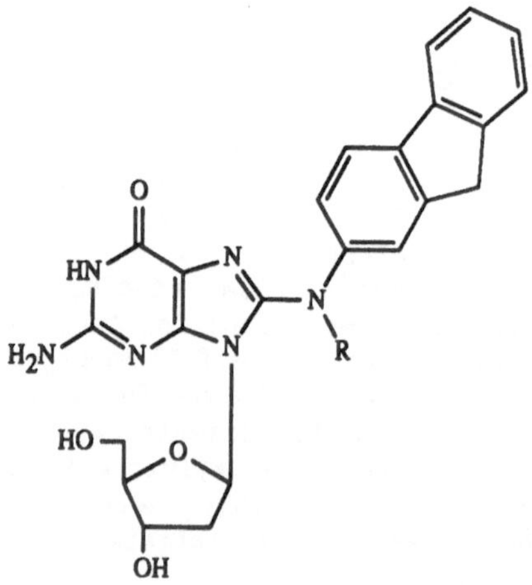

3   R = Ac ; dG(C8)AAF
4   R = H ; dG(C8)AF

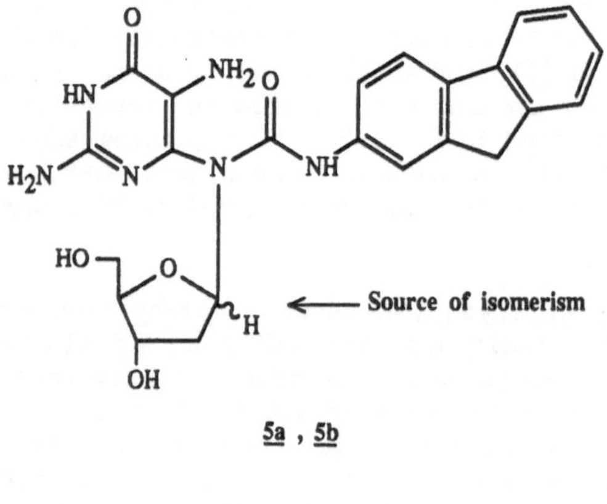

← Source of isomerism

5a , 5b

quantitative yields of the desired dG(C-8)AF-containing oligomers, with essentially no degradation. This gave credibility to the suggestion by Stohrer that an aerial oxidation was compromising the system (12,13). Because this type of oxidation appeared to be little recognized and its mechanism not understood, we undertook an examination of the reaction at the level of the nucleosides dG(C-8)AAF (3), and dG(C-8)AF (4). We have also made extensive studies at the oligomer level and although these results are not reported here, they parallel completely those obtained with the monomers.

Results: When dG(C-8)AAF (3) is treated in air at room temperature with aqueous alkali, it suffers a rapid loss of the acetyl group giving rise to dG(C-8)AF (4). The latter however, is itself rapidly destroyed and three new major products designated I, II, and III, can be observed (Fig. 1A) by HPLC analysis together with traces of two other materials IV and V.

When the reaction is conducted anerobically (Fig. 1B) or in the presence of an antioxidant (Fig. 1C) such as 2-mercaptoethanol (MSH) or ascorbic acid (data not shown) the dominant product is the deacetylated compound dG(C-8)AF (4) and only traces of I, II, and III are produced, while IV, and V are virtually absent. When the hydrolysis reaction is conducted in air at 75° (Kriek and Westra conditions), the small amount of compound III which is produced initially, suffers further degradation and quickly vanishes from the system (Fig. 2A). In fact a temperature study revealed the curious phenomenon that at higher temperatures compounds I and II are the dominant products whereas at 10°C, III is the main product (Fig.3). Under all hydrolysis conditions tested it was further found that compounds I and II are not interconvertible and do not give rise to III, neither does III give rise to I or II. However, when the hydrolysis of 4 is conducted at pH 8 over a period of 5 days at 37°C the major product becomes IV (Fig. 4A) at the expense of III. Morever IV, at pH 13, is rapidly converted to V which is stable under all pH conditions. All of these compounds (I-V) have been isolated and studied, but it was obvious at an early stage that the overall degradative pathway is similar to the much-studied (15-18) alkaline oxidation of uric acid.

Structure Studies on Compounds I-V - Compounds I and II gave [1]HNMR spectra which quickly identified them as the two compounds isolated by Kriek and Westra (9) from the alkaline hydrolysis of 3. However, the FAB positive and negative ion mass spectra of these substances show that each has a molecular ion at m/z 462 rather than the m/z 464 previously reported (9). This leaves no doubt that these materials are products that arise via an oxidation mechanism rather than by simple alkaline hydrolysis. Although not completely conclusive, the mass spectral degradation patterns suggest that these substances are epimers represented by the spiro structures 6a and 6b and with which the [1]HNMR spectral data is still in agreement. The isomeric relationship of I and II is now much better ascribed to differences at the central spiro carbon atom rather than at the anomeric C-1' atom. The latter assignment (9) suffered from the lack of an obvious mechanistic pathway to explain a base-induced epimerization at this site. It now appears that I and II (6a and 6b) are cyclic reaction-path analogs of 7 a skeletal-rearrangement intermediate postulated to occur alóng the uric acid-allantoin-uroxanate oxidative pathway (16).

The third compound III according to [1]HNMR data appears to be a 4:1 mixture of inseparable isomers to which we assign structures 8a and 8b although the isomeric structures 9a and 9b cannot really be excluded at this point. The epimeric hydrogen atoms at C-5 appear as non-exchangeable ($D_2O$) singlets at 5.44 and 5.48 respectively. The positive and negative ion FAB-mass spectra show good correlation exhibiting parent ions at m/z 437 and 435 respectively, identifying the molecular weight as 436 daltons. Significantly,

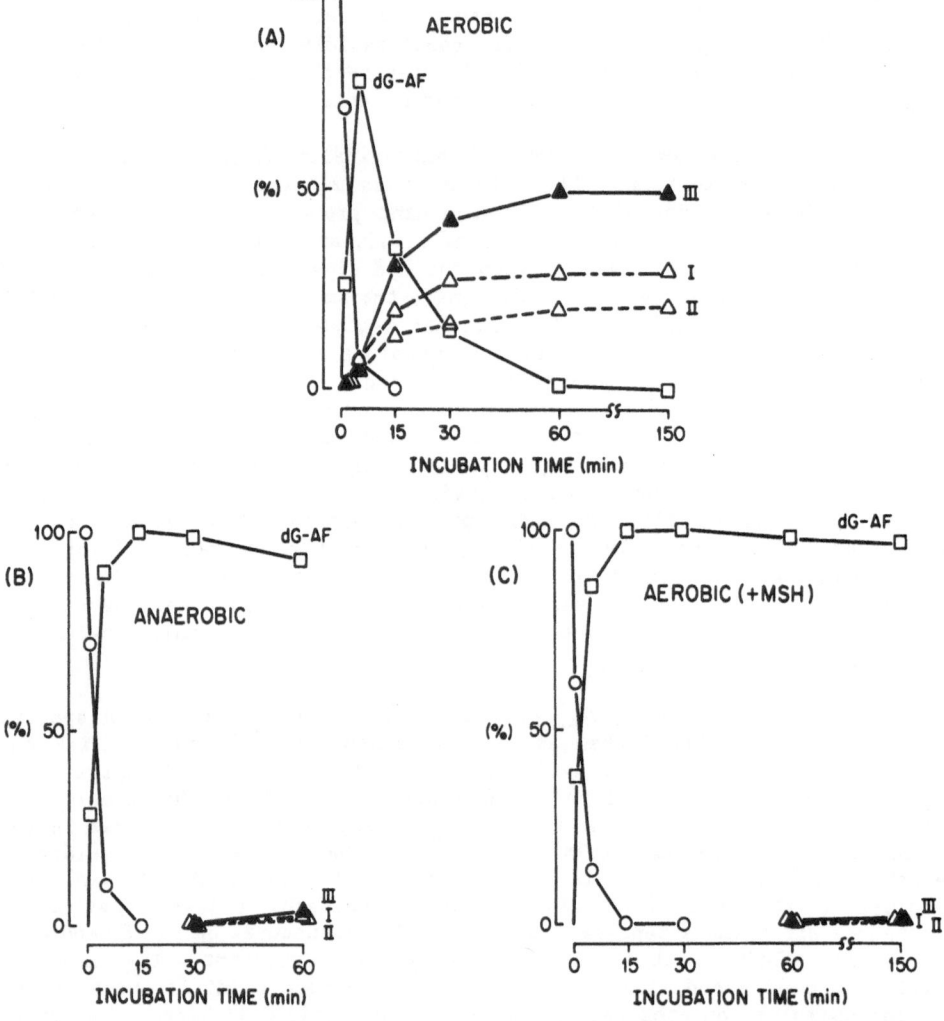

Figure 1. (A) A sample of dG(C-8)AAF (5 µg) was incubated with 1 N NaOH (100 µl) at 37°C (pH 13). (B) Conditions were identical to those used in A except that He was bubbled through the NaOH solution prior to the addition of dG(C-8)AAF and also during the reaction. (C) Conditions were identical to those used in A except that the solution was made 0.25 N in MSH (pH 12.93).

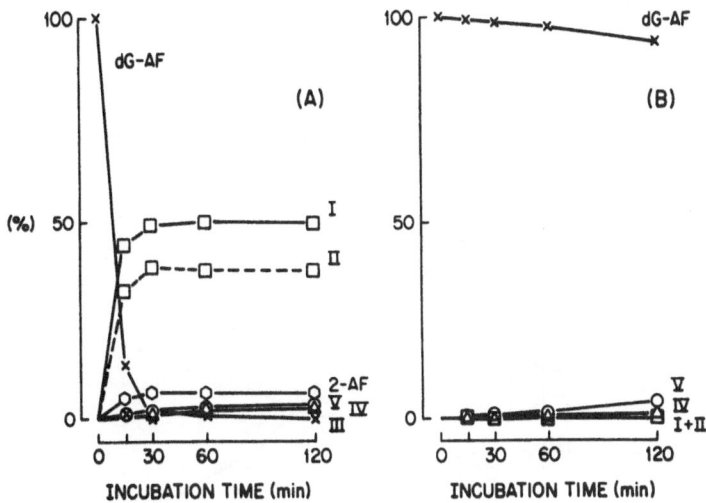

Figure 2. (A) The conditions used were identical to those described by Kriek and Westra (9) at 75°C except that dG(C-8)AF was used as the substrate instead of dG(C-8)AAF. (b) The same reaction conditions in the presence of 0.1 N MSH.

**6a** and **6b** (I and II)

**7**

**8a** and **8b** (III)

**10**

**9a** and **9b**

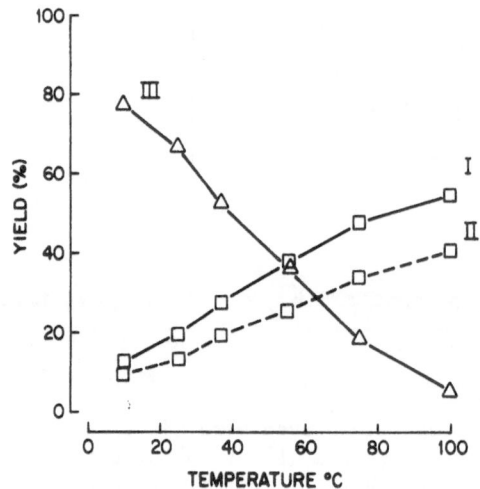

Figure 3. A sample of dG(C-8)AAF (5 μg) was
incubated with 1 N NaOH at various
temperatures (°C) and times, namely:
10° (8 h), 25° (4 h), 37° (2.5 h),
56° (15 min), 75° (10 min), m 100°
(5 min). These times were chosen to
minimize the degradation of Compound
III.

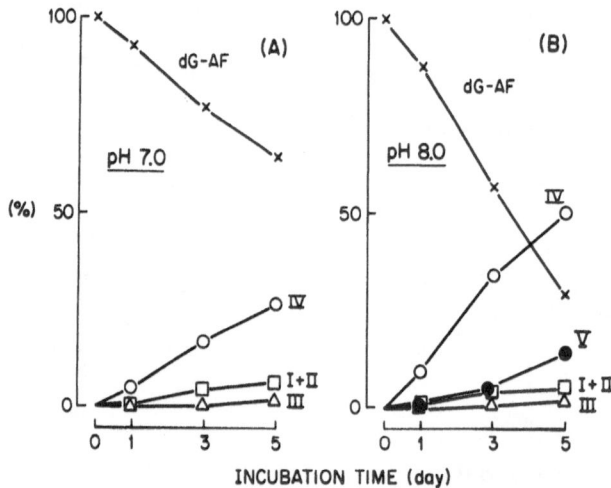

Figure 4. In both (A) and (B) a sample of dG(C-8)AF
(5 μg) was incubated at 37° with 200 μL of
100 mM buffer, adjusted to the appropriate
pH with HCl. The same results were obtained
in 50 mM sodium phosphate buffer.

there is a positive ion peak at m/z 340 (negative ion peak at m/z 338) corresponding to the loss of the imidazolone ring. Subsequent loss of the sugar residue to give the fluorenylguanidine ion is indicated by peaks at m/z 207 (+ ve ion) and m/z 205 (-ve ion). The structure 8 postulated for III is analogous to allantoin (10) again a well-established oxidative degradation product of uric acid (17,18).

Compound V was easily identified as N-(2-fluorenyl)guanidine by comparison with an authentic sample. Its isolation as an end product confirms that the overall reaction involves an oxidative mechanism, otherwise the end product should have been the corresponding N-(2-fluorenyl)urea, no trace of which could be found in the reaction mixtures. There remains only compound IV which lacks a purine residue and whose UV spectrum is identical to that of V. Accordingly it is assigned structure 11 being the hydrolytic intermediate between III and V. It is interesting to note also that we have found that III (8a, 8b) is unstable in aqueous buffer at pH 7.0 and has a half-life of 6.9 days, giving rise almost exclusively to IV (11). However, above neutral pH, compound V (12) rapidly appears also (Fig.4B) and this may be the material that was isolated in small yield by Tarpley et al. (11) from an "hydrolysis" at pH 7.4 of an oligomer containing dG(C-8)AAF. This was characterized as containing no purine residue but having displayed a high aromatic hydrogen (to other hydrogen) ratio in its $^1$HNMR spectrum. In addition, the quoted (11) UV spectrum is in agreement with that found for V (12).

The Mechanism of Oxidation and the Role of the 2-Amino Group in the Chemical Reactivity of Guanine-An Implication for the Determinant of Hot Spots in DNA Damage and Mutation.- The mechanistic pathways that we consider to be operating in the oxidative degradation of 4 are shown in Schemes 1-4. Under basic conditions the 1-position is deprotonated (Scheme 1) and the redistribution of the resulting anion increases the electron density at the 5-position a process undoubtedly aided by the presence of the arylimino group at C-8. Subsequent oxidation at C-5 generates a radical which combines with the oxygen radical anion prduced concomitantly. Exchange of the hydroperoxy group with water to generate 13 then might be expected to occur easily by an elimination-addition mechanism in view of the amine hemicetal nature of the system and the presence of an adjacent carboxyl group. The further transformations of 4 to produce 6 and/or 9 are illustrated in Scheme 2. These are highly temperature dependant. At higher temperatures the benzilic acid-like rearrangement (17) dominates (pathway A) producing the pair of spiro-diastereomers 6. At lower temperatures the rate of the rearrangement diminishes and the hydrolytic route (pathway B) is dominant. The latter proceeds through a $\beta$-imino carboxylic acid which then undergoes spontaneous decarboxylation. Facile decarboxylations of this type of system are well-documented in the chemical literature. The tautomeric nature of the enolic intermediate in this process leads to the diastereomers 9a and 9b. To this point the pathways find parallel in the recent oxidation studies (18, 19) on uric acid, except that in the two series the stabilities of the related intermediates towards alkali, are quite different. Whereas 6a and 6b (I and II) are stable, 7 is not. On the other hand, allantoin (10) is stable to base, but 9a and 9b are not. The latter appear to undergo an isomerization to the diasteromers 8a and 8b (Scheme 3). Although these isomers can be isolated as a mixture from low temperature basic reactions, they suffer further degradation, first loss of the imidazolidine ring to give IV, then the sugar group to produce the stable end product 12. Of great interest is the fact that the oxidative degradation of 4 takes place albeit slowly, under essentially neutral pH conditions. Here anion formation at N-1 cannot be a factor and it seems highly likely that under these circumstances, it is the 2-amino group that is the source of the increased electron density at the C-5 position as noted in 14.

SCHEME 1

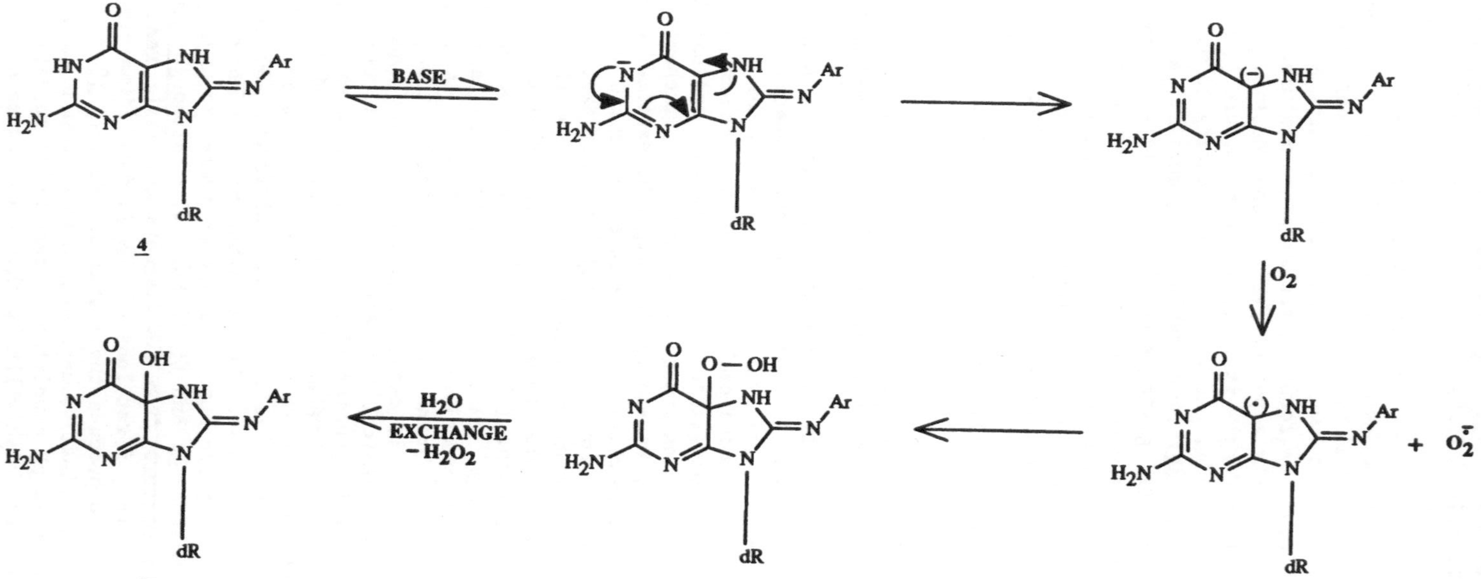

SCHEME 2

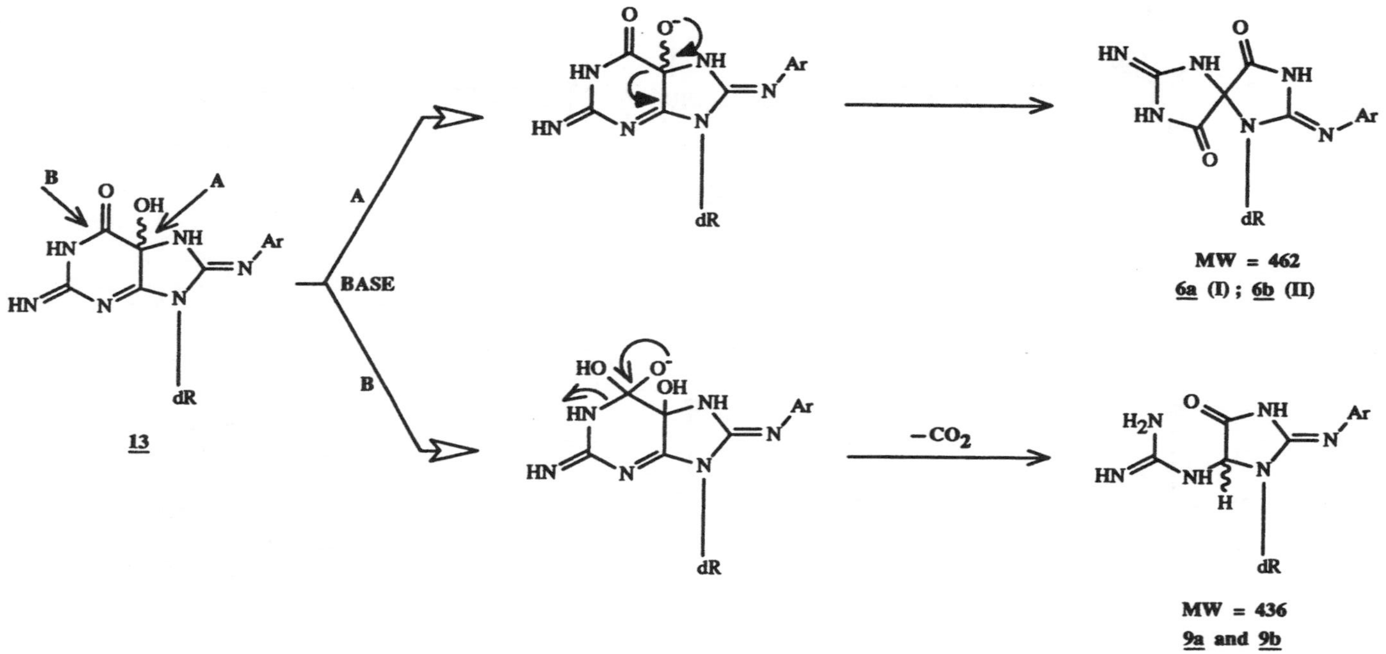

MW = 462
**6a** (I) ; **6b** (II)

MW = 436
**9a** and **9b**

$-CO_2$

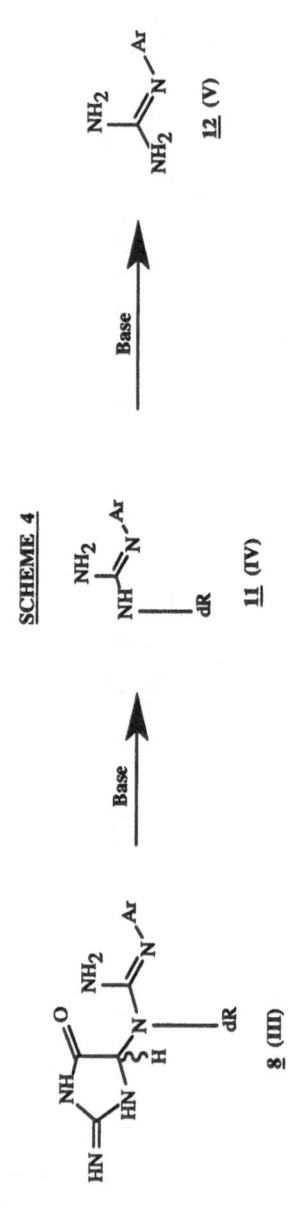

SCHEME 3

Base - catalysed equilibration of **8** and **9**

8a and 8b

9a and 9b

SCHEME 4

8 (III)

Base

11 (IV)

Base

12 (V)

14                    15                    16

17a                              17b

The control that the 2-amino group exerts over the reactivity of the guanine nucleus is highly significant from a chemical point of view. Reactions at the C-8 position such as oxidation (20) and arylamination (5-8) are undoubtedly facilitated by the electronic redistribution (see 15) that can occur at the approach of an appropriate electrophile. For different electrophiles, such as alkylating agents, the 2-amino group again influences the reactivity both at C-5 (21) at the 6-oxygen directly (22, 23) (see 16; although a nucleophilic contribution from N-1 cannot be discounted), and indirectly at N-7 (22) by increasing its nucleophilicity through its induction of higher electron density at C-5 (as in 14). However, if the electrons on the 2-amino group were not readily polarizable then reactivity at all of these positions would be diminished. Such a situation would be expected to exist when the amino group is acylated or more importantly when it is involved in hydrogen bonding where the resonance canonical form 17b is dominant. The latter situation affords an explanation for the diminished reactivity of dG residues in double-stranded DNA as opposed to single-stranded DNA. This in turn offers a chemical basis for the presence of mutational "hot spots" in double-stranded DNA, whose origins Kohwi Shigamatsu and her co-workers (24,25) have suggested are associated with improperly-paired DNA bases. Broadly speaking, the chemical argument should apply to dG, dC and dA residues because in all of these heterocyclic bases, the amino groups are the prime regulators of electron distribution and therefore of base reactivity. This can be generalized by stating that at least one major class of "hot spots" in DNA appears to be associated with points at which the interstrand hydrogen bonding of one or more base pairs has been attenuated or even abolished by such internal factors as conformational strain (super-coiled DNA), spontaneous strand separation ("breathing") or changes in helicity (B to Z DNA transition points). External factors that might compromise hydrogen bonding and thus lead to an increase in base reactivity towards electrophiles, include intercalating groups, simple binding agents or polymerizing enzymes (strand separation at the replication fork).

Further discussion of this concept will be deferred to a future publication. However, with regard to the present work one might expect that once a dG-8-AF residue has been formed at a "hot spot", the lack of

145

interstrand H-bonding at such a site should then also allow slow oxidation to occur at physiological pH. Thus the <u>oxidative degradation products of dG-8-AF residues may have real biological significance with regard to the induction of mutation during DNA replication</u>. Experiments designed to explore these ideas are in progress.

<u>Conclusions</u> - Previous hydrolytic studies on carcinogenic amine adducts of DNA had identified only two types of product, both of which were considered to be simple derivatives formed by alkali-mediated opening of the imidazole ring of guanosine residues. The current studies conducted on guanosine derivatives having an aminofluorene residue at the 8-position, show that all products, other than that derived by simple hydrolytic removal of an acetyl group, are the result of an aerial oxidation reaction that was virtually unrecognized previously. This has important implications for all guanosine compounds having different'amine substituents at the 8-position such as the 2-naphthylamine adduct (26) or those derived from the food mutagens (27). Consideration of the reactivity of dG and other DNA bases towards electrophiles allows a broad chemical explanation of the action of these agents on DNA bases and of the occurrence of mutational "hot spots" in DNA.

<u>Acknowledgement</u>: The authors would like to thank Dr. Charles R. Iden and Mr. Robert Rieger for their expert assistance in obtaining and interpreting the mass spectral data. The authors are pleased also to acknowledge the interest and helpful discussions provided by Drs. Arthur P. Grollman and M. Takeshita. This research was supported by a much appreciated grant (No. ESO4068) from the National Institute of Environmental Health Sciences.

# References

1.  Yoshida, T. <u>Proc. Imp. Acad.</u> (Tokyo) <u>1932</u>, <u>8</u>, 464; Sasaki, T., Yoshida, T. <u>Virchows Arch.</u> <u>1935</u>, <u>295</u>, 175.

2.  Heuper, W.C., Wiley, F.H., Nolfe, H.D. <u>J. Ind. Hyg.</u> and <u>Toxicol.</u> <u>1938</u>, <u>20.</u> 46

3.  Wilson, R. H., DeEds, F., Cox, A. J. <u>Cancer Res.</u> <u>1941</u>, <u>1</u>, 595; see also

    Bielschowsky, F. <u>Brit. J. Exptl. Path.</u> <u>1944</u>, <u>25</u>, 1.

4.  For recent reviews see King, C. M. in Nicolini, C. (ed), <u>Chemical Carcinogenesis</u> Plenum Press, New York, pp.25-46; Gardner, R. C., Martin, C. N. Clayson, D. B. In Searle, C. E. (ed.) <u>Chemical Carcinogenesis</u> A.C.S. Monograph 182 American Chemical Society, Washington, D. C., pp. 175-276; Osborne, M.R. <u>ibid</u> pp. 785-524.

5.  Kriek, E., Miller, J. A., Juhl, U., Miller, E. C. <u>1967</u>, <u>Biochemistry</u> <u>1967</u>, <u>6</u>, 177; Kriek, E. <u>Cancer Res.</u> <u>1972</u>, <u>32</u>, 2042.

6.  Westra, J. G. Kriek, E., Hittenhausen, H. <u>Chem. Biol. Interactions</u> <u>1976</u>, <u>15</u>, 149.

7.  Beland, F.A. Dooley, K. L., Jackson, C. D. <u>Proc Am. Assoc. Cancer Res.</u> <u>1979</u>, <u>20</u>, 518.

8.  Beland, F. A. Allaben, W. T., Evans, F. E. <u>Cancer Res.</u> <u>1980</u>, <u>40</u>, 751.

9.  Kriek, E., Westra, J.G. <u>Carcinogenesis</u> <u>1980</u>, <u>1</u>, 459.

10. Johnson, D. L., Reid, T. M., Lee, M-S, King, C. M., Romano, L. J. Carcinogenesis 1987, 8, 619.

11. Tarpley, W.G., Miller, J. A, Miller, E. C. Carcinogenesis 1982, 3, 81.

12. Stohrer, G., Osband, J. A., Alvarado-Urbina, G. Nucl. Acids. Res. 1983, 11, 5093.

13. O'Connor, D., Stohrer, G. Proc. Natl. Acad. Sci. 1985, 82, 2325.

14. We also had attempted to synthesize dG(C-8)AF-containing oligomers using the resin-based automated approach but were discouraged by the poor yields and the problems of purification that were encountered using the procedures recommended by Stohrer (12).

15. Brandenberger, H. Biochem. Biophys. Acta 1952, 15, 108; ibid. Experientia 1936, 12,208, ibid. Helv. Chim. Acta. 1954, 37, 641.

16. Brandenberger, H., Brandenberger, R. H. Helv. Chim. Acta 1954, 37, 2207.

17. For a review of this type of reaction see Acheson, R.M. Account of Chem. Res. 1971, 4, 177. Specifically references 12 and 13 contained therein.

18. Poje, M., Sokolic-Maravic, L. Tetrahedron 1986, 42, 747.

19. Poje, M., Sokolic-Maravic, L. Tetrahedron 1988, 44, 6723.

20. Richter, C., Park, J.-W., Ames, B. N.. Proc. Natl. Acad. Sci. 1988, 85, 6465 and references 14-18 contained therein.

21. Moschel, R. C., Hudgins, R. W. Dipple, A. J. Org. Chem. 1986,51,4180.

22. Singer, B. Cancer Investigation 1984, 2, 233.

23. Pegg, A. E. Cancer Investigation 1984, 2 223.

24. Kohwi-Shigematsu,T., Gelinas, R., Weintraub, H. Proc. Natl. Acad. Sci. U.S.A. 1983, 80, 4389. See also: Kohwi-Shigematsu, T., Enomoto, T., Yamada, M. Nakanishi, M., Tsuboi, M. ibid 1978, 75, 4689.

25. Kohwi-Shigematsu, T., Scribner, N., Kohwi, Y., Carcinogenesis 1988, 9 457. See also: Kimura, K., Nakanishi, M., Yamamoto, T., Tsuboi, M.J Biochem.. (Tokyo) 1977, 81, 1699.

26. Kadlubar, F.F., Unruch, L. E., Beland, F.A., Straub, K.M., Evans, F.E. Carcinogenesis 1980, 1, 139.

27. A recent review of these substances is given by Sugimura, T. Science 1986, 233, 32; Sugimura, T., Nagao, M., Wakabayashi, K. in "Environmental Carcingonesis: Selected Methods of Analysis" (ed. Egan, H.) IARC Publication No. 40, IARC, Lyon 1982, pp. 251-267.

# MUTATIONS AND HOMOLOGOUS RECOMBINATION INDUCED BY *N*-SUBSTITUTED ARYL COMPOUNDS IN MAMMALIAN CELLS

Veronica M. Maher, M. Chia-Miao Mah, Jia-Ling Yang,
Nitai P. Bhattacharyya, and J. Justin McCormick

Carcinogenesis Laboratory - Fee Hall
Department of Microbiology and Department of Biochemistry
Michigan State University, East Lansing, Michigan  48824-1316

## INTRODUCTION

It is widely recognized that the transformation of normal cells into tumorigenic cells is a multistep process, and there is substantial evidence that mutations and gene rearrangements are involved in causing one or more of the required changes. However, the molecular mechanisms by which carcinogens induce such changes in mammalian cells are not well understood. We are making use of a shuttle vector, pZ189, containing a target gene, *supF*, to examine the frequency and kinds of mutations induced by a series of *N*-substituted aryl compounds and structurally-related carcinogens, as well as their specific location in the target gene (spectrum of mutations) (Yang et al., 1987; 1988; Mah et al., 1989). In these experiments, the plasmid is treated with radiolabeled carcinogens and allowed to replicate in the human cell line 293 where the mutations are introduced. The progeny plasmids are rescued, analyzed for mutated *supF* genes, and the kinds and spectra of mutations are determined in order to identify common features in the modes of action of the various agents and determine how they differ from each other.

The *supF* gene codes for a tRNA, and because the structure of the tRNA is essential for its purpose, a single base pair (bp) change at almost any site of the 85 bp structure of the tRNA results in a mutant phenotype, making this gene exceptionally responsive to mutagens and allowing for very few silent mutations. The small size of the target gene greatly facilitates sequence analysis and determination of "hot spots" and "cold spots" for mutation induction by particular carcinogens.

We and our colleagues are also investigating the ability of carcinogens to induce homologous recombination between genes in mammalian cells, including human cells (Wang et al., 1988; Bhattacharyya et al., 1989, 1990; Tsujimura et al., 1990). Genetic recombination may represent an important mechanism for rendering a cell homozygous for a recessive allele which would permit expression of a recessive phenotype. Evidence that such recombination can play a role in the development of tumors comes from studies of retinoblastoma patients (Cavanee et al., 1983). However, the role of DNA damage and repair in the induction of intrachromosomal homologous recombination is less understood. Therefore, we have made use of an assay system originally designed by Liskay et al. (1984) in which a thymidine kinase (*tk*)-deficient mouse L cell line contains a single integrated plasmid with duplicated copies of the Herpes simplex *tk* (H*tk*) gene, each containing an 8 bp *Xho*I site inserted in a different place. Only by undergoing a productive recombinational event between the two nonfunctional H*tk* genes can a functional gene product be made and the recombinant survive selection. The design of the plasmid allows one to deduce the kinds of recombinational events that occurred from the types of products obtained as analyzed by Southern blotting (Liskay et al., 1984; Wang et al., 1988; Bhattacharyya et al., 1989).

We have used these two assay systems, one detecting mutational changes, the other detecting recombinational events, to gain insight into the mechanism of action of 1-nitrosopyrene (1-NOP) and *N*-acetoxy-2-acetylaminofluorene (N-AcO-AAF) in causing changes in the genetic material of mammalian cells, including human cells.

RESULTS

## Studies of Mutation Induction

The plasmid, pZ189, was treated with tritiated *N*-acetoxy-*N*-trifluoroacetyl-2-aminofluorene (N-AcO-TFA-AF) at neutral pH or with 1-NOP in the presence of ascorbic acid pH 5.0 as a reducing agent. The DNA was analyzed by HPLC and found to contain only one major adduct, *N*-(deoxyguanosin-8-yl)-2-aminofluorene (dG-C8-AF) and *N*-(deoxyguanosin-8-yl)-1-aminopyrene (dG-C8-AP), respectively. The plasmids were assayed for the number of residues bound per plasmid and for loss of ability to transform bacteria to ampicillin resistance. As shown in Figures 1A and 2A, 1-NOP was twice as reactive as N-AcO-TFA-AF, but this merely reflects the ability of ascorbic acid to activate 1-NOP. Figures 1B and 2B show that 1-NOP adducts are three times more active than AF adducts in interfering with bacterial transformation. Figures 1C and 2C show that when plasmids treated with various doses of N-AcO-TFA-AF or 1-NOP were transfected into 293 cells and allowed to replicate and the progeny plasmids were rescued and assayed for the frequency of *supF* mutants as a function of the number of adducts per plasmid, the frequency of plasmids with a mutated *supF* was approximately the same for both agents. For example, plasmids carrying ~40 AF adducts per plasmid increased the frequency of *supF* mutants to $19.4 \times 10^{-4}$, whereas ~40 1-NOP adducts raised the frequency to $23 \times 10^{-4}$. The background frequency was $1.3 \times 10^{-4}$. These results suggest that the mechanisms of mutation induction by these compounds is similar.

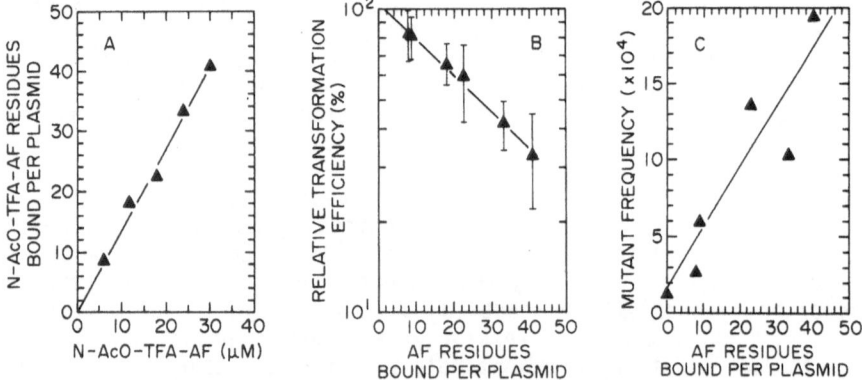

Fig. 1. Number of AF adducts per plasmid as a function of concentration of N-AcO-TFA-AF (A); frequency of transformation of bacteria to ampicillin resistance as a function of the number of adducts per plasmid (B); frequency of *supF* mutants as a function of the number of AF adducts per plasmid (C). (From Mah et al., 1989.)

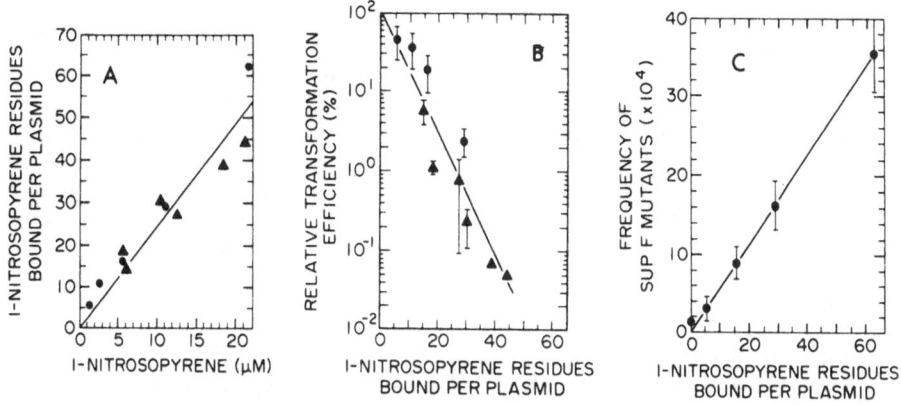

Fig. 2. Number of 1-NOP adducts per plasmid as a function of concentration 1-NOP in the presence of ascorbic acid (A); frequency of transformation of bacteria as a function of the number of adducts per plasmid (B); frequency of *supF* mutants as a function of the number of adducts per plasmid (C). (From Yang et al., 1988.)


Characterization of *supF* Mutants Induced. This conclusion was supported by the results of our analysis of the kinds of mutations induced by these structurally-related agents. Less than 13% of the *supF* mutants obtained with N-AcO-TFA-AF-treated or 1-NOP-treated plasmids that were examined exhibited gross rearrangements or contained deletions consisting of nine or more nucleotides. In contrast, at least 60% of the mutants obtained with untreated plasmids did so. DNA sequencing analysis of 50 *supF* genes from N-AcO-TFA-AF-treated plasmids that did not show gross alterations in gel patterns and 60 from 1-NOP-treated plasmids indicated that the majority consisted of base substitutions. With N-AcO-TFA-AF, all of these base substitutions involved G·C; with 1-NOP, 87% did so. The majority were G·C --> T·A transversions.

Figure 3 compares the specific location of the base substitutions in the *supF* gene. With N-AcO-TFA-AF treated plasmids there were four prominent "hot spots," positions 123, 133, 159, and 169. With 1-NOP, the prominant "hot spots" were located at positions 109, 123, 127, and 159.

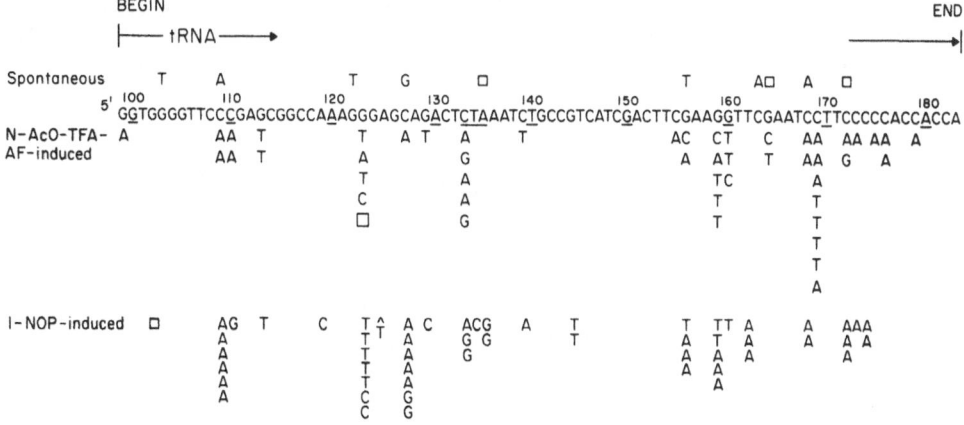

Fig. 3. Location of independent point mutations in the *supF* gene. The DNA strand shown is the 5' to 3' strand. The rectangle represents a deleted guanine; the caret shows the location of an inserted thymine. Every 10th residues and the anticodon triplet are underlined. (Data from Mah et al., 1989 and Yang et al., 1988.)

Correlation Between Sites of AF- or 1-NOP-Adducts and Locations of AF- or 1-NOP-Induced Mutations. To determine if the "hot spots" for mutation corresponded to "hot spots" for formation of dG-C8-AF adducts, we carried out the polymerase-stop assay of Moore and Strauss (1979) using the Klenow fragment of *E. coli* polymerase I. This method of estimating the percentage of carcinogen adducts formed at particular sites on a gene assumes that the density of the bands in a sequencing gel is proportional to the number of DNA molecules of a particular length and that the length of the fragments reflects the chance of adduct-induced premature termination of the polymerization. With this assay we found that with N-AcO-TFA-AF-treated plasmids, termination occurred opposite or 1 base prior to every cytosine (Mah et al., 1989); with 1-NOP-treated plasmids, termination occurred 1 base prior to every cytosine except that at position 131 (Yang et al., 1988). No bands corresponding to positions 1 nucleotide away from any base other than cytosine were seen, and there was no evidence of any interference with polymerization when untreated template was used. The pattern of bands did not vary significantly with the number of AF or 1-NOP adducts per molecule of plasmid. In some cases there was good correlation between the extent of binding and the frequency of mutation induction, but the correlation was not perfect. The lack of correlation between the results could not be explained simply by so-called "silent mutations".

## Studies of Homologous Recombination

To prepare an assay for studying intrachromosomal homologous recombination, Liskay et al. (1984) transfected *tk⁻* mouse L cells with pJS-3 (Fig. 4) that had been linearized at the unique *Cla*I site to increase the chances of the H*tk* genes being intact. pJS-3 also carries the *neo* gene, which codes for resistance to Geneticin, located between the H*tk* genes. This is useful for identifying cells that have taken up the plasmid and also for determining if the sequence of DNA between the two H*tk* genes has been eliminated during a recombination event. Such elimination will occur if a single reciprocal exchange takes place. Since each H*tk* gene has been inactivated, expression of a functional H*tk* enzyme requires a productive recombinational event. Such recombination can be detected because it allows the recombinants to be selected for their ability to form colonies in medium containing deoxycytidine, hypoxanthine, aminopterin, and thymidine (CHAT medium). A cell line with a single, stably-integrated, intact copy of the plasmid and a low background rate of spontaneous recombination (~3 per $10^6$ cells per cell generation) was identified by Liskay et al. (1984) and subsequently used to study of induction of recombination by various carcinogens (Wang et al., 1988; Bhattacharyya et al., 1989).

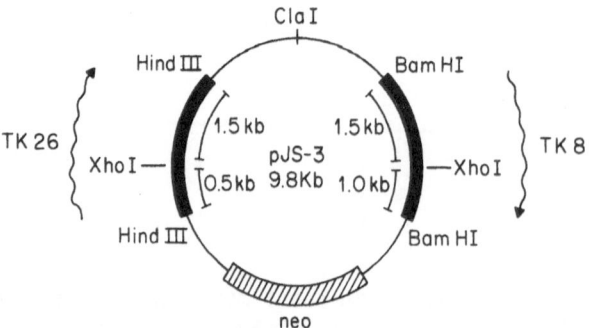

Fig. 4. Structure of plasmid pJS-3. (From Bhattacharyya et al., 1989.)

Exponentially growing cells were plated and exposed to such agents as N-AcO-AAF or 1-NOP for 1 hour. As noted above, these two carcinogens predominantly form closely-related guanine adducts. The doses were adjusted to obtain a survival level of between 90% and 20% of the untreated control population, as determined by loss of colony-forming ability. Sufficient target cells were used for each dose to have at least $2 \times 10^6$ survivors. Figure 5 shows the frequency of recombination induced by these agents above the background frequency. As a function of applied concentration, 1-NOP was twice as active as N-AcO-AAF. But when the two agents were compared on the basis of equal cytotoxicity, i.e., recombina-"recombination efficiency", 1-NOP was found to be only slightly more active than N-AcO-AAF (Bhattacharyya et al., 1989). When compared on the basis of equal numbers of DNA adducts initially formed in the mouse L cells, 1-NOP proved to be 2.5-fold more active than N-AcO-AAF (data not shown). This difference could not be explained by a difference in overall rate of removal of the adducts from bulk DNA (Bhattacharyya et al., 1989). However, there might be detectable differences in rate of repair of the two kinds of adducts from the DNA of actively transcribed genes, such as the Htk gene in these mouse cells. This question is being investigated. In addition, Bhattacharyya et al. (1990) has recently introduced the recombination assay plasmid into a series of human cells which differ in their ability to remove such DNA damage. Use of these strains will allow us to compare the two agents for the ability to induce recombination in the total absence of nucleotide excision repair.

Types of recombination events by these agents. The presence of the neo gene in the sequence between the two Htk genes in pJS-3 facilitates inference of the kinds of recombination events that have occurred from the types of products obtained. If the recombination event involves a single reciprocal exchange within a chromatid, or a single unequal exchange between chromatids, only a single wild-type copy of the Htk gene will be present and the neo gene will be lost. Such CHAT-resistant recombinants

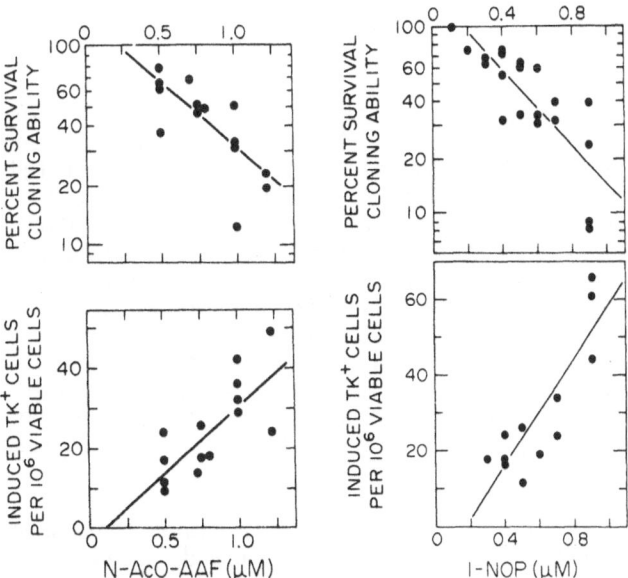

Fig. 5. Cell killing and induction of intrachromosomal homologous recombination in the 333M mouse L cell line by N-AcO-AAF and 1-NOP as a function of the applied concentration. The background frequencies, which averaged $19 \times 10^{-6}$ for N-AcO-AAF and $16 \times 10^6$ for 1-NOP, have been subtracted. (Taken from Bhattacharyya et al., 1989.)

will be sensitive to Geneticin. In contrast, if the event consists of a nonreciprocal transfer of wild-type information, gene conversion, the H*tk* gene duplication with the *neo* gene will be retained in the recombinant, and the cells will be Geneticin resistant. CHAT-resistant colonies were isolated and tested for resistance to Geneticin. 85 to 90% of the recombinants from each group retained the *neo* gene, consistent with gene conversions (Bhattacharyya et al., 1989). This ratio is the same as that found for untreated cells (Liskay et al., 1984), suggesting that the carcinogens increased the chances of an event taking place, but did not affect the mechanism involved.

DNA from representative Geneticin-resistant or sensitive colonies was analyzed by Southern blot hybridization, using the H*tk* gene as a probe. The results confirmed that the Geneticin-resistant recombinants retained the H*tk* gene duplication, with one H*tk* gene being wild type, i.e., lacking an *Xho*I restriction site. As predicted, the Geneticin-sensitive recombinants analyzed contained only a single (wild-type) H*tk* gene.


DISCUSSION

Studies of Mutation Induction

We conclude from our results that the mutations obtained with the N-AcO-AAF-treated plasmids and 1-NOP-treated plasmids were targeted to the adducts formed. The increase in frequency of mutations was linearly correlated with the number of adducts per plasmid; the frequency of those containing gross rearrangements, or deletions or insertions greater than 1 to 3 bp was approximately equal to the background frequency; the rest carried point mutations exclusively or predominantly involving G·C base pairs; none of the mutations in the 5' to 3' strand were located at sites where termination of polymerization on a 3' to 5' template by the Klenow fragment was not detected; and the two mutational "hot spots" for which the dG-C8-AF or dG-C8-AP adduct would be located in the 3' to 5' strand were also prominent positions for termination of polymerization in the stop assay.

The majority of the base substitutions observed with either carcinogen were G·C --> T·A transversions (65% for N-AcO-TFA-AF-treated plasmids and 61% for 1-NOP-treated plasmids). This may have resulted from a DNA polymerase preferentially inserting adenosine triphosphate opposite the adduct. However, recent studies by Norman et al. (1989) support the hypothesis that adenosine triphosphate might be accommodated opposite dG-C8-AF during replication of the plasmid. This would result in the type of transversion that we observed most frequently with N-AcO-TFA-AF-treated plasmid. It is possible that stable base pairing of adenosine triphosphate opposite dG-C8-AP adducts also occurs since the two C8 quanine adducts are so similar. Our results with pZ189, showing that 98% of the dG-C8-AF- mutations consisted of G·C base substitutions, agree with what was recently found by Carothers et al. (1989) in an endogenous gene of CHO cells treated with N-AcO-AAF.

Both N-AcO-TFA-AF- and 1-NOP-treated plasmids contain adducts at the C8 position of guanine and the covalently bound residues are structurally quite similar. Although the dG-C8-AF adduct was not as effective as the dG-C8-AP adduct in interfering with bacterial transformation to ampicillin resistance (cf. Fig. 1B with Fig. 2B), the frequency of *supF* mutants when the plasmids replicated in human cells, as a function of the number of adducts per plasmid, was very similar (cf. Fig. 1C and Fig. 2C) and the kinds of mutations were very similar. Although N-AcO-TFA-AF- and 1-NOP-treated plasmids gave one prominent hot spot in common (at position 123), each carcinogen produced its own set of unique "hot spots." The spectra of "hot spots" also differed from the spectra found with $(\pm)7\beta,8\alpha$-dihydroxy-$9\alpha,10\alpha$,epoxy-7,8,9,10-tetrahydro-benzo[a]pyrene (BPDE) (Yang et al., 1988) or with benzo(c)phenanthrene(4R,3S)-dihydrodiol-(2S,1R)-epoxide (Bigger et al., 1989).

## Studies of Recombination Induction

Mitotic recombination is considered to play a causal role in certain human cancers. Our finding that $N$-substituted aryl carcinogens cause a dose-dependent increase in the frequency of intrachromosomal homologous recombination between duplicated genes in the mouse L cell line indicates that these agents may be capable of inducing similar changes in human cells so that they become tumorigenic. The types of recombinational events by these carcinogens were similar to what was reported earlier by Liskay et al. (1984) for spontaneous recombinational events in this 333M cell line, suggesting that the mechanism involved in carcinogen-recombination is similar to that operating in untreated cells.

The recombinational events may be stimulated by bulky adducts present in the DNA, perhaps because the adducts interrupt DNA synthesis. Alternatively, recombination may be stimulated by the cellular DNA excision repair processes that remove such adducts. The study in the mouse L cell line sheds light on these questions, but cannot answer them definitely. The finding that the relationship between the frequency of recombination and the number of DNA adducts differed for these two carcinogens and also for several other structurally-related compounds (Bhattacharyya et al., 1988) suggests that the mechanism of recombination recognizes intrinsic differences in the nature of the adducts.

The fact that we observed no significant difference in the rate of removal of N-AcO-AAF and 1-NOP adducts from the DNA of the 333M cells during the 24 hour prior post-treatment does not rule out the possibility that the observed differences in recombinagenic effectiveness result not from the intrinsic nature of the adducts, but from differential rates of excision repair. This is because we measured the rate of removal of the total number of adducts, but certain rodent cell lines in culture have been shown to preferentially remove lesions from actively transcribed genes (Bohr et al., 1985). If this is the case with these 333M cells, they might actually remove specific adducts from the region of interest, i.e., the actively transcribed, duplicated H$tk$ genes, at different rates, even though the observed overall rates of loss were equal. Circumstantial evidence that the cells preferentially repair certain genes is the fact that their sensitivity to the cytotoxic effect of these carcinogens was similar to that found previously for diploid human fibroblast cell lines, and yet the latter cell lines remove adducts formed by these carcinogens at a much faster rate than do the mouse L cells.

ACKNOWLEDGEMENTS

We thank our colleagues Dr. Frederick A. Beland of the National Center for Toxicological Research for providing us with the tritiated 1-NOP and for HPLC analysis of the adducts formed on the plasmid and Dr. Charles M. King and Dr. Thomas M. Reid of the Michigan Cancer Foundation for reacting the plasmid with N-AcO-TFA-AF and for HPLC analysis of the adducts formed. The excellent technical assistant of Neil Davis and Paul Ferguson is gratefully acknowledged. The research was supported by DHHS grant CA21253 from the National Cancer Institute and by Contract 87-2 from the Health Effects Institute, an organization jointly funded by the U.S. EPA and automotive manufacturers, is currently under review by the Institute. The contents of this article do not necessarily reflect the policies of EPA, or automotive manufacturers.

# REFERENCES

Bhattacharyya, N. P., Maher, V. M., and McCormick, J. J., 1989, Ability of structurally related polycyclic aromatic carcinogens to induce homologous recombination between duplicated chromosomal sequences in mouse L cells, Mutat. Res., 211:205.

Bhattacharyya, B. P., Maher, V. M., and McCormick, J. J., 1990, Intrachromosomal homologous recombination in human cells which differ in nucleotide excision repair capacity, Mutat. Res., 234:31.

Bigger, C. A. H., Strandberg, J, Yagi, H., Jerina, D. M., and Dipple, A., 1989, Mutagenic specificity of a potent carcinogen, benzo[c]phenanthrene(4$R$,3)-dihydroliol (2$S$,1$R$)-epoxide, which reacts with adenine and guanine in DNA, Proc. Natl. Acad. Sci. (U.S.A.), 86:2291.

Bohr, V. A., Smith, C. A., Okumoto, D. S., and Hanawalt, P. C. 1985, DNA repair in an active gene: removal of pyrimidine dimers from the DHFR gene of CHO cells is much more efficient than in the genome overall, Cell, 40:359.

Carothers, A. M., Steigerwalt, R.W., Urlaub, G., Chasin, L. A., and Grunberger, D., 1989, DNA base changes and RNA levels in N-acetoxy-2acetylaminofluorene- dihydrofolate reductase mutants of CHO cells. J. Mol. Biol., 208:417.

Cavanee, W. K., Dryja, T. P., Phillips, R. A., Benedict, W. R., Godbout, R., Gallie, B. L., Morphree, A. C., Strong, L. C., and White, R. L., 1983, Expression of recessive alleles by chromosomal mechanisms of retinoblastoma, Nature, 305:779.

Liskay, R. M., Stachelek, J. L., and Letsou, A., 1984, Homologous recombination between repeated chromosomal sequences in mouse cells, Cold Spring Harbor Symp. Quant. Biol., 49:183.

Mah, M. C.-M., Maher, V. M., Thomas, H., Reid, T. M., King, C. M., and McCormick, J. J., 1989, Mutations by aminofluorene-DNA adducts during replication in human cells, Carcinogenesis, 10:2321.

Moore, P., and Strauss, S., 1979, Sites of inhibition of in vitro DNA synthesis in carcinogen- and UV-treated OX174 DNA, Nature, 278:664.

Norman, D., Abauf, P., Hingerty, B. E., Live, D., Grunberger, D., Broyde, S., and Patel, D. J., 1989, NMR and computational characterization of the N-(deoxyguanosin-8-yl) aminofluorene adduct [(AF)G] opposite adenosine in DNA: (AF)G[syn]-A[anti] pair formation and its pH dependence, Biochemistry, 28:7462.

Tsujimura, T., Maher, V. M., Godwin, A. R., Liskay, R. M, and McCormick, J. J., 1990, Frequency of intrachromosomal homologous recombination induced by UV radiation in normally repairing and excision repair-deficient human cells, Proc. Natl. Acad. Sci., U.S.A., 87:1566.

Wang, Y., Maher, V. M., Liskay, R. M., and McCormick, J. J., 1988, Carcinogens can induce homologous recombination between duplicated chromosomal sequences in mouse L cells, Mol. Cell. Biol., 8:196.

Yang, J.-L., Maher, V. M., and McCormick, J. J., 1987, Kinds of mutations formed when a shuttle vector containing adducts of ($\pm$)-7$\beta$,8$\alpha$-dihydroxy-9$\alpha$,10$\alpha$-epoxy-7,8,9,10-tetrahydrobenzo[a]pyrene replicates in human cells, Proc. Natl. Acad. Sci. (U.S.A.), 84:3787.

Yang, J.-L., Maher, V. M., and McCormick, J. J., 1988, Kinds and spectrum of mutations by 1-nitrosopyrene adducts during plasmid replication in human cells, Mol. Cell. Biol., 8:3364.

COMPARISON OF THE MUTAGENIC POTENCY OF DNA ADDUCTS FORMED BY REACTIVE

DERIVATIVES OF AFLATOXIN, BENZIDINE AND 1-NITROPYRENE IN A PLASMID SYSTEM

Carl N. Martin and Gary S. Jennings

Cancer Research Unit
University of York
York YO15DU, England

INTRODUCTION

1-Nitropyrene is an ubiquitous environmental pollutant which has been detected in the urban atmosphere in many areas of the world (1). It is generated primarily by incomplete combustion of fossil fuels and thus is present in relatively high concentrations in emissions such as diesel engine exhaust fumes (2,3) and power station fly-ash (4,5). This compound has been shown to be biologically active in a number of in vitro genotoxicity assay systems, inducing mutation in Salmonella typhimurium (6) and inducing mutations, unscheduled DNA synthesis, transformation or chromosomal rearrangements in a variety of mammalian cell lines (7-10). 1-nitropyrene has also been shown to be carcinogenic in laboratory animals (11,12). The activation of 1-nitropyrene has been shown to involve a sequential two-electron reduction leading to the formation of 1-NOP, $\underline{N}$-OH-1-aminopyrene and 1-aminopyrene (13,14). The $\underline{N}$-hydroxy derivative has been shown to be an ultimate DNA-reactive species (13). The adduct produced in DNA by the reactive metabolite of 1-nitropyrene has been identified in bacterial systems (15) and in rat liver (16) as $\underline{N}$-(deoxyguanosin-8-yl)-1-aminopyrene. Thus, 1-nitropyrene is a compound to which the human population is exposed and is biologically active in a number of experimental systems. However, very little information is available as to the relative carcinogenic risk to humans chronically exposed to this compound. One approach to the assessment of such risk at the level of carcinogenicity initiation potential is by comparison of the mutagenic potential of 1-nitropyrene-DNA adducts with that of adducts generated by compounds of which human carcinogenicity has been established by epidemiological study.

The present investigation is an attempt to determine the mutagenic potency of the major DNA-adduct of 1-nitropyrene in an in vitro plasmid system and compare this with the mutagenic potency of DNA adducts generated by benzidine, a known animal and human carcinogen and aflatoxin, a potent animal and suspected human carcinogen. The structures of the DNA adducts of these compounds have previously been elucidated (17-19), and the nature of their reactive intermediates has been identified. The effect of bacterial post-replication repair systems on these relative mutagenicities has also been measured.

## MATERIALS AND METHODS

### Chemicals, enzymes and biological materials

DNA hydrolysing enzymes were obtained from Sigma Chemical Company, Poole, Dorset, England; [³H] aflatoxin (sp.rad.act. 15 Ci/mmol) was obtained from Moravek Biochemicals Inc. Brea, California, USA; [³H] acetic anhydride (500 mCi/mmol) and [³H] 1-aminopyrene (sp.rad.act. 1.0 Ci/mmol) were obtained from The Radiochemical Centre, Amersham, England. Plasmid pDS40A was a kind gift from Dr D Schumperli, University of Zurich, Switzerland. E. coli strains AB1886 (uvrA6) and AB2480 (uvrA6, recA13; no inducible recombinant repair, no SOS repair) were strains maintained in our laboratory. E. coli strain AB6002 (uvrA6, recB22; no recombinational repair pathway, SOS repair present, recE,F present but may be dependent on recB) was a kind gift from Dr P Strike, Dept of Genetics, University of Liverpool, Liverpool, England. Bacteriological media was obtained from Difco Laboratories Inc, Detroit, Michigan, USA and Oxoid Ltd, Wade Road, Basingstoke, Hants, UK. All other reagents used in these studies were of analytical reagent grade.

8,9-diCl-AFB1 and [³H] 8,9-diCl-AFB1 (sp.rad.act. 49.68 - 439.16 mCi/mmol) were synthesised as described by Swenson et al (20).

### Synthesis of 1-NOP and [3H] 1-NOP (sp.rad.act. 11.68 mCi/mmol)

m-Chloroperoxybenzoic acid in chloroform was added in stoichiometric amounts to 1-aminopyrene also in chloroform. The reaction was continued for 1h at 22°C. The mixture was extracted three times with 0.1 M sodium hydroxide, washed twice and dried. Chloroform was removed in vacuo and the residue redissolved in benzene. The solution was applied to a neutral alumina column developed in benzene. The first band to elute was collected and benzene removed in vacuo. The orange-brown residue was recrystallized from ethanol and stored at -20°C. Tlc and hplc analyses showed the presence of a single species co-eluting with an authentic 1-NOP standard.

### Synthesis of N-OH-N'-AcBZ and [³H] N-OH-N'-AcBZ (sp.rad.act. 2.115 mCi/mmol)

Biphenyl was nitrated by addition to nitric acid at 35°C. The temperature was allowed to rise to 50°C and then the solution was stirred until it cooled to 20°C. Pale yellow 4,4'-dinitrobiphenyl crystallized out and was then recrystallized from ethanol.

A solution of 4,4'-dinitrobiphenyl in 50% ethanol was stirred vigourously and refluxed whilst a solution of sodium sulphide nonahydrate and sulphur in water was added over 15 mins. Refluxing was continued for 30 min and then orange-red 4-amino-4'-nitrobyphenyl was crystallized out. The product was recrystallized from 10% hydrochloric acid.

To a solution of 4-amino-4'-nitrobiphenyl in ether, was added excess acetic anhydride in ether and the mixture stirred at 22°C for 16h. Etheral suspension was washed with aqueous sodium carbonate and water before removal of ether in vacuo. Mustard yellow 4-acetylamino-4'-nitrobiphenyl was recrystallized from 50% ethanol.

A solution of 4-acetylamino-4'-nitrobiphenyl in dimethylformamide was mixed with an excess of 1 M ammonium chloride in water. The mixture was cooled on ice and argon bubbled through whilst zinc powder was added. The reaction was continued for up to 5h and monitored by hplc. When all of the starting material was converted to the N-hydroxy derivative, cold water was added to precipitate the pale yellow product. The precipitate was centrifuged and washed again leaving a pale yellow powder. N-OH-N'-AcBZ was stored under argon at -90°C. Synthesis of the [³H] product was as described above except that 1 molar equivalent of [³H] acetic anhydride was added to 4-amino-4'-nitrobiphenyl.

## Preparation of plasmid DNA

Plasmid DNA was prepared by the modified cleared-lysate technique of Mizusawa et al (21) Routinely, 4.0L plasmid host (E. coli N100) were grown up in ampicillin containing medium. Bacteria were harvested and treated with lysozyme in isotonic sucrose solution. Bacterial spheroplasts were then treated sequentially with EDTA and Triton X-100 solutions. Lysate was 'cleared' by centrifugation and the bacteriosol extracted three times with phenol (TEN saturated) and three times with diethylether to remove proteins. DNA was precipitated from the aqueous phase at -20°C by making it 300 mM in sodium acetate, pH 6.0 and adding 3 volumes of ethanol. DNA was resuspended in TEN, pH 8.0 and banded by equilibrium centrifugation on a caesium chloride gradient using ethidium bromide to visualise bands under UV-light. The lower, supercoiled DNA, band was removed and the ethidium bromide extracted with butan-1-ol. DNA was precipitated again, centrifuged, washed and dried before resuspension in TEN, pH 8.0. DNA was stored in 200 μg aliquots at -90°C. The integrity of plasmid DNA was checked on 0.8% agarose gel. Typically a yield of 3-4 mg plasmid DNA was obtained.

## Reaction of test chemicals with plasmid DNA

20 μg Plasmid DNA was dispersed into reaction buffer in a 2.0 ml reactivial (Alltech associates). For reaction with 8,9-diCl-AFB1, the buffer used was 25 mM sodium acetate, pH 5.5; for 1-NOP the buffer was made 0.5 mM with ascorbic acid pH 5.5; for N-OH-N'-AcBZ the buffer used was 10 mM potassium citrate, 1 mM EDTA, pH 4.6. Chemicals were added in 20 μl dry DMSO to make a total volume of 300 μl. Vials were rotated on an end-over-end mixer at 37°C for 2h (4h in the case of N-OH-N'-AcBZ). DNAs were reprecipitated at -90°C after making the mixtures 300 mM with respect to sodium acetate, pH 6.0 and addition of 3 volumes of cold ethanol. DNAs were centrifuged, washed, dried and re-dissolved in 50 mM calcium chloride, 25 mM sodium acetate buffer, pH 6.0 (Tfb). Concentrations of DNA were determined by UV-spectroscopy.

For reaction of radiolabelled chemicals to plasmid DNA, 100 μg DNA was dispersed in a total of 1500 μl reaction mixture. Radiochemicals were added in 100 μl dry DMSO and the whole incubated in 50 ml-Falcon tubes at 37°C as described above. Non-covalently bound chemical was removed by extraction with butan-1-ol (or chloroform for 8,9-diCl-AFB1) and the organic phase counted for radioactivity until this fell to zero. The aqueous layer was extracted three times with phenol:chloroform:isoamyl alcohol (25:24:1) DNA was then precipitated by ice cold ethanol, washed with acetone dried.

## Mutation experiments

Three strains of E. coli were separately transformed with chemically modified plasmid pDS40A by calcium phosphate coprecipitation. Bacteria grown to early log phase were harvested and resuspended in 50 mM calcium chloride, 25 mM sodium acetate, pH 6.0 (TfB). After incubation on ice for 20 min cells were harvested and resuspended in TfB at approx 10⁸ cells/ml. Cells were maintained on ice for 1 hr. 500 ng chemically modified plasmid DNA in 100 μl TfB was added to 10⁷ competent cells and incubation on ice was continued for 30 min. Bacterial cells were heat shocked at 42°C for 2 mins then returned to ice where 800 μl recovery medium was added. After incubation at 37°C for 90 min transformants were plated on MacConkey agar containing 1% galactose and 50 μg/ml ampicillin. Bacteria were grown overnight at 37°C and total ampicillin resistant colonies were counted by automatic image analysis. White (unable to utilise galactose) mutants were scored by hand.

Determination of DNA binding

Plasmid DNA modified by radiolabelled chemicals was re-dissolved in Bis Tris buffer pH 7.1 or 1.5mM NaCl for aflatoxin-DNA samples. An aliquot was taken for colourimetric assay to determine DNA concentration and the remainder was enzymically hydrolysed to nucleosides as previously described (19). Aflatoxin DNA samples were acid hydrolysed to purine bases by 100 mM HCl at 70°C. Adducted nucleosides or bases were extracted into n-butanol and prepared for hplc chromatography (19). The chromatographic conditions were as follows: -for all runs, elution was at a flow rate of 2 ml/min; for 8,9-diCl-AFB1 reacted DNA - 10% to 100% methanol over 40 min and eluent monitored at 360 nm; for 1-NOP reacted DNA - 30% methanol for 10 min, 30% to 70% over 30 min, eluent monitored at 280 nm; for N-OH-N'-AcBZ reacted DNA - 30% methanol for 10 min, 30% to 70% over 10 min and 70% to 100% over 5 min, eluent monitored at 280 nm.

In each case samples were co-injected with authentic standards of the relevant major adduct. 30 second fractions were collected and counted for radioactivity. Counts associated with the major adduct peak were summated and the amount of adduct was calculated from the known specific radioactivity of the starting material. The number of adducts present per plasmid were calculated and, assuming a random distribution, the number associated with the galactose gene and its promoter was deduced.

RESULTS

When measured as a function of dose, the highest mutagenicity was induced by plasmid treated with 8,9-diCl-AFB1 and transformed in E.coli AB1886. Table 1 shows the number of mutants induced in plasmid pDS40 after treatment with 500 pmol of the three agents under study and transformed into the different E.coli strains.

Figure 1 presents an example of an hplc analysis of enzymically hydrolysed 1-nitrosopyrene treated plasmid DNA. 8,9-diCl-AFB1 and N-OH-N'-AcBZ treated plasmid DNAs were similarly treated and counts present in the major adduct peak were measured. Using these data, the binding of each compound was determined over a dose range and found to be linear in each case (Figure 2). At a dose of 1 nmol the binding values, expressed as nmol compound/mg DNA were for 8,9-diCl-AFB1, 1, for N-OH-N'-AcBZ, 0.2 and for 1-NOP, 2.5. Figure 3 shows the mutation induced in plasmid pDS40 treated by the three test agents and transformed into the E.coli strains of differing post-replication repair capacities.

Table 1

| Test agent (Dose 500 pmol) | E.coli strain | Mutants per $10^6$ transformants | | |
|---|---|---|---|---|
| | | AB1886 | AB6002 | AB2480 |
| 8,9-dichloro-aflatoxin | | 90,000 | 26,400 | 11,000 |
| N-OH-N'-acetylbenzidine | | 114 | 6 | 3 |
| 1-nitrosopyrene | | 37,500 | 32,700 | 41,400 |

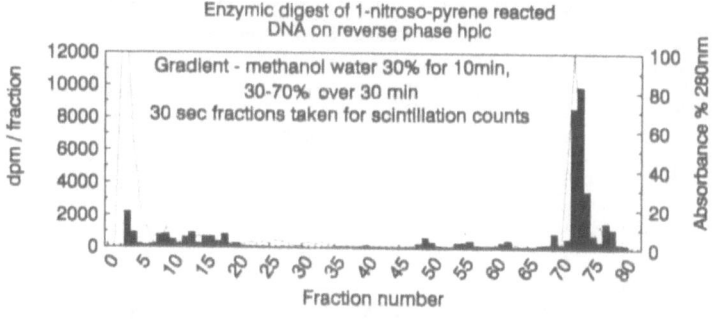

Figure 1

The induced mutation is expressed as a function of the number of DNA adducts present in the target gene. Table 2 shows the data expressed as the number of mutants/$10^6$ transformants induced by the presence of 1 adduct in the target gene.


DISCUSSION

A great deal of scientific effort is being directed toward population monitoring to determine the extent to which populations exposed to potentially carcinogenic agents have suffered damage to their hereditary material. The techniques of $P^{32}$ post-labelling and the use of adduct-directed monoclonal antibodies have been of significant value in the advances made in this area. However, the detection of DNA adducts in, for example, the circulating lymphocytes of an exposed individual has little value unless the quantification of such adducts can be expressed in terms of that individual's increased risk of resultant carcinogenicity. Since the early 1970's the use of in vitro genetic toxicology assays as predictors of chemical carcinogenicity has become widely accepted. The value of these assays is still the subject of debate but the correlation between mutation induction and the initiation stage of the carcinogenic process in now established at least in part. It is generally recognised however that such assays are poor predictors of carcinogenic potency. Clearly an important factor is the difficulty to imitate in in vitro systems many of the complex factors important in chemical carcinogenesis for example metabolism, pharmacodynamics and extracellular control mechanisms.

One approach to obtaining information on the carcinogenic risk of individuals exposed to chemicals of unknown human carcinogenicity, is to determine the relative mutagenic potency of DNA adducts generated by such

Table 2

| Chemical | Mutants/$10^6$ transformants induced per DNA adduct in target gene E.coli strain | | |
|---|---|---|---|
| | AB1886 | AB6002 | AB2480 |
| 8,9-dichloro-aflatoxin | 57,800 | 18,200 | 7,500 |
| N-OH-N'-acetylbenzidine | 62 | 2 | 0.2 |
| 1-nitrosopyrene | 1,900 | 1,300 | 1,400 |

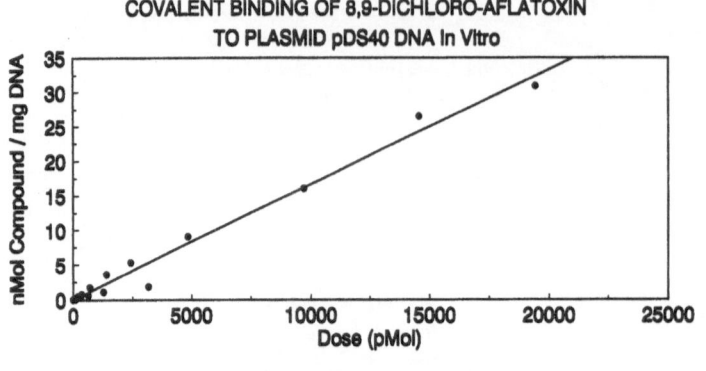

COVALENT BINDING OF 8,9-DICHLORO-AFLATOXIN
TO PLASMID pDS40 DNA In Vitro

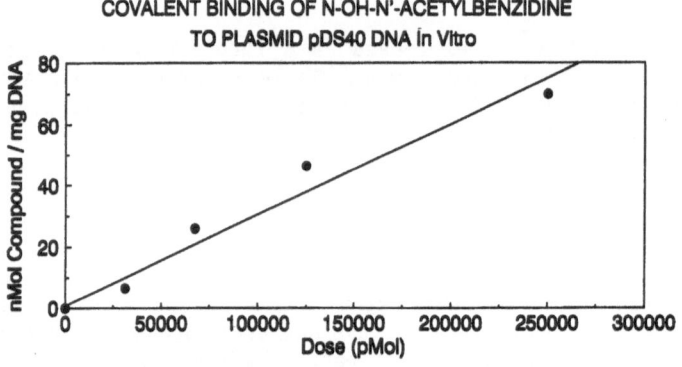

COVALENT BINDING OF N-OH-N'-ACETYLBENZIDINE
TO PLASMID pDS40 DNA In Vitro

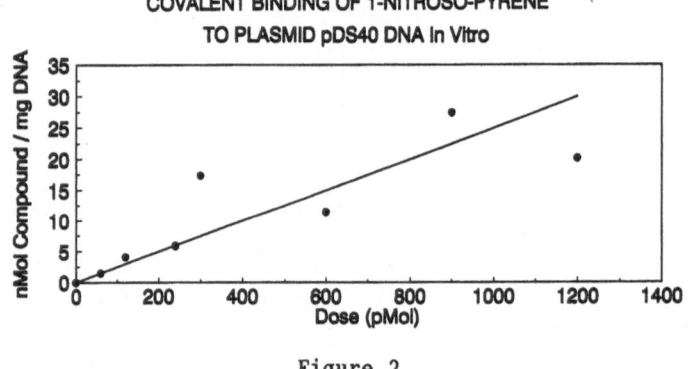

COVALENT BINDING OF 1-NITROSO-PYRENE
TO PLASMID pDS40 DNA In Vitro

Figure 2

compounds in a variety of in vitro assay systems. This is then compared
with the mutagenic potency of DNA adducts generated by known human
carcinogens in the same assay systems. Clearly, absolute mutagenic
potential will vary between assay systems but data being generated in our
laboratory is beginning to suggest that the relativity of mutagenic
potential of adducts of different chemical compounds within any given
system is maintained when other systems are investigated. The use of
synthetic activated intermediate compounds or their analogues removes the
uncertainty inherent in in vitro metabolising systems.

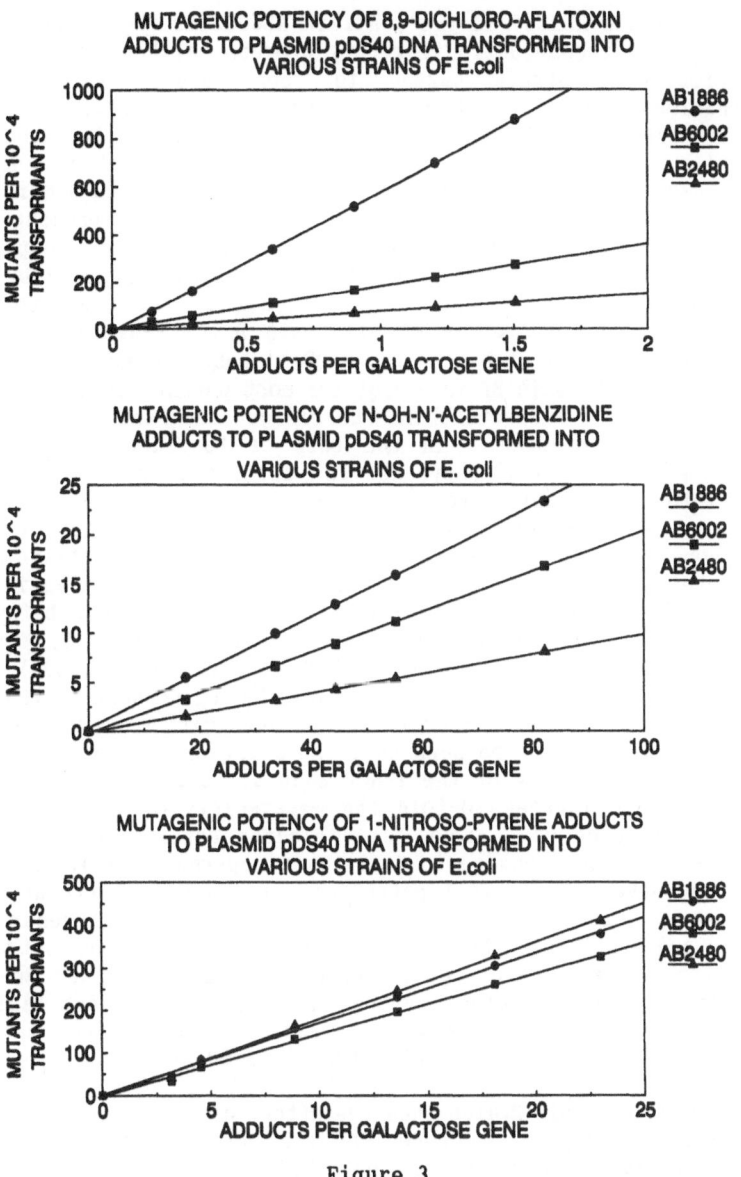

Figure 3

The present study was initiated to investigate the relative mutagenic potential of adducts generated by 1-NOP in an _in vitro_ plasmid system compared with the mutagenic potential of DNA adducts generated by 8,9-diCl-AFB1 and N-OH-N'-AcBZ. In the present study, the relative mutagenic potential of adducts generated by 1-NOP, an activated intermediate of 1-nitropyrene, was measured in an _in vitro_ plasmid system. Plasmid pDS40 containing an active galK gene was treated with test chemical and then transformed into three different strains of _E.coli_ unable to utilise galactose as an energy source. The transferance of the active galK into the bacteria conveyed an ability to utilise galactose. All three strains are uvrA⁻; strain AB1886 is rec⁺; strain AB6002 is recB⁻ and hence has no active major recombinational repair process, although SOS repair is still present.  Strain AB2480 is recA⁻ and therefore neither recombinational nor SOS repair systems are active. Mutations, resulting from treatment of the plasmid with test agent leading to the inactivation of the plasmid galactose gene  or its promoter, were detected as white colonies growing among red colonies on McConkey agar.  Mutagenicity was expressed as a function of the number of DNA adducts present in the plasmid's target gene determined from at least three seperate experiments using radiolabelled test compound.  The mutagenicity induced by 1-NOP-DNA adducts was compared with those induced by the analogue of the ultimate carcinogenic species of aflatoxin $B_1$, 8,9-diCl-AFB1 and the activated intermediate of benzidine, N-OH-N'-AcBZ.  These compounds were chosen because aflatoxin $B_1$ is one of the most potent animal carcinogens known and is also a suspected human carcinogen and benzidine is a known human bladder carcinogen.  In the case of benzidine it is further relevant as, like 1-nitropyrene is activated via an N-hydroxy intermediate.  From Table 1 it can be seen that for the same dose of each test agent 8,9-diCl-AFB1 displays some 2-5 fold the mutagenicity of 1-NOP and some 800-fold the mutagenicity of N-OH-N'-AcBZ in strain AB1886. From Figure 2 it can be seen that the binding of 1-NOP to plasmid DNA is greater than the binding of 8,9-diCl-AFB1 and both are substantially higher than the binding of N-OH-N'-AcBZ (1-NOP slope; 0.0194 nmol/mg DNA/pmol 8,9-diCl-AFB1 slope 0.0017; N-OH-N'-AcBZ slope 0.0003).  Thus, when mutation is related to the number of adducts present in the target gene as in Table 2, the true relative mutagenicities in strain AB1886 become apparent.  It can be seen that the aflatoxin adduct induces some 30-fold the mutagenicity of the 1-NOP adduct.  Significantly, however, the 1-NOP adduct induces 30-fold the mutagenicty of the N-OH-N'-AcBZ adduct, the product of a known human bladder carcinogen.  These data suggest that if N-(deoxyguanosin-8-yl)-1-aminopyrene adducts are induced in DNA they have a biological potential, at least in a prokaryotic in vitro system, somewhere intermediate between the adducts generated by aflatoxin and benzidine.  As 1-nitropyrene has not been demonstrated to be a potent animal carcinogen, these data indicate, as has been previously suggested in other systems (14,22), that potent biological activity may be prevented by the inadequacy of mammalian systems in reducing this compound to its activated intermediate.

Data presented in Table 2 also show the effect of error-prone, bacterial post-replication repair processes on the mutagenicity of the three compounds studied.  In the case of both 8,9-diCl-AFB1 and N-OH-N'-AcBZ, removal of the recombinational repair pathway, as in strain AB6002, significantly reduces the induced mutation.  In the case of 1-NOP, however, a very much lower reduction in mutation was observed.  Mutation in strain AB6002 may be more representative of true targeted, induced mutation than that observed in strain AB1886.  Data exists in the literature to suggest that in AB1886 recombination can take place between the galactose gene in the plasmid and its equivalent in the host DNA. Therefore, if an adduct were present in a part of the plasmid gene which corresponded to an undamaged section of the deficient host gene, a recombinational event could result in an active gene without mutation having occurred.  In strain AB2480, which is completely deficient in post-replicational repair activity, the mutagenicity of 8,9-diCl-AFB1 and N-OH-N'-AcBZ adducts is again significantly reduced whereas the mutagenicty of 1-NOP adducts is unaffected.  These data suggest that in the case of 1-NOP, the majority of induced mutation occurs as a direct result of the presence of the adduct in the target gene and that the mechanism is base-substitution and frameshift mutation occuring before or during normal DNA synthesis, as we have shown in other systems (23). In the case of the other two compounds under investigation, the vast majority of apparent mutation induction appears to occur as a result of recombinational events. If these involve the host genome as in strain AB1886, then the numbers of mutants scored may be an overestimate of the actual mutagenicity of these compounds. For this reason data indicating the relative mutagenic potential of these adducts in this system may best be derived from strain AB2480, a strain in which it is believed bacterial post-replicational repair processes are playing little or no part in the observed mutagenicity.

Studies in our laboratory attempting to measure the relative mutagenic potential of adducts generated by these and other related compounds including methylene-bis-o-chloroaniline, are ongoing in the NIH 3T3 protooncogene activation assay; mouse lymphoma L1578Y HGPRT mutation assay; Salmonella typhimurium mutation assay and prophage lambda induction assay. Data being generated from these systems indicate that the relative order of biological activity induced by the DNA adducts observed in the present study is maintained between systems.

In vitro studies on the biological activity of chemical carcinogens are a long way from the goal of replicating the conditions in the intact animal. However, by combining pharmacodynamic and pharmacokinetic data from in vitro studies with data on the potential of specific DNA adducts to induce mutation in in vitro systems; and comparing these data with data from compounds of known carcinogenic activity; it may be possible to derive valuable information on the likely hazard to man. The results generated in this study suggest that if N-(deoxyguanosin-8-yl)-1-aminopyrene adducts are found in DNA, then the risk of mutation may be far higher than the presence of N-(deoxyguanosin-8-yl)-N'-acetylbenzidine. However, a great deal more data on relative metabolism and relative extent of DNA repair in vivo will be required before extrapolation of carcinogen initiation risk to man can be attempted.

REFERENCES

1  Gibson T, (1982) Nitroderives of polynuclear aromatic hydrocarbons in airborne and source particulate. Atmos. Environ. 16, 2037-2040.
2  Pederson T and Siak J S (1981) The role of nitroaromatic compounds in the direct-acting mutagenicity of diesel particle extracts. J. Appl. Toxicol. 1, 54-60.
3  Salmeen I, Durisin A M, Prater T J, Riley T and Schuetzle D (1982) Contribution of 1-nitropyrene to direct-acting Ames assay mutagenicities of diesel particulate extracts. Mutat. Res. 104, 17-23.
4  Wei C, Raobe O G and Rosenblatt L S (1982) Microbial detection of mutagenic nitro-organic compounds in filtrates of coal fly ash. Environ. Mutagenesis 4, 249-258.
5  Li A P, Clark C R, Hanson R L, Henderson T R and Hobbs C H (1982) Coal combustion fly ash extract as direct-acting mutagen in Salmonella and promutagen in Chinese hamster ovary cells. Environ. Mutagenesis 3, 407-408.
6  Mermelstein R, Kiriazides, D K, Butler M, McCoy E C and Rosenkranz H S (1981) The extraordinary mutagenicity of nitropyrenes in bacteria. Mutat. Res 89, 187-196.
7  Nakayasu M, Sakamoto H, Wakabayashi K, Terada M, Sugimura T and Rosenkranz H S (1982) Potent mutagenic activity of nitropyrenes on Chinese hamster lung cells with diphtheria toxin resistance as a selective marker. Carcinogenesis 3, 917-922.
8  Li A P and Dutcher J S (1983) Mutagenicity of mono-, di- and tri-nitropyrenes in Chimese hamster ovary cells. Mutat. Res. Lett. 119, 387-392.

9  Kawachi T (1982)  Mutagenicity and carcinogenicity of nitropyrene. In: Rickert D (ed), The Toxicity of Nitroaromatic Compounds. Hemisphere, New York.

10 Natchman J P and Wolff S (1982)  Activity of nitropolynuclear aromatic hydrocarbons in the sister chromatid exchange assay with and without metabolic activation.  Environ. Mutageneses 4, 1-5.

11 Hirose M, Lee M S, Wang C Y and King C M (1984)  Induction of rat mammary gland tumours by 1-nitropyrene, a recently recognised environmental mutagen.  Cancer Res. 44, 1158-1162.

12 El-Bayoumi K, Hecht S S, Sackl T and Stoner G D (1984)  Tumorigenicity and metabolism of 1-nitropyrene in A/J mice.  Carcinogenesis 5, 1499-1452.

13 Heflich R H, Howard P C and Beland F A (1985)  1-Nitrocopyrene : an intermediate in the metabolic activation of 1-nitropyrene to a mutagen in Salmonella typhimurium TA1538.  Mutat. Res. 149, 25-32.

14 Beland F A, Ribovich M, Howard P C, Heflich R H, Kurian P and Milo G E (1986)  Cytotoxicity, cellular tranformation and DNA adducts in normal human diploid fibroblats exposed to 1-nitropyrene, a reduced derivative of the environmental contaminant, 1-nitropyrene. Carcinogenesis 7, 1279-1283.

15 Howard P C, Heflich R H, Evans F E and Beland F A (1983)  Formation of DNA adducts in vitro and in Salmonella typhimurium upon metabolic reduction of the environmental mutagen 1-nitropyrene.  Cancer Res. 43, 2052-2058.

16 Stanton C A, Chow F L, Phillips D H, Grover P L, Garner R C and Martin C N (1985)  Evidence of N-(deoxyguanosin-8-yl)-1-aminopyrene as a major DNA adduct in female rats treated with 1-nitropyrene. Carcinogenesis 6, 535-538.

17 Martin C N and Garner R C (1977) Aflatoxin B-oxide generated by chemical or enzymic oxidation of aflatoxin B$^1$ causes guanine substitution in nucleic acids. Nature 267, 863-865.

18 Essigmann J M, Croy R G, Nadyan A M, Busby W F, Reinhold V N, Buchi G and Wogan G N (1977)  Structural identification of the major DNA adduct formed by aflatoxin B$_1$ in vitro.  Proc. Natl. Acad. Sci. 74, 1870-1874.

19 Martin C N, Beland, F A, Roth, R W and Kadlubar, F F (1982) Covalent binding of benzidine and N-acetylbenzidine to DNA at the C-8 atom of deoxyguanosine in vivo and in vitro. Cancer Res. 42, 2678-2686.

20 Swenson D H, Miller J A and Miller E C (1975)  The reactivity and carcinogenicity of aflatoxin B$_1$-2,3-dichloride, a model for the putative 2,3-oxide metabolite of aflatoxin B$_1$. Cancer Res. 35, 3811-3823

21 Mizusawa H, Lee C-H and Kakefuda T (1981) Alteration of plasmid DNA-mediated transformation and mutation induced by covalent binding of benzo(a)pyrene-7,8-dihydrodiol-9,10,-epoxide in Escherichia coli. Mutation Res. 82, 47-57.

22 McCoy E C, Anders M, McCartney M, Howard P C, Beland F A and Rosenkranz H S (1984). The recombinogenic inactivity of 1-nitropyrene for yeast is due to a deficiency in a functional nitroreductase. Mutation Res. 139, 115-118.

23 Stanton C A, Garner RC, and Martin C N (1988).  The mutagenicity and DNA base sequence changes induced by 1-nitroso-and 1-nitropyrene in the -cI gene of lambda prophage. Carcinogenesis 9, 1153-1157.

# MUTATIONS INDUCED IN THE lacI GENE OF *E. COLI* BY 1-NITROSO-8-NITROPYRENE AND FURYLFURAMIDE: THE INFLUENCE OF PLASMID pKM101 AND EXCISION REPAIR ON THE MUTATIONAL SPECTRUM

I.B. Lambert[1], A.J.E. Gordon[2], T.A. Chin[1], D.W. Bryant[1], B.W. Glickman[2], and D.R. McCalla[1]

[1]McMaster University, Hamilton, Ontario
[2]York University, Toronto, Ontario (Canada)

## INTRODUCTION

Nitrated polyaromatic hydrocarbons ($NO_2PAH$) and nitroheterocyclic compounds are of potential human health importance. $NO_2PAH$ are emitted into the environment as byproducts of combustion processes (1), while nitroheterocyclic compounds such as 5-nitrofuran derivatives have been widely used in both human and veterinary medicine, and as food additives and preservatives. Representatives of each of these chemical classes are potent mutagens in bacteria, and induce tumors in rodents (1,2).

Our recent studies have examined the mutational specificity of the $NO_2PAH$, 1,8-dinitropyrene (1,8-DNP), and a 5-nitrofuran derivative, 2-furylfuramide (AF2). In bacteria both of these compounds are metabolically activated by nitroreductases to species capable of reacting with DNA (2,3). In the case of 1,8-DNP, acetylation of the partially reduced hydroxylamine intermediate is required for maximal mutation induction. The major DNA adduct formed by 1,8-DNP is the C(8) adduct of guanine, 1-N-(2'-deoxyguanosin-8-yl)-amino-8-nitropyrene (4). In the *S. typhimurium* tester strains developed by Ames, maximal mutagenicity is observed at the frameshift loci hisD3052 (TA1538, TA98) and hisC3076 (TA1537) (1). In strains harboring the plasmid pKM101 1,8-DNP also exhibits significant mutagenicity at the base substitution locus hisG46 (i.e. in TA100 but not in TA1535) (1).

The DNA adducts(s) formed by AF2 are unstable and subsequently have not been characterized. AF2 is mutagenic in assays measuring base substitution activity (2) but requires the induction of error-prone repair functions such as those encoded by the mucAB locus of plasmid pKM101 or the endogenous umuCD locus in *E. coli* (2,5).

In order to understand the mutagenic mechanisms of $NO_2PAH$ and 5-nitrofurans better, we have determined the DNA sequence of a large number of forward mutations induced in the lacI gene or *E. coli* by either 1-nitroso-8-nitropyrene (1,8-NONP), an intermediate in the metabolic activation of 1,8-DNP, or AF2. In this paper we summarize some of the results of these studies and propose models which accommodate our observations.

*Nitroarenes*, Edited by P. C. Howard *et al.*
Plenum Press, New York, 1990

## The Mutagenic System

The <u>lac</u> repressor is a tetrameric protein consisting of 4 identical subunits of 360 amino acids each. The 59 amino acids at the amino terminus form the DNA-binding domain, while the remainder of the protein forms the domain responsible for subunit aggregation and inducer binding (Reviewed in Ref 6). As a target for mutagenesis, the <u>lacI</u> gene contains about 1000 base pairs which can be altered by point mutations to produce a phenotypically selectable mutant. All frameshift or nonsense mutations within this sequence will be selected without bias. In contrast, missense mutations tend to cluster in defined regions, the most sensitive of which is the DNA-binding domain (7).

In our studies we have taken advantage of these properties to obtain as comprehensive a description of 1,8-NONP and AF2 mutagenesis as possible. Since nitroarenes are known to be capable of generating both frameshift and base substitution mutations (1), it was necessary to examine the mutational specificity of 1,8-NONP in the entire gene. In contrast, our study of AF2-induced mutations was restricted to the DNA binding domain of the <u>lacI</u> gene which has been extremely well characterized as a base substitution target.

## METHODS

### Selection and Characterization of LacI⁻ Mutants

LacI⁻ mutants were selected from five different *E. coli* strains. Excision repair proficient strains were NR6112 [F'<u>lacpro</u>; <u>ara</u> <u>thi</u> <u>rfa</u> Δ(<u>lacpro</u>)] and EE125 [NR6112 / pKM101]. Strains deficient in excision repair were NR6113 [F'<u>lacpro</u>; <u>ara</u> <u>thi</u> <u>rfa</u> Δ(<u>uvrB-bioFCD-chlA</u>) Δ(<u>lacpro</u>)]; CM6114 [NR6113 / pKM101]; and TC3960 [F'<u>lacpro</u>; <u>ara</u> <u>thi</u> <u>trpE9777</u> Δ(<u>uvrB-bioFCD-chlA</u>) Δ(<u>lacpro</u>) / pKM101]. Multiple cultures of the <u>rfa</u> strains were treated with 25 nmoles of 1,8-NONP for 15 minutes at 37°C following which the bacteria were pelleted by centrifugation, washed and immediately plated onto agar plates containing PGal (phenyl-β-D-galactoside), a noninducing substrate for β-galactosidase. With PGal as the sole carbon source, only cells constitutively expressing β-galactoside (<u>lacI</u>⁻ or <u>lacO</u>ᶜ mutants) will form colonies (8). Mutants selected in this manner were transferred into phage mRS81 (9), the approximate location of the mutation determined by deletion mapping (10), and the DNA sequence of the mutated region determined by the dideoxy chain termination method (11,12).

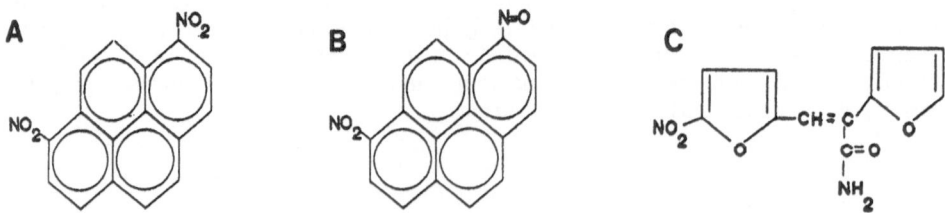

Figure 1. Structures of (A) 1,8-DNP; (B) 1,8-NONP; and (C) AF2.

AF2-induced mutants were selected by treating cultures of *E. coli* TC3960 with 1 nmole of AF2 for 15 minutes at 37°C prior to selection on PGal. LacI⁻ mutations located in the amino terminal portion of the gene encoding the DNA binding domain of the repressor were identified by means of a complementation test (13), and the DNA sequence determined as described (14).

## $^{32}$P-postlabelling of DNA Adducts

Following treatment with 1,8-NONP (as above) some *E. coli* cultures were centrifuged, and the DNA extracted with phenol/chloroform/isoamyl alcohol. The DNA was examined for adducts by $^{32}$P-postlabelling analysis following enrichment of the adducts from the normal nucleotides in a butanol extraction step (15). Solvents for the TLC procedure (16) were as follows: $D_1$- 1.1 M LiCl; $D_2$- omitted; $D_3$- 3.0 M lithium formate, 7.0 M urea, pH 3.5; $D_4$- 0.8 LiCl, 0.5 M tris-HCl, 7.0 M urea, pH 8.0; $D_5$- 1 M sodium phosphate, pH 6.8.

## RESULTS AND DISCUSSION

### DNA Adduct Formation with 1,8-NONP

The $^{32}$P-postlabelling procedure revealed one major DNA adduct and two minor adducts in the DNA of cells treated with 1,8-NONP. The fingerprint pattern is shown in Figure 2. The adduct represented by spot A has previously been identified as the C-8 adduct of guanine, 1-N-(2'-deoxyguanosin-8-yl)-amino-8-nitropyrene (4). It accounts for about 95% of the DNA adduct formed in *E. coli* under these conditions and is also the major adduct observed in: 1) rabbit tracheal epithelial cells treated with either 1,8-DNP or 1,8-NONP (17); 2) rat mammary gland, lung, mesentery tissue, and liver following intraperitoneal injection with 1,8-DNP (18); and 3) calf thymus DNA or alternating (dG·dC) copolymer treated *in vitro* with 1,8-NONP in the presence of ascorbate at pH 5.0. The minor adducts B and C each accounted for about 2.5% of the total adduct formed in *E. coli* cells treated with 1,8-NONP. We have found that adduct B was the major adduct formed in alternating (dA·dT) copolymer treated *in vitro* with 1,8-NONP, while adduct C comigrated with a minor product of the *in vitro* reaction between 1,8-NONP and alternating (dG·dC) copolymer.

### LacI⁻ mutations recovered following mutagen treatment

a) 1,8-NONP. LacI⁻ mutants were selected from the four rfa strains (NR6112, EE125, NR6113, and CM6114) following treatment with 1,8-NONP (average survival > 70%). Relative to the spontaneous frequency, the mutation frequency in identically treated cultures was increased 2-fold in NR6112 (wild-type), 5 fold in EE125 (pKM101), 33-fold in NR6113 (ΔuvrB),and 88-fold in CM6114 (pKM101, ΔuvrB). The precise DNA sequence of 743 mutants (248 mutants in NR6112, 129 in EE125, 148 in NR6113, and 218 in CM6114) has been determined.

Several classes of mutation were recovered from *E. coli* strains following 1,8-NONP treatment. In the excision repair proficient strains NR6112 and EE125, which were relatively insensitive to 1,8-NONP mutagenesis, a frequently observed type of mutation was the gain or loss of the sequence 5'-TGGC-3' from a frameshift hot spot at positions 621-632. These specific mutations comprise about 70% of spontaneous lacI⁻ mutants (12,19) and therefore their presence is probably due in part to the inclusion of mutants of spontaneous origin in the mutant collections. This type of mutation accounted for a very small proportion of the mutational spectrum in NR6113 and CM6114, which had a much higher level of induced mutagenesis.

The most commonly observed mutations other than those due to (+/-) TGGC events were frameshift mutations, base substitutions and deletions. The degree to which each of these mutational classes was induced by 1,8-NONP in our studies is shown in Table 1. The level of induction was estimated by comparing the frequency

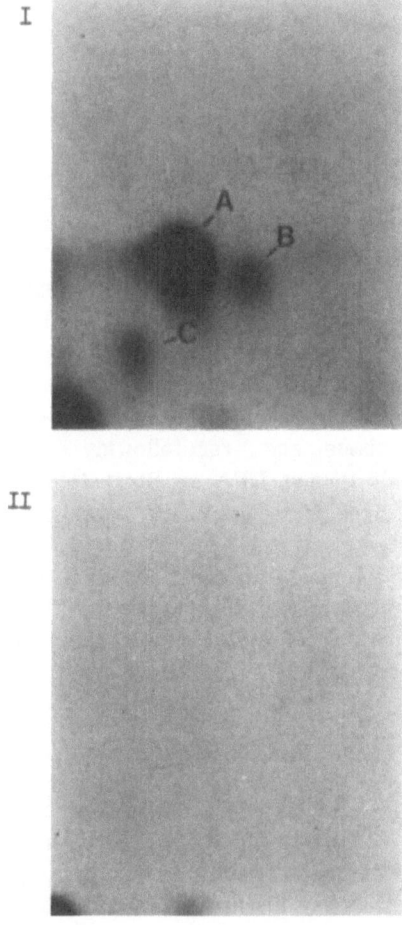

Figure 2. ³²P-postlabelling analysis of adducts generated by treatment of *E. coli* with 1,8-NONP. (I) *E. coli* + 1,8-NONP; (II) *E. coli* + DMSO.

**Table 1.** Induction (fold increase over the spontaneous frequency) of mutational classes following 1,8-NONP treatment.

| Strain | NR6112 (Wild-type) | EE125 (pKM101) | NR6113 (ΔuvrB) | CM6114 (pKM101, ΔuvrB) |
|---|---|---|---|---|
| **Class of Mutation** | | | | |
| Frameshifts | 8 | 15 | 630 | 850 |
| Base Subst. | no increase | 8 | 24 | 180 |
| Deletions | 2 | 2 | 4 | 43 |

The frequency of mutational events in a given strain was calculated by considering the contribution of each mutational class (based on the distribution of mutants in the mutational spectra) to the spontaneous or 1,8-NONP-induced mutation frequency. It was assumed that the spontaneous mutational spectra in uvrB- strains is similar to that of isogenic uvrB+ strains (34). Therefore for NR6112 and NR6113, the spontaneous mutation spectra used in these calculations was from wild-type bacteria (12; Halliday and Glickman, in preparation), while for EE125 and CM6114 a spontaneous pKM101 spectrum (Gordon and Glickman, in preparation) was used.

of each mutational class in the presence and absence of 1,8-NONP treatment. In addition, a small number of duplications, tandem base substitutions, and complex mutations were recovered from each strain following 1,8-NONP treatment.

The genetic dependence of frameshift and base substitution mutations induced by 1,8-DNP has previously been demonstrated using Ames tester strains (1). The trends obtained from the *S. typhimurium* reversion tests closely parallel those obtained from our studies in *E.coli* using a forward mutation system. However, DNA sequencing studies allow us to extend these observations by elucidating the precise nature of the mutagenic events (see below).

Although deletion mutations have been recovered from all strains treated with 1,8-NONP, the frequency of this mutation is highest in the strain which is both deficient in excision repair and contains the plasmid pKM101. The endpoints of most deletions contain regions of extensive homology. This is consistent with models involving slippage of the template strand during DNA synthesis stabilized by base pairing between the misaligned DNA strands (19). In some cases, the slipped intermediates are stabilized by palindromic or quasipalindromic sequences (20) within the deleted section.

**Table 2.** Summary of base changes induced in the lacI gene by AF2 in pKM101, ΔuvrB *E. coli*.

| Class of Mutation | | # of Occurences |
|---|---|---|
| Base | substitutions | 145 |
| | transversions | 95 |
| | transitions | 50 |
| Tandem Base Substitutions | | 2 |
| Frameshifts | | 11 |
| Complex mutations | | 5 |
| Deletions | | 2 |
| Total | | 165 |

b) AF2. Treatment of strain TC3960 (pKM101, ΔuvrB) with AF2 led to a large increase in the mutation frequency relative to untreated cultures (>300-fold at 20% survival). One hundred and sixty-five lacI⁻ mutants defective in DNA binding were identified by means of a genetic test, and their DNA sequence determined. A summary of the mutations recovered is given in Table 2.

Almost 90% of the mutations recovered following AF2 treatment were base substitutions, of which transversions outnumbered transitions by a ratio of 2:1. A more detailed descriptions of AF2-induced base substitutions is given below. Small numbers of other mutational classes were also recovered, the most abundant of which were frameshift mutations. Interestingly, none of these frameshift mutations involved loss of bases from repeated sequences. This is in contrast to frameshift mutations induced by 1,8-NONP which tend to occur within runs of guanine.

## Mutational Specificity of 1,8-NONP and AF2 and Possible Mechanisms of Mutation

Base Substitution Mutations. A total of 142 base substitution mutations were recovered from the 4 strains which had been treated with 1,8-NONP. As shown in Table 3, the vast majority of these (84%) were induced in strains harboring the plasmid pKM101. More than 92% of the base substitutions were at G:C sites giving strong evidence that they were targeted by the major DNA adduct. G:C=>T:A transversions were by far the most common form of base substitution suggesting that adenine is incorporated almost exclusively opposite the site of the lesion during mutagenic bypass.

The specificity of AF2-induced base substitution mutations is very different from that seen following 1,8-NONP treatment. Although G:C=>T:A transversions were still the most common base substitution event at G:C sites, a significant number of G:C=>A:T transitions and G:C=>C:G transversions were also observed. The distribution of AF2-induced base substitutions is given in Table 4. For comparative purposes, those base substitution mutations induced by 1,8-NONP in the ΔuvrB, pKM101 background which occur in the DNA binding domain of the lacI gene are also shown. The distribution of these mutations is clearly different. While G:C=>T:A transversions are the prominent base substitution induced by both agents, AF2 induces significantly higher levels of additional base substitution mutations, particularly G:C=>A:T transitions.

**Table 3.** Distribution of base substitutions in *E. coli* strains treated with 1,8-NONP.

| Strain | NR6112 (Wild-type) | EE125 (pKM101) | NR6113 (ΔuvrB) | CM6114 (pKM101, ΔuvrB) |
|---|---|---|---|---|
| **Class of Mutation** | | | | |
| **Transversions** | | | | |
| G:C=>T:A | 5[a] | 31 | 4 | 67 |
| G:C=>C:G | 1 | 1 | 1 | 2 |
| A:T=>T:A | 0 | 1 | 0 | 7 |
| A:T=>C:G | 0 | 2 | 0 | 0 |
| **Transitions** | | | | |
| G:C=>A:T | 5 | 3 | 7 | 4 |
| A:T=>G:C | 0 | 1 | 0 | 0 |
| Total | 11 | 39 | 12 | 80 |
| % at G:C sites[b] | 100% | 90% | 100% | 91% |

[a] values are the number of independent occurrences.
[b] the percent of base substitution mutations occurring at G:C base pairs.

**Table 4.** Specificity of base substitution matations induced by 1,8-NONP and AF2 in the DNA binding domain of ΔuvrB, pKM101 *E. coli*.

| Strain | TC3960 + AF2 (ΔuvrB, pKM101) | CM6114 + 1,8-NONP (rfa, ΔuvrB, pKM101) |
|---|---|---|
| **Class of Mutation** | | |
| **Transversions** | | |
| G:C=>T:A | 76 | 30 |
| G:C=>C:G | 10 | 2 |
| A:T=>T:A | 9 | 4 |
| A:T=>C:G | 0 | 0 |
| **Transitions** | | |
| G:C=>A:T | 49 | 1 |
| A:T=>G:C | 1 | 0 |
| Total | 145 | 37 |

The very different specificities observed in the 1,8-NONP and AF2 base substitution spectra probably reflect different mutational mechanisms. Since base substitutions induced by both chemicals are influenced by pKM101 encoded functions, the mechanistic differences are likely to be a direct consequence of the templating properties of the respective lesions. One explanation that has frequently been invoked to account for the common tendency of bulky lesions to induce G:C=>T:A transversions is the formation of apurinic (AP) sites followed by preferential incorporation of adenine opposite the noncoding site during DNA synthesis (21). The mutational specificity of depurination in lac DNA of phage M13mp2 has been determined by Kunkel (22) and suggests that while incorporation of adenine opposite the AP site is favored, significant incorporation of thymine and guanine also occurs. Table 5 shows a comparison between the ratio of incorporated adenine, thymine, or guanine expected opposite an AP site (22), and that actually observed in the 1,8-NONP and AF2 base substitution spectra. While the distribution of AF2 induced base substitutions is not statistically different from those formed at putative AP sites, the distribution of 1,8-NONP induced mutation is quite distinct.

Table 5.  Base incorporation opposite apurinic sites (22), or opposite guanine in base substitutions induced by 1,8-NONP and AF2.

| Incorporation opposite guanine | Apurinic Sites[a] | 1,8-NONP | AF2 |
|---|---|---|---|
| Adenine | 59% | 91% | 56% |
| Thymine | 28% | 3% | 36% |
| Guanine | 11% | 6% | 8% |

[a] as determined by Kunkel (22).

If base substitution mutations induced by 1,8-NONP are not due to the formation of AP sites, at least two alternative mechanisms might account for the predominance of G:C=>T:A transversions. 1) It is possible that the C(8) adduct is a stable, noninformational lesion and that DNA polymerase almost exclusively incorporates adenine opposite such sites. If this were the case, our data suggest that the specificity of base incorporation opposite a stable noninformational lesion is different from that opposite an AP site. 2) The modified bases might direct the incorporation of bases during DNA synthesis. If the adducted base maintained the anti conformation during DNA synthesis, the templating face would be similar to that normally presented to the incoming base and should facilitate incorporation of cytidine. Recent modelling studies with 2- acetylaminofluorene (2-AAF) adducts formed at the C(8) position of guanine suggest that these lesions may also be capable of shifting the position of the modified base within the helix to allow direct pairing with adenine (23). Structures for the resultant G:A base pairs have been reported

(24). This model is supported by the observation that both cytidine and adenine are incorporated across from the C(8) adduct of 2-AAF by both prokaryotic and mammalian polymerases *in vitro* (25). However, only adenine incorporation would yield a base substitution mutation.

In either case, the continued presence of the adduct may be expected to represent a physical impediment to progression of the replicational complex. Therefore, the functions encoded by the <u>mucAB</u> genes of plasmid pKM101 or the endogenous <u>umuCD</u> locus of *E. coli*, which facilitate resumption of DNA synthesis following incorporation of nucleotides opposite the lesion (26), would increase the efficiency of base substitution formation. This is entirely consistent with the genetic dependence of base substitution mutation observed in our studies.

The specificity of base substitutions induced by AF2 is very similar to the specificity of AP site mutagenesis. Knowledge of the DNA adducts formed by nitrofurans would help to clarify whether this association is purely fortuitous or a reflection of the underlying mechanism of mutagenesis. However, despite considerable effort on the part of a number of laboratories, the putative DNA adducts have eluded characterization. Because nitrofurans, like nitropyrenes, require reduction of the nitro moiety in order to produce the reactive species, hypothetical schemes for adduct formation have tended to emphasis the role of the hydroxylamino derivative. However, an intriguing alternative possibility is that the hydroxylamine derived form nitroreduction of AF2 undergoes ring opening to form a substituted acrylonitrile (Figure 3). The reactive centre in the intermediate would be carbon 3 (*). With another 5-nitrofuran derivative, furazolidone (27), this sequence of reactions leads to the formation of protein adducts and would provide a route for adduction at the N(7) position of guanine, the product of which would be predicted to be extremely labile. We note that base substitutions induced by two agents which are known to react at the N(7) position of guanine, aflatoxin $B_1$-2,3-dichloride (28), and benzo(a)pyrene-7,8-diol-9,10-epoxide (Gordon, Bernelot-Moens and Glickman, in preparation), include a considerable number of G:C=>A:T transitions and G:C=>C:G transversions in addition to the major base substitution, the G:C=>T:A transversions.

<u>Frameshift Mutations</u>. Frameshift mutations resulting from the gain or loss of 1 or 2 bases were the class of mutation induced to the greatest extent by 1,8- NONP treatment. This was expected based on the potent mutagenic effect of 1,8- DNP in Ames tester strains TA1538, TA98, and TA1537 (1). Table 6 shows that the vast majority of detected frameshift events are due to the loss of a G:C base pair. Close examination of the sequences at which these mutations occur reveal that 1,8-NONP-induced frameshift mutagenesis exhibits remarkable specificity: 1) most frameshift events occur at sequences of contiguous guanines and the probability of such an event occurring increases with the length of the reiterated sequence; 2) frameshift events at lone G:C base pairs or runs of 2 guanines are highly favoured at sites with a 5' flanking thymine; 3) most frameshifts involve the loss of a G:C base pair suggesting that they are targeted by the major adduct G-C(8)-ANP (the possibility that a minor adduct is the mutagenic lesion cannot be ruled out, but is unlikely); and 4) losses of G:C base pairs are dominant in contrast to intercalating agents which induce both additions and losses of base pairs at reiterated sequences.

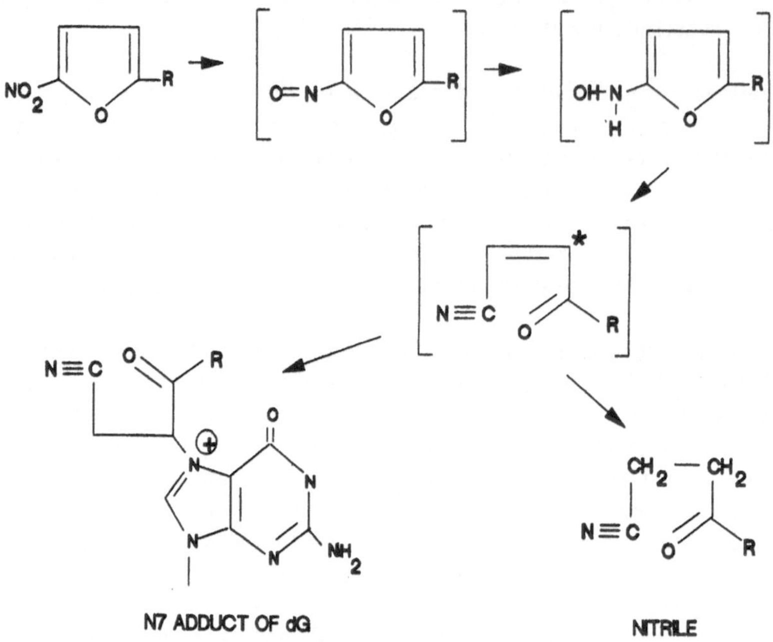

Figure 3. Hypothetical pathway for adduct formation with nitrofurans.

Table 6. Distribution of frameshift mutations collected from *E. coli* strains following 1,8-NONP.

| Strain | NR6112 (Wild-type) | EE125 (pKM101) | NR6113 (ΔuvrB) | CM6114 (pKM101, ΔuvrB) |
|---|---|---|---|---|
| Class of Mutation | | | | |
| −1 Frameshifts | 34[a] | 21 | 111 | 115 |
| −2 Frameshifts(−GC) | 3 | 0 | 2 | 3 |
| +1 Frameshifts | 3 | 2 | 1 | 1 |
| Total | 40 | 23 | 114 | 119 |
| % at G:C sites[b] | 87% | 87% | 99% | 99% |

[a] values are the number of independent occurrences

[b] the percent of frameshift mutations involving the gain or loss of G:C base pairs.

176

The tendency for frameshift mutations to occur in sequences of reiterated bases is common for frameshift events of spontaneous origin (29,30) as well as those induced by chemicals which are intercalators or which form bulky adducts (30,31,32,33). Reiterated sequences are believed to stabilize replication intermediates derived from slippage of one strand relative to the other such that one or more extrahelical bases bulge from the duplex (30). If the extrahelical base occurs on the template strand during DNA synthesis then loss of base(s) is predicted. The frameshift mutations characterized here are distinctive in that virtually all result from the loss of G:C base pairs. This is similar to frameshift events induced by 2-AAF (32). In contrast, frameshifts of spontaneous origin or those which are induced by intercalating agents include a substantial number of additions (29,30,31).

The presence of bulky lesions during DNA synthesis pose a fundamental challenge to slippage models because such models require that a portion of the reiterated sequence be copied accurately onto the nascent strand prior to slippage. But how does the nascent strand get faithfully copied in the presence of a replicational block? One solution to this problem is to postulate that cytidine can be incorporated opposite the C(8) adduct (as described above for base substitution mechanisms) but that replication is stalled immediately subsequent to the incorporation event. Halting of the replication complex, perhaps accompanied by dissociation of the polymerase from the replication fork, would increase the time available for slippage events to occur (28). If the lesion existed in a reiterated sequence the slippage would be stabilized by the formation of a G:C base pair at the base 5' to the adduct. Figure 4 (pathway A) shows that resumption of DNA synthesis from this point would produce a -1 frameshift. If the lesion was at a base without a 5'-guanine, then the slippage would not be stabilized and resumption of DNA synthesis would finally occur from opposite the lesion with no resultant mutation. A similar series of events would be possible if misincorporation of adenine occurred (Figure 4: pathway B). In this case, slippage would be stabilized if the 5'-base were a thymine. This is, in fact, entirely consistent with the specificity of the observed 1,8-NONP induced frameshift mutations which show a significant bias towards the loss of guanines adjacent to a 5' flanking thymine.

In this model the DNA adduct plays two roles in the formation of a frameshift mutation: 1) it directs incorporation of cytidine or adenine; and 2) it blocks the progression of the DNA polymerase subsequent to misincorporation allowing misalignment to occur. The model is a direct extension of the concepts previously discussed regarding base substitution mutation, and stresses the role of DNA sequence in the mutagenic outcome.

The frequency of frameshift mutations is elevated by almost 2 orders of magnitude in strains deficient in excision repair. However the DNA sequence analysis suggests that the specificity of the frameshift mutations induced in strains which are uvrB- is similar to those recovered from isogenic uvrB+ strains. It is likely therefore that the principal effect of reduced excision repair is to elevate both the total adduct burden and the mean adduct residence time, thereby increasing the probability that the appropriate replication (or possibly repair) enzymes might encounter a lesion in a sequence capable of giving rise to a frameshift event.

|                     | Pathway A                          |            | Pathway B                          |
|---|---|---|---|

## Pathway A                                                    Pathway B

1)    5'-A-T-G-G*G-A-C-3'                                5'-A-T-G*-G-A-C-A-3'

      a)  Replication up to lesion
      b)  Incorporation of nucleotide
          across from lesion
      c)  Replication halted by
          bulky adduct

2)    5'-A-T-G-G*G-A-C-3'                                5'-A-T-G*-G-A-C-A-3'
        C-C-T-G-5'                            A - C-T-G-T-5'

      d)  Misalignment (formation
          of extrahelical base)

               G*  ·                                           G*
3)    5'-A-T-G  G-A-C-3'                        5'-A-T  G-A-C-A-3'
        C - C-T-G-5'                          A - C-T-G-T-5'

      e)  Progression of replication
          at base 5' - to the lesion

               G*                                                 G*
4)    5'-A-T-G  G-A-C-3'                        5'-A-T  G-A-C-A-3'
        3'-T-A-C - C-T-G-5'                  3'-T-A - C-T- G-T-5'

5)    3'-T-A-C-C-T-G-5'     Final Product          3'-T-A-C-T-G-5'
                    -(G:C)

**Figure 4.** A model for 1,8-NONP induced frameshift mutagenesis. 1) Adduct formation occurs at the C-8 position of guanine. DNA polymerase attempts to incorporate bases across from the lesion during DNA synthesis. The adducted base is predicted to be capable of coding properly for cytosine (pathway A) or miscoding for adenine (pathway B). Following incorporation of the base, the bulky adduct interacts with the replicational complex to halt its progression. 2) Stalling at the site of the lesion allows time for slippage to occur (possibly aided by a conformational change facilitating the formation of an extrahelical base). 3) If the base 5' to the lesion pairs properly with the newly incorporated base then the stability of this structure will be enhanced. Pathway A predicts that this would occur only in runs of G:C bases. Misincorporation of adenine (pathway B) could lead to a -1 frameshift at guanines with a 5' flanking thymine. 4) Progression of the replicational complex would resume at the base 5' to the lesion leading to the final product (5) which is a -1 frameshift.

The influence of plasmid pKM101 on 1,8-NONP induced frameshift mutations is modest. The frequency of frameshifts is increased approximately 2-fold in strains harboring pKM101 (EE125 or CM6114) relative to isogenic strains not containing the plasmid (NR6112 or NR6113). It is possible that frameshifts are enhanced by the presence of the error prone repair capability encoded by the mucAB locus of plasmid pKM101 (or the endogenous umuCD locus), but not absolutely dependent on the presence of these functions. This is supported by the large reversion frequency observed in S. typhimurium TA1537 or TA1538 (which do not contain pKM101) following treatment with 1,8-DNP. S. typhimurium does not contain genes homologous to umuCD or mucAB and therefore frameshift mutations occur in these strains independent of any known error prone repair functions.

The spectrum of mutations recovered from 1,8-NONP treated bacteria provides a vivid illustration of how the nature of the local DNA sequence influences the process of mutagenesis. The mechanisms postulated above emphasize different outcomes of stalled replication following incorporation of bases opposite a bulky lesion. Resolution of this intermediate would be modulated by the nature of the surrounding sequence as well as the presence of error prone repair functions. A number of additional factors are likely to influence whether a mutation is observed at a particular site. These include the relative reactivity of bases within a given sequence; the ability of certain sequences to efficiently stabilize extrahelical structures; the preference of base incorporation by various polymerases; and the susceptibility of bases incorporated opposite the lesion to the exonucleolytic proofreading activity of polymerases. Development of sophisticated techniques for measuring DNA damage distribution, modelling the influence of adducts on DNA structure, and measuring the activities of DNA polymerases during replication of mutagen-modified templates will more precisely define those factors and bring evidence to bear on hypotheses formulated solely on the basis of the observed mutational specificity.

## ACKNOWLEDGEMENTS

We thank E. Eisenstadt for providing some of the E. coli strains used in this study. IBL was supported by an Ontario Graduate Scholarship. Financial support for these studies was provided by a grant from the National Cancer Institute of Canada.

## REFERENCES

1. H. Tokiwa, and Y. Ohnishi, CRC Crit. Rev. Toxicol. 17, 23-60 (1986).
2. D.R. McCalla, Env. Mutagenesis, 5, 745-765 (1983).
3. P.C. Howard, E.C. McCoy, and H.S. Rosenkranz, Mutagenesis, 2, 431-432 (1983).
4. P.J. Andrews, M.A. Quilliam, B.E. McCarry, D.W. Bryant, and D.R. McCalla, Carcinogenesis, 7, 105-110 (1986).
5. D.W. Bryant, D.R. McCalla, Chem.-Biol. Interactions, 31, 151-166 (1980).
6. B. Muller-Hill, Prog. Biophys. Molec. Biol. 30 227-252 (1975).

7. A.J.E. Gordon, P.A. Burns, D.F. Fix, F. Yatagai, F.L. Allen, M.J. Horsfall, J.A. Halliday, J. Gray, C. Bernelot-Moens, and B.W. Glickman. J. Mol. Biol. 200, 239-251 (1988).

8. J.H. Miller, D. Ganem, P. Lu, and A. Schmitz, J. Mol Biol. 109, 275-302 (1977).

9. R.M. Schaaper, B.N. Danforth, and B.W. Glickman, Gene, 39, 181-189 (1985).

10. U. Schmeissner, D. Ganem, and J.H. Miller, J. Mol. Biol. 109, 303-326 (1977).

11. F. Sanger, S. Nicklen, and A.R. Coulsen, Proc. Natl. Acad. Sci. (USA), 74, 5463-5467 (1977).

12. R.M. Schaaper, B.N. Danforth, and B.W. Glickman, J. Mol. Biol. 189, 273- 284 (1986).

13. J.H. Miller, Experiments in Molecular Genetics, Cold Spring Harbor Laboratory Press, Cold Spring Harbor, New York (1972).

14. P.A. Burns, A.J.E. Gordon, and B.W. Glickman, Carcinogenesis, 9, 1607-1610 (1988).

15. R.C. Gupta, Cancer Res. 45, 5656-5662 (1985).

16. R.C. Gupta, M.V. Reddy, and K. Randerath, Carcinogenesis, 3, 1081-1092 (1982).

17. C.A. Norman, I.B. Lambert, L.M. Davison, D.W. Bryant, and D.R. McCalla, Carcinogenesis, in press (1989).

18. C.A. Norman, MSc. Thesis, McMaster University, Hamilton, Canada (1988).

19. P.J. Farabaugh, U. Schmeissner, M. Hofer, and J.H. Miller, J. Mol. Biol. 126, 847-857 (1978).

20. B.W. Glickman, and L.S. Ripley, Proc. Natl. Acad. Sci. (USA), 81, 512- 516 (1984).

21. J.H. Miller, Ann. Rev. Genetics, 17, 215-238 (1983).

22. T.A. Kunkel, Proc. Natl. Acad. Sci. (USA), 81, 1494-1498 (1984).

23. E.L. Loechler, Biopolymers, 28, 909-927 (1989).

24. L.S. Kan, S. Chandrasegarian, S.M. Pulford, and P.S. Miller, Proc. Natl. Acad. Sci. (USA), 80, 4263-4265 (1983).

25. S.D. Rabkin, and B.S. Strauss, J. Mol. Biol. 178, 569-594 (1984).

26. B.A. Bridges, and R. Woodgate, Proc. Natl. Acad. Sci. (USA), 82, 4193-4197 (1985).

27. L.H.M. Vroomen, M.C.J. Berghmans, J.P. Groten, J.H. Koeman, and P.J. Van Bladeren, Tox. Appl. Pharmacol. 95, 53-60 (1988).

28. K. Sambamurti, J. Callahan, X. Luo, C.P. Perkins, J.S. Jacobsen, and M.Z. Humayun, Genetics, 120, 863-873 (1988).

29. L.S. Ripley, A. Clark, and J.G. deBoer, J. Mol. Biol. 191, 601-613 (1986).

30. G. Streisinger, and J. Owen, Genetics, 109, 633-659 (1985).

31. M. Calos, and J.H. Miller, J. Mol. Biol. 153, 39-66 (1981).

32. N. Koffel-Schwartz, J.-M. Verdier, M. Bichara, A.-M. Freund, M.P. Duane, and R.P.P. Fuchs, J. Mol. Biol. 177, 33-51 (1984).

33. L.M. Refolo, C.B. Bennett, and M.Z. Humayun, J. Mol. Biol. 193, 609-636 (1987).

34. R.M. Schaaper, R.L. Dunn, and B.W. Glickman, J. Mol. Biol. 198, 187-202 (1987).

# DNA ADDUCT FORMATION BY 1-NITROPYRENE 4,5- AND 9,10-OXIDE

Beverly A. Smith,[1] Robert H. Heflich,[1] Yoshinari Ohnishi,[2]
Akinobu Ohuchida,[3] Takemi Kinouchi,[2]
Janice R. Thorton-Manning,[1] and Frederick A. Beland[1,4]

[1]National Center for Toxicological Research,
Jefferson, Arkansas 72079, USA
[2]University of Tokushima, Tokushima, Japan
[3]Taiho Pharmaceutical Co., Tokushima, Japan
[4]University of Arkansas for Medical Sciences,
Little Rock, Arkansas 72205, USA

## INTRODUCTION

1-Nitropyrene is an environmental pollutant with both mutagenic and tumorigenic properties (reviewed in reference 1). This nitropolycyclic aromatic hydrocarbon is metabolized in vivo by either oxidative or reductive pathways, or by a combination of the two. Reduction of 1-nitropyrene leads to the formation of N-hydroxy-1-aminopyrene, an electrophilic metabolite that reacts with DNA to give N-(deoxyguanosin-8-yl)-1-aminopyrene (dG-C8-AP). This adduct is associated with the induction of mutations in Salmonella typhimurium and Chinese hamster ovary (CHO) cells (2-6), and evidence has been presented for its formation in vivo (7-11). Although 1-nitropyrene undergoes nitro-reduction in vivo, the major metabolic pathways appear to be oxidative (reviewed in reference 1), and these may lead to the DNA adducts that have been detected in addition to dG-C8-AP (8-11). 1-Nitropyrene 4,5-oxide and 1-nitropyrene 9,10-oxide are prime candidates for the ring-oxidized 1-nitropyrene metabolites that produce some of these unidentified DNA adducts. Both oxides display strong mutagenicity in Salmonella typhimurium (12) and CHO cells (13). In addition, Djurić et al. (14) demonstrated their binding to DNA by direct reaction and after nitroreduction by the mammalian nitroreductase, xanthine oxidase. In this study, we have examined the DNA adducts formed by 1-nitropyrene 4,5- and 9,10-oxide following reaction with calf thymus DNA, both in the presence and absence of xanthine oxidase and hypoxanthine, and have compared these adducts to those responsible for the mutagenicity of the oxides in CHO cells.

## MATERIALS AND METHODS

### Reaction of 1-Nitropyrene 4,5- and 9,10-Oxide with DNA and CHO Cells

1-Nitropyrene 4,5- and 9,10-oxide were synthesized as described in Fifer et al. (12). Calf thymus DNA was incubated with the oxides in the presence and absence of xanthine oxidase and hypoxanthine as outlined in

Djurić et al. (14). CHO cells were exposed to the oxides as reported in Heflich et al. (13).

DNA Adduct Analysis

DNA modified with 1-nitropyrene 4,5-oxide in the presence and absence of xanthine oxidase and hypoxanthine was hydrolyzed with DNase (0.1 mg per mg DNA) for 3 hr at 37°, and then with a combination of snake venom phosphodiesterase (0.3 U per mg DNA) and alkaline phosphatase (1 U per mg DNA) for 18 hr at 37°. The samples were adjusted to 50% methanol, centrifuged and analyzed by HPLC on a Waters system equipped with a Hewlett Packard 1040A detector adjusted to 280 nm. The adducts were separated with a C18 µBondapak reversed-phase column using a 30-min nonlinear (Waters #2) gradient of 20-65% methanol at a flow rate of 2 ml per min (2). Additional samples of adducted DNA were hydrolyzed using nuclease P1 (15 µg per mg DNA) instead of snake venom phosphodiesterase.

CHO cell DNA was isolated as described in Beland et al. (15). This DNA, as well as the in vitro modified DNA, was analyzed by $^{32}$P-post-labeling using the contact transfer method of Lu et al. (16), following enrichment of the adducts by extraction with n-butanol as described by Gupta (17). Adducts were resolved by thin-layer chromatography with the following solvents: D1 and D5, 0.9 M sodium phosphate, pH 6.8; D3, 3.6 M lithium formate, 8.5 M urea, pH 8.0; D4, 1.2 M lithium chloride, 0.5 M Tris-HCl, 8 M urea, pH 8.0. The adduct concentrations were quantified by comparison with a DNA standard substituted with dG-C8-AP at a level of 1.8 fmol per µg DNA (3). This reference was run as a control in each experiment.

RESULTS

HPLC analysis of calf thymus DNA that had been reacted with 1-nitro-pyrene 4,5-oxide indicated the presence of three adducts (Figure 1) having phenanthrene-type UV spectra (not shown). When the DNA was hydrolyzed with nuclease P1 instead of snake venom phosphodiesterase, the concentration of these adducts decreased by >70%. Incubations were also conducted with 1-nitropyrene 4,5-oxide and calf thymus DNA in the presence of xanthine oxidase and hypoxanthine. In addition to the three "direct-reaction" adducts, an adduct was observed with a HPLC retention time and UV spectrum identical to dG-C8-AP (not shown). The structure of this adduct was confirmed by NMR spectroscopy through comparison to a previously reported spectrum (2).

$^{32}$P-Postlabeling analysis of the 1-nitropyrene 4,5-oxide-modified calf thymus DNA indicated either one or possibly several closely related adducts (Figure 2A). DNA reacted with 1-nitropyrene 4,5-oxide in the presence of xanthine oxidase and hypoxanthine also had these adducts along with an additional adduct having identical migration characteris-tics to the bis-phosphate of dG-C8-AP (Figure 2B). In CHO cells incubated with 1-nitropyrene 4,5-oxide, the only adducts detected corresponded to those observed by the direct reaction of the oxide with DNA (Figure 2C).

Calf thymus DNA was also reacted with 1-nitropyrene 9,10-oxide and analyzed by $^{32}$P-postlabeling. Two adducts were observed that migrated in the vicinity of the "direct-reaction" adducts obtained from 1-nitropyrene 4,5-oxide (Figure 3A). The same adduct pattern was obtained when calf thymus DNA was incubated with 1-nitropyrene 9,10-oxide in the presence of xanthine oxidase and hypoxanthine (Figure 3B). CHO cells also had these adducts; in addition, there appeared to be a small amount of the bis-phosphate of dG-C8-AP (Figure 3C).

182

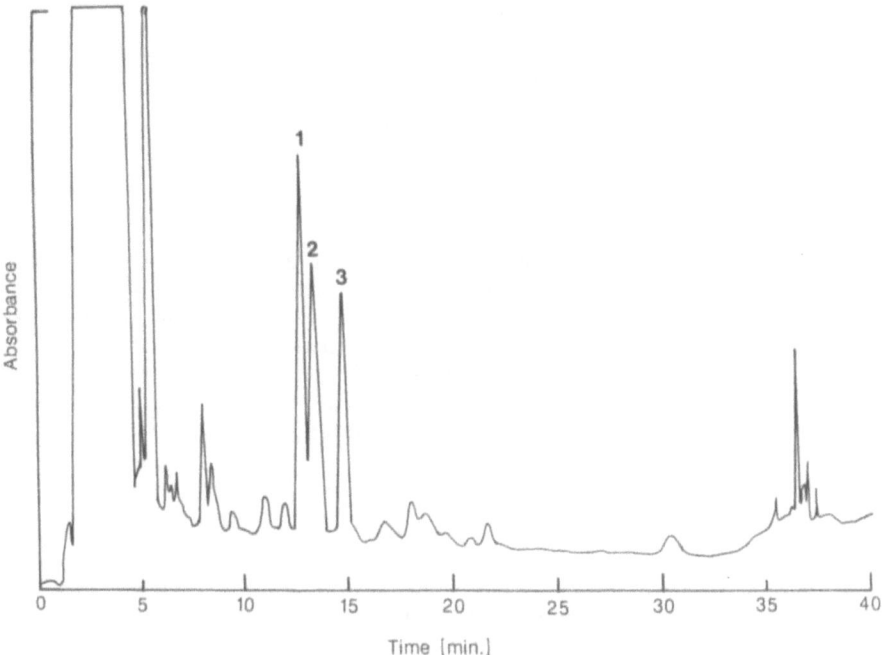

Fig. 1. HPLC elution profile of calf thymus DNA reacted with 1-nitro-
pyrene 4,5-oxide. The DNA was hydrolyzed with DNase, snake
venom phosphodiesterase and alkaline phosphatase. Adducts were
separated with a 30-min nonlinear gradient of 20-65% methanol at
a flow rate of 2 ml per min. The peaks labeled 1, 2 and 3 had
phenanthrene-type UV spectra.

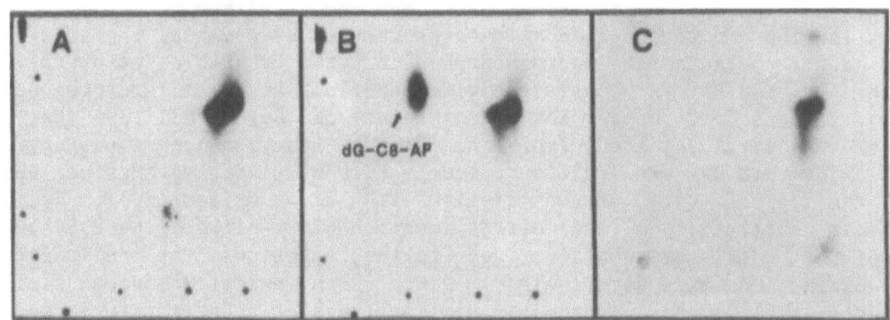

Fig. 2. $^{32}$P-Postlabeling autoradiograms of adducts obtained from: A.
calf thymus DNA reacted with 1-nitropyrene 4,5-oxide; B. calf
thymus DNA reacted with 1-nitropyrene 4,5-oxide in the presence
of xanthine oxidase and hypoxanthine; C. DNA from CHO cells
incubated with 1-nitropyrene 4,5-oxide. $^{32}$P-Postlabeling was
conducted as described in Materials and Methods, and the adducts
were visualized by autoradiography.

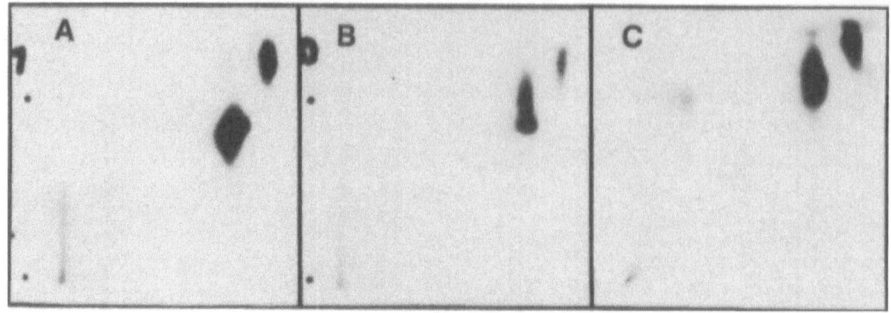

Fig. 3. $^{32}$P-Postlabeling autoradiograms of adducts obtained from: A. calf thymus DNA reacted with 1-nitropyrene 9,10-oxide; B. calf thymus DNA reacted with 1-nitropyrene 9,10-oxide in the presence of xanthine oxidase and hypoxanthine; C. DNA from CHO cells incubated with 1-nitropyrene 9,10-oxide. $^{32}$P-Postlabeling was conducted as described in Materials and Methods, and the adducts were visualized by autoradiography.

Table 1 compares mutation induction by 1-nitropyrene 4,5- and 9,10-oxide (13) with the concentration of DNA adducts produced by these oxides. At a 10 μM dose, 1-nitropyrene 9,10-oxide produced approximately two times as many mutations as 1-nitropyrene 4,5-oxide. Likewise, 1-nitropyrene 9,10-oxide bound to the CHO cell DNA approximately four-fold more than 1-nitropyrene 4,5-oxide.

DISCUSSION

1-Nitropyrene 4,5-oxide reacted with calf thymus DNA to produce adducts having phenanthrene-type UV spectra, which indicates that addition to the oxide ring had occurred. Although the site of substitution on the nucleic acid bases was not determined, the fact that the adduct concentration decreased when analyzed by HPLC after hydrolysis with nuclease P1 is consistent with reaction at $N^2$ of deoxyguanosine (18). These data support the recent report of Roy et al. (19) that the major adducts obtained from the direct reaction of 1-nitropyrene 4,5-oxide with DNA are addition products with $N^2$ of deoxyguanosine, which is the typical site of substitution with arene oxides (20). The DNA adducts obtained from the direct reaction of 1-nitropyrene 9,10-oxide with DNA were not characterized to the extent of the 1-nitropyrene 4,5-oxide adducts; nevertheless, since both oxides produced similar $^{32}$P-postlabeling maps, presumably similar types of adducts were formed.

Reaction of the 1-nitropyrene oxides with DNA in the presence of the mammalian nitroreductase xanthine oxidase and hypoxanthine had no obvious effect on the adduct pattern produced by 1-nitropyrene 9,10-oxide, but an additional adduct with properties consistent with dG-C8-AP was observed in reactions conducted with 1-nitropyrene 4,5-oxide. It appears that dG-C8-AP was produced from the reduction of 1-nitropyrene 4,5-oxide to 1-nitropyrene followed by nitroreduction to N-hydroxy-1-aminopyrene.

Table 1.  Mutations and DNA adducts produced in CHO cells by
10 µM 1-nitropyrene 4,5-oxide and 1-nitropyrene
9,10-oxide.

| Compound | Mutants per $10^6$ clonable CHO cells[a] | fmol adduct per µg DNA[b] |
|---|---|---|
| 1-Nitropyrene 4,5-oxide | 10 | 12 |
| 1-Nitropyrene 9,10-oxide | 21 | 46 |

[a]Mutations were measured at the hypoxanthine-guanine phosphoribosyl transferase locus and are from reference 13.

[b]The concentration of DNA adducts was determined by $^{32}$P-postlabeling through comparison to a standard modified with dG-C8-AP at a level of 1.8 fmol adduct per µg DNA.

Sugiura et al. (21) have shown that xanthine oxidase and hypoxanthine will reduce arene oxides, while we (2) have demonstrated that the same system will catalyze the reduction of 1-nitropyrene to N-hydroxy-1-aminopyrene, which will react with DNA to give dG-C8-AP. When the adducts produced by xanthine oxidase-catalyzed reduction of the 1-nitro-pyrene oxides were previously examined by HPLC (14), both compounds produced a major adduct (about 35% of total DNA binding) that coeluted with dG-C8-AP. In the present study, about 40% of the xanthine oxidase-mediated binding of 1-nitropyrene 4,5-oxide to calf thymus DNA was identified as dG-C8-AP; however, no dG-C8-AP was detected in incubations with 1-nitropyrene 9,10-oxide. This observation suggests that some of the 1-nitropyrene 9,10-oxide adducts detected by $^{32}$P-postlabeling analysis must have HPLC chromatographic properties similar to dG-C8-AP.

1-Nitropyrene 4,5-oxide produced a DNA adduct profile in CHO cells similar to that found by reacting this compound directly with DNA. The adduct profile produced in CHO cells by 1-nitropyrene 9,10-oxide also included a small amount (~5%) of an adduct that migrated in the vicinity of dG-C8-AP. This adduct was not seen with the 9,10-oxide when incubated with calf thymus DNA either in the presence or absence of xanthine oxidase and hypoxanthine. That the CHO adduct profiles for the 1-nitro-pyrene oxides differ from those found with xanthine oxidase is not surprising since xanthine oxidase is a one-electron nitroreductase (22) and CHO cells are thought to contain only two-electron nitroreductase activity (23). 1-Nitropyrene 9,10-oxide also bound to CHO cell DNA to a much greater extent and was more mutagenic than 1-nitropyrene 4,5-oxide, while binding of the two oxides to calf thymus DNA is nearly equal (14). The differential activities of these compounds in CHO cells could reflect the fact that 1-nitropyrene 4,5-oxide is more readily detoxified than 1-nitropyrene 9,10-oxide through epoxide hydrase-catalyzed hydration (24).

ACKNOWLEDGEMENTS

Research described in this article was conducted, in part, under contract to the Health Effects Institute (HEI), an organization jointly funded by the United States Environmental Protection Agency (EPA) (Assistance Agreement X-812059) and automotive manufacturers. The contents of this article do not necessarily reflect the views of the HEI, nor do they necessarily reflect the policies of the EPA, or automotive manufacturers. We thank Cindy Hartwick for helping prepare this manuscript.


REFERENCES

1.  Beland, F.A. and F.F. Kadlubar. Metabolic activation and DNA adducts of aromatic amines and nitroaromatic hydrocarbons. In: C.S. Cooper and P.L. Grover (eds.), Handbook of Experimental Pharmacology: Chemical Carcinogenesis and Mutagenesis, Springer-Verlag, Heidelberg, in press.

2.  Howard, P.C., R.H. Heflich, F.E. Evans, and F.A. Beland (1983). Formation of DNA adducts in vitro and in Salmonella typhimurium upon metabolic reduction of the environmental mutagen 1-nitropyrene. Cancer Res. 43:2052-2058.

3.  Heflich, R.H., P.C. Howard, and F.A. Beland (1985). 1-Nitroso-pyrene: an intermediate in the metabolic activation of 1-nitropyrene to a mutagen in Salmonella typhimurium TA1538. Mutation Res. 149:25-32.

4.  Heflich, R.H., E.K. Fifer, Z. Djurić, and F.A. Beland (1985). DNA adduct formation and mutation induction by nitropyrenes in Salmonella and Chinese hamster ovary cells: relationships with nitroreduction and acetylation. Environ. Health Persp. 62:135-143.

5.  Heflich, R.H., N.F. Fullerton, and F.A. Beland (1986). An examination of the weak mutagenic response of 1-nitropyrene in Chinese hamster ovary cells. Mutation Res. 161:99-108.

6.  Heflich, R.H., S.M. Morris, D.T. Beranek, L.J. McGarrity, J.J. Chen, and F.A. Beland (1986). Relationships between the DNA adducts and the mutations and sister-chromatid exchanges produced in Chinese hamster ovary cells by N-hydroxy-2-aminofluorene, N-hydroxy-N'-acetylbenzidine and 1-nitrosopyrene. Mutagenesis 1:201-206.

7.  Hashimoto, Y. and K. Shudo (1985). Modification of nucleic acids with 1-nitropyrene in the rat: identification of the modified nucleic acid base. Jpn. J. Cancer Res. (Gann) 76:253-256.

8.  Stanton, C.A., F.L. Chow, D.H. Phillips, P.L. Grover, R.C. Garner, and C.N. Martin (1985). Evidence for N-(deoxyguanosin-8-yl)-1-aminopyrene as a major DNA adduct in female rats treated with 1-nitropyrene. Carcinogenesis 6:535-538.

9.  Mitchell, C.E. (1988). Formation of DNA adducts in mouse tissues after intratracheal instillation of 1-nitropyrene. Carcinogenesis 9:857-860.

10. El-Bayoumy, K., G.-H. Shiue, and S.S. Hecht (1988). Metabolism and DNA binding of 1-nitropyrene and 1-nitrosopyrene in newborn mice. Chem. Res. Toxicol. 1:243-247.

11. Roy, A.K., K. El-Bayoumy, and S.S. Hecht (1989). $^{32}$P-postlabeling analysis of 1-nitropyrene-DNA adducts in female Sprague-Dawley rats. Carcinogenesis 10:195-198.

12. Fifer, E.K., P.C. Howard, R.H. Heflich, and F.A. Beland (1986). Synthesis and mutagenicity of 1-nitropyrene 4,5-oxide and 1-nitropyrene 9,10-oxide, microsomal metabolites of 1-nitropyrene. Mutagenesis 1:433-438.

13. Heflich, R.H., J.R. Thornton-Manning, T. Kinouchi, and F.A. Beland. Mutagenicity of oxidized microsomal metabolites of 1-nitropyrene in Chinese hamster ovary cells. Mutagenesis, in press.

14. Djuric, Z., E.K. Fifer, P.C. Howard, and F.A. Beland (1986). Oxidative microsomal metabolism of 1-nitropyrene and DNA-binding of oxidized metabolites following nitroreduction. Carcinogenesis 7:1073-1079.

15. Beland, F.A., N.F. Fullerton, and R.H. Heflich (1984). Rapid isolation, hydrolysis and chromatography of formaldehyde-modified DNA. J. Chromatography 308:121-131.

16. Lu, L.-J.W., R.M. Disher, M.V. Reddy, and K. Randerath (1986). [32]P-Postlabeling assay in mice of transplacental DNA damage induced by the environmental carcinogens safrole, 4-aminobiphenyl, and benzo(a)pyrene. Cancer Res. 46:3046-3054.

17. Gupta, R.C. (1985). Enhanced sensitivity of [32]P-postlabeling analysis of aromatic carcinogen:DNA adducts. Cancer Res. 45:5656-5662.

18. Gupta, R.C., K. Earley, N.F. Fullerton, and F.A. Beland (1989). Formation and removal of DNA adducts in target and nontarget tissues of rats administered multiple doses of 2-acetylaminophenanthrene. Carcinogenesis 10:2025-2033.

19. Roy, A.K., K. El-Bayoumy, and S.S. Hecht (1989). [1]H NMR analysis of adducts of 1-nitropyrene-4,5-oxide with DNA. Proc. Amer. Assoc. Cancer Res. 30:127.

20. Beland, F.A. and M.C. Poirier (1989). DNA adducts and carcinogenesis. In: A.E. Sirica (ed.), The Pathobiology of Neoplasia, Plenum Publishing Corp., New York, pp. 57-80.

21. Sugiura, M., Y. Yamazoe, T. Kamataki, and R. Kato (1980). Reduction of epoxy derivatives of benzo(a)pyrene by microsomal cytochrome P-450. Cancer Res. 40:2910-2914.

22. Orna, M.V. and R.P. Mason (1989). Correlation of kinetic parameters of nitroreductase enzymes with redox properties of nitroaromatic compounds. J. Biol. Chem. 264:12379-12384.

23. Eddy, E.P., E.C. McCoy, H.S. Rosenkranz, and R. Mermelstein (1986). Dichotomy in the mutagenicity and genotoxicity of nitropyrenes: apparent effect of the number of electrons involved in nitroreduction. Mutation Res. 161:109-111.

24. Djuric, Z., B. Coles, E.K. Fifer, B. Ketterer, and F.A. Beland (1987). In vivo and in vitro formation of glutathione conjugates from the K-region epoxides of 1-nitropyrene. Carcinogenesis 8:1781-1786.

# NITROPOLYCYCLIC AROMATIC HYDROCARBONS AND DIESEL EXHAUST:

# POTENTIAL ROLE OF DNA BINDING IN CARCINOGENICITY

James A. Bond[1], Charles E. Mitchell, Joe L. Mauderly, and
Ronald K. Wolff[2]

Lovelace Inhalation Toxicology Research Institute, P.O. Box
5890, Albuquerque, New Mexico 87185;
[1]Present address:  CIIT, P.O. Box 12137, Research Triangle
Park, NC 27709
[2]Present address:  Lilly Research Laboratories, P.O. Box 708
Greenfield, IN 46140

## INTRODUCTION

The potential health effects of exposure to diesel exhaust (DE) have
been a topic of intensive research during the last decade.  DE consists of
a mixture of gases, vapors, and soot particles of a respirable size (Cheng
et al., 1984).  Recent results from several studies (Mauderly et al., 1987;
Stöber, 1986; Ishinishi et al., 1986; Brightwell et al., 1986) indicate
that DE, when inhaled chronically at high concentrations, is a pulmonary
carcinogen in laboratory rats.  The carcinogenicity of inhaled DE was shown
to be both dose- and time-dependent (Mauderly et al., 1987).

It is not clear at this time whether genetic or epigenetic mechanisms,
or a combination of the two, are responsible for the carcinogenic action of
inhaled DE.  The results of several studies indicate that the organic
chemicals associated with DE soot have a role as causative factors in the
carcinogenic response of rodents (reviewed in McClellan, 1987).  One
hypothesis of a genetic mechanism for the tumorigenic response is that the
organic chemicals, desorbed from DE particles, must interact with lung DNA
to initiate carcinogenesis.  DNA adducts have been identified in the rat
lung after both short-term (i.e., 12 wk; Bond et al., 1988, 1990) and long-
term exposures (i.e., 30 month; Wong et al., 1986) to DE.  After 12 wk of
exposure to DE, we found relatively higher levels of total DNA adducts in
those same tissues in which tumors developed as a result of chronic
exposure to DE (Bond et al., 1988, 1989).

Nitropolycyclic aromatic hydrocarbons (nitro-PAH) have been implicated
as one class of DE-associated chemicals (Schuetzle, 1983) that may be
responsible, in part, for the carcinogenic action of DE.  Nitro-PAH are
mutagenic in both bacterial and mammalian systems (Beland et al., 1985).
Of the nitro-PAH, 1-nitropyrene (NP) is one of those found most abundantly
in the environment (Schuetzle, 1983).  Concern about NP has been raised,
because several studies have indicated that it might be a carcinogen in
animals (El-Bayoumy et al., 1984; Hirose et al., 1984; Moon 1986), although
other studies have produced negative results for carcinogenicity (Tokiwa et
al., 1984; Ohgaki et al., 1985; Maeda et al., 1986).

Our laboratory is exploring the molecular basis for DE-induced pulmonary carcinogenesis in rats by investigating the molecular dosimetry of inhaled DE. The goal of our research is to provide insight into the potential roles of DNA adducts in the carcinogenic response to DE and the role of NP, one of the key DE-associated chemicals, in adduct formation. Thus, studies at this Institute have focused on the abilities of DE and NP, either in pure form or associated with particulate matter, to induce DNA adducts in the lungs of animals exposed by inhalation to these materials.

## METHODS

### Animals

In all studies, male F344/N rats (11-15 wk old) or male B6C3F$_1$ mice (8-12 wk old) were used. The rats were born and raised in the Institute's barrier-maintained colony and were housed two per polycarbonate cage supplied with sterilized hardwood-chip bedding and filter tops. The mice were purchased from Charles River Laboratories, Inc. (Kingston, MA). Animal rooms were maintained at 20 to 22°C, with a relative humidity of 20 to 50% and a 12-h, light-dark cycle starting at 0600. Feed (Lab Blox, Allied Mills, Chicago, IL) and water were provided ad libitum.

For the inhalation exposures, animals were transferred from the animal housing quarters to inhalation exposure chambers (Hazleton Systems, Aberdeen, MD). Animals were housed in individual wire cages within the chamber, and were acclimated for 2 to 3 wk before initiation of exposures. Feed and water were supplied ad libitum (automated water system) during the acclimation.

### Experimental Design

For the DE exposure studies, rats were exposed for 7 h/day, 5 days/wk, for up to 12 wk to filtered air (controls) or to diluted DE (0.35-10 mg soot/m$^3$). For the NP studies, rats were exposed nose-only, 4 h/day, 1 day/wk, for up to 12 wk to filtered air (controls), NP (2 mg/m$^3$), or NP adsorbed on carbon black (NP/CB; 2 mg NP + 98 mg CB/m$^3$). Rats were euthanized by CO$_2$ asphyxiation and lungs were removed for analysis of DNA adducts. For both the DE and NP studies, rats were sacrificed at 2 (DE only), 4, 8, 12, 14, 16, and 20 wk after the start of exposure.

Additional studies, in which mice were intratracheally instilled with $^3$H-NP (4,5,9,10-$^3$H-NP; 1.8 Ci/mmole; > 99% radiochemical purity), were conducted . The purpose of these studies was to separate and identify specific NP adducts in mouse lungs.

### Exposure Systems

Diesel Exhaust. Details of the DE exposure system have been reported (Mokler et al., 1984). Briefly, exhaust was generated by 1980 Model, 5.7 L, Oldsmobile V-8 engines mounted on test stands and connected by automatic transmissions to dynamometers and flywheels to simulate a mid-sized passenger car operated on continuously repeating, U.S. Federal Test Procedure urban cycles. Fuel meeting USEPA certification standards was used (D-2 diesel control fuel, Phillips Chemical Co., Borger, TX) (Cheng et al., 1984). The exhaust was routed through a standard automotive muffler and tailpipe, diluted 10:1 with filtered air, and then diluted to the final concentration. Exposure chamber atmospheres were measured for particles by daily filter samples and for carbon monoxide, carbon dioxide, hydrocarbon vapors, nitrogen oxides, and ammonia by weekly bag samples. The particle size distribution was measured by using a Lovelace multijet cascade impactor and a parallel flow diffusion battery operated in series (Cheng et al., 1984).

1-Nitropyrene. The aerosols of pure NP were generated by methods described previously (Sun et al., 1983; Bond et al., 1986). NP was placed in a quartz vessel inside a tube furnace and vaporized at temperatures of 180-200°C, with a $N_2$ flow of 0.4 L/min. Air flow at 2.0 L/min was used to carry the vapor to a 2-liter cooling chamber for condensation. The condensed particles were then delivered to an exposure chamber, with an additional 25 L/min of diluting air. Inhalation exposures were conducted using an 80-port, nose-only exposure chamber (Raabe et al., 1973).

1-Nitropyrene Adsorbed to Carbon Black. Aerosols of NP adsorbed to carbon black were produced as follows. NP was added as a dry powder to carbon black (Cabot Corp.; Elftex 12), at a mass ratio of 2:98, to provide a total mass of 10 g of material. The powders were mixed for 6 h in a jar mill, and the mixture was then compressed into a Wright dust feeder cup for 2 h at 1500 psi using a hydraulic press. The mixed powder was dispersed by using a Wright dust feeder (BGI Inc., Waltham, MA), as previously described (Wolff et al., 1989a).

## DNA Isolation and $^{32}$P-Postlabeling Assay

DNA was isolated from lung tissue by a modification (Gupta, 1985) of the Marmur procedure (Marmur, 1961). The $^{32}$P-postlabeling assay (Randerath et al., 1981) was used to separate DNA adducts. Two enrichment procedures were used in these studies. For analyses of DNA adducts induced by DE (Bond et al., 1988), the nuclease $P_1$ enrichment procedure, as described in detail by Reddy and Randerath (1986), was used to separate DNA adducts. We have previously demonstrated (unpublished observations) that the nuclease $P_1$ and butanol extraction enrichment procedures yield similar patterns of adducts derived from DE exposures. Recent reports (Gallagher et al., 1989; Gupta and Earley, 1988) have indicated that C8-dG-arylamine adducts are poorly recovered when the nuclease $P_1$-mediated procedure is used. Thus, for analysis of DNA adducts derived from NP studies, the butanol extraction enrichment procedure, as described by Gupta (1985), was used.

DNA adducts were separated chromatographically by applying the $^{32}$P-postlabeled solutions to PEI-cellulose sheets. The sheets were developed according to published methods (Gupta et al., 1983; Reddy et al., 1984; Reddy and Randerath, 1986; Bond et al., 1988). To separate the DE-induced DNA adducts the following solvent conditions were used: D1, 1 M sodium phosphate buffer, pH 6.8; D3, 4.2 M lithium formate, 7.5 M urea, pH 3.5; D4, 0.8 M lithium chloride, 8.5 M urea, 0.5 M Tris-HCl, pH 8.0; D5, 1 M sodium phosphate buffer, pH 6.0. To separate NP-induced DNA adducts the following solvent conditions were used: D1, 1 M sodium phosphate buffer, pH 6.8; D3, 3 M lithium formate, 7 M urea, pH 3.5; D4, 0.6 M lithium chloride, 0.5 M Tris-HCl, 7.5 M urea, pH 8.0; D5, 1.7 M sodium phosphate buffer, pH 6.0.

## RESULTS AND DISCUSSION

### Diesel Exhaust

The exposures did not cause overt signs of toxicity in the animals and there were no significant exposure-related differences in body weight (data not shown). Gross observations of the lungs of animals exposed to DE showed that these lungs were darkly pigmented, while the lungs from sham-exposed animals were not.

Autoradiograms of DNA adducts in lung tissue from control and DE-exposed (7 mg soot/m$^3$) rats showed many more adducts in the DNA from rats exposed to DE than in DNA from control rats, and many of these adducts were unique to DE-exposed rats (Fig. 1). Levels of DNA adducts in lungs of rats exposed to the different exhaust concentrations for 12 wk were similar (Table 1). The overall mean level of 14 adducts per 10$^9$ bases in exhaust-exposed rats was significantly ($p < 0.05$; Student's t-test) higher than the level of adducts in sham-exposed rats. These data demonstrated that, while DNA adduct formation in lung tissue from DE-exposed rats was significantly elevated compared to adduct formation in lung tissue from control rats, adduct formation was independent of exhaust exposure concentration at the exhaust exposure levels tested. Bond and Mauderly (1984) demonstrated that the capacity of isolated, perfused rat lungs to metabolize chemicals associated with DE could be saturated at high concentrations. One explanation for the results reported here is that, at the exhaust concentrations tested, the lung enzymes responsible for formation of metabolites that bind to DNA were saturated. Furthermore, the lack of an exposure-response relationship for DNA adduct formation as reported here, and the earlier finding that DNA adducts were increased at an exposure level that did not significantly increase lung tumor incidence (0.35 mg/m$^3$; Mauderly et al., 1987), suggest that additional factors are also important in exhaust-induced carcinogenicity.

We designed experiments to determine whether adducts increased linearly with time during the 12-wk exposure, and whether steady-state

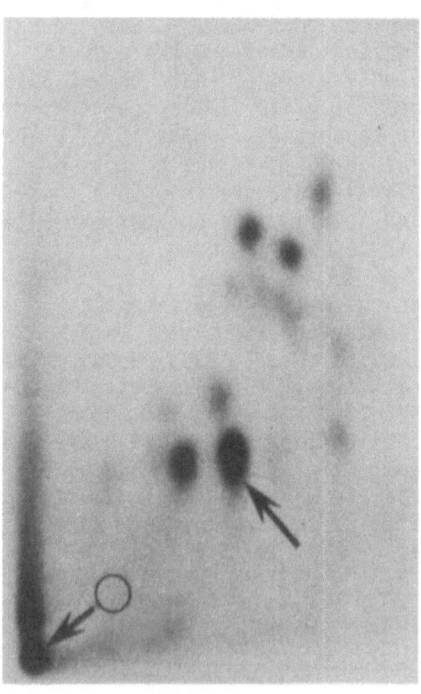 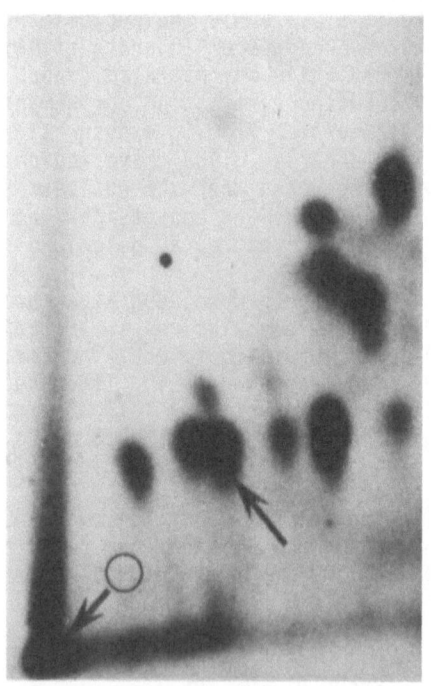

CONTROL                      DIESEL EXHAUST

Fig. 1.   Autoradiograms of chromatograms of DNA from lungs of control and diesel exhaust-exposed rats (7 mg soot/m$^3$) after 12 wk of exposure. The origin (0) is indicated on each chromotagram. The arrow points to the location where the BPDE-DNA adduct migrates.

Table 1.  Effect of Diesel Exhaust Concentration on
         DNA Adduct Levels in Rat Lung[a]

| Exposure Concentration $(mg/m^3)$ | Adducts per $10^9$ Bases |
|---|---|
| 0 (sham) | 7.3 ± 2.7 |
| 0.35 | 13.6 ± 1.7 |
| 3.5 | 13.1 ± 1.9 |
| 7.0 | 14.0 ± 3.9 |
| 10 | 14.0 ± 1.9 |

[a]Values are the mean ± SE; n = 6.

levels of adducts could be predicted after a chronic exposure to DE.  Lung
DNA adduct levels in rats exposed to 7 mg/m$^3$ of DE soot were less than
those in the lungs of controls at 4 wk of exposure, after which levels of
adducts accumulated slowly for the remainder of the 12-wk exposure (Fig.
2).  By the end of the exposure, adduct levels in exposed rat lungs were
about 160% of control levels.  DNA adduct levels declined rapidly after the
termination of exposure; by 4 wk after the end of exposure, adduct levels
were not significantly different from those of controls.  The observation
that adduct repair was faster than adduct formation suggests that
steady-state levels of DNA adducts would be reached during long-term
exposure to DE.  Wong et al. (1986) reported adduct levels of 20 ± 8
adducts per 10$^9$ bases ($\bar{X}$ ± SD; n = 6) in lungs of rats exposed for 30

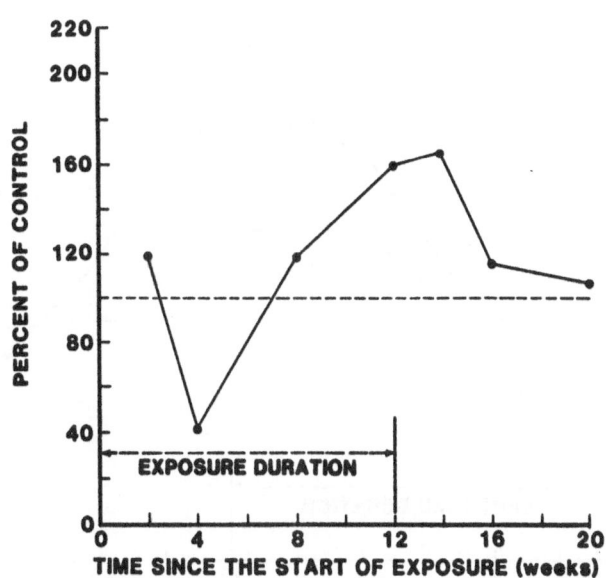

Fig. 2.  Time course for the formation and persistence
         of DNA adducts in lungs of rats exposed for
         up to 12 wk to 7 mg diesel exhaust soot/m$^3$.

months to 7 mg soot/$m^3$. This level of adducts is not significantly different from the 12 ± 4 adducts per $10^9$ bases present at 12 wk of exposure to 7 mg soot/$m^3$ in the present study. These data suggest that adduct levels were at, or near, steady-state by 12 wk of exposure.

## 1-Nitropyrene

The identities of the chemicals in DE that are responsible for DNA adduct formation have not been determined. As noted earlier, NP has been detected in DE and may be a chemical responsible for adduct formation. Therefore, studies were initiated to assess whether inhaled NP, either in pure form or adsorbed onto carbon black (a surrogate particle for DE soot), could induce DNA adduct formation in rat lungs and whether the pattern of adducts was similar to the patterns observed after exposure to DE.

Our preliminary results on the time course for formation and persistence of NP-induced adducts is shown in Fig. 3. Autoradiograms of DNA adducts in lung tissue from control, NP, and NP/CB exposures are shown in Fig. 4. Adduct levels increased in both the NP and NP/CB exposure groups, reaching a maximum by 12 wk of exposure (~ 160% of control values). There were no significant differences between the levels of adducts in the NP and NP/CB exposure groups. These results are consistent with the results of previous studies conducted in our laboratory. In the latter studies, we observed no significant differences in lung DNA adduct levels in rats exposed by inhalation to benzo[a]pyrene (BaP) or to BaP/CB (Wolff et al., 1989b). Because of the small quantities of adducts, it was not possible to specifically identify the chemical nature of the adducts. However, one of the spots noted on the chromatograms migrated to the same area as did the synthetic standard, N-(deoxyguanosin-8-yl)-1-aminopyrene (C8-dG-AP).

More detailed studies were initiated to determine whether lung tissue could metabolize NP to a reactive metabolite that would form the C8-dG-AP

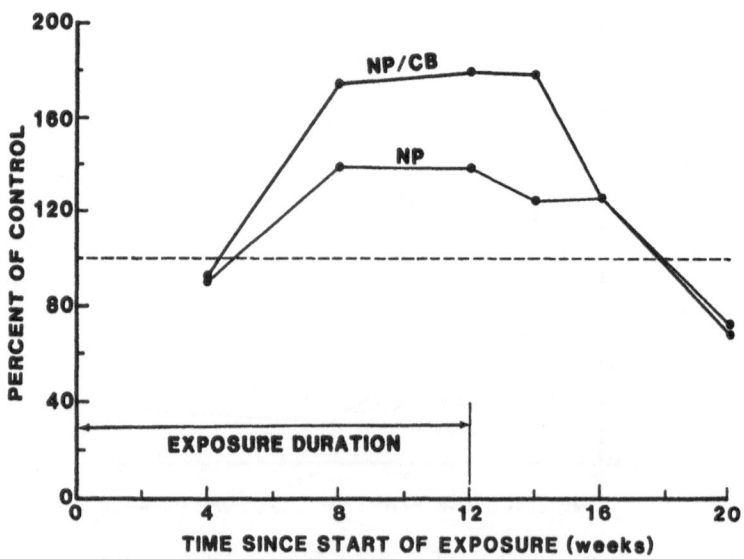

Fig. 3. Time course for the formation and persistence of DNA adducts in lungs of rats exposed for up to 12 wk to NP or NP/CB.

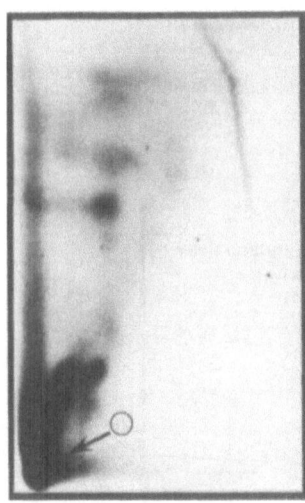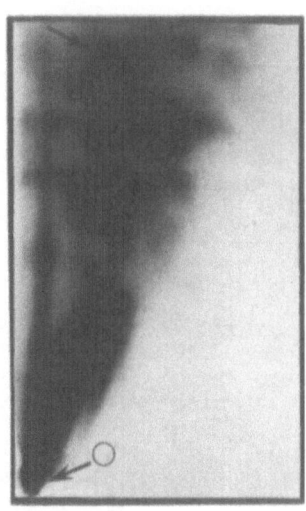

**CONTROL**     **PURE NP**     **NP/CB**

Fig. 4.  Autoradiograms of chromatograms of DNA from lungs of
control, NP-, and NP/CB-exposed rats.  The origin (O)
is indicated on each chromatogram.  The arrow points
to the location where the C8-dG-AP DNA adduct migrates.


adduct.  For these studies, NP was administered intratracheally.  DNA
obtained from mouse lungs at 1 day and 28 days after administration of
[$^3$H]-1-NP was enzymatically hydrolyzed and the adducts were analyzed by
HPLC as described previously (Mitchell, 1988).  Multiple peaks of
radioactivity were observed in the DNA hydrolysates at 1 day post-treatment
(Fig. 5A).  At least five distinct peaks of radioactivity were observed in
the lung hydrolysates.  These were, in most cases, broad bands of radioactivity that could represent several adducts.  This is particularly true
for the bands of radioactivity that eluted in fractions 42-50.  One major
band of radioactivity was observed in fractions 24-25 of the lung
hydrolysates.  These fractions represented approximately 20% of the total
eluted radioactivity.  This major band of radioactivity coeluted with the
synthetic marker, C8-dG-AP.

The HPLC profile of the lung hydrolysates from mice at 28 days post-
treatment is shown in Fig. 5B.  The major peaks of radioactivity after the
void volume were in fractions 23-24 and in fractions 44-48.  The peak of
radioactivity in fractions 23-24 coeluted with the synthetic marker,
C8-dG-AP, and was identical to the one that was observed in the lung DNA
hydrolysates at 1 day after administration.  The adduct that eluted in
fractions 23-24 at 28 days post-treatment did, however, comprise a lower
percentage of the total eluted radioactivity than was observed at 1 day
post-treatment; this adduct comprised approximately 10% of the total eluted
radioactivity at 28 days after treatment; whereas, it represented
approximately 20% of the total eluted radioactivity at 1 day post-treatment.

Previous studies have identified NP DNA adducts in rat liver, kidney,
and mammary glands following administration of 1-NP (Stanton et al., 1985;
Hashimoto and Shudo, 1985).  In in vitro studies, NP adducts have also been
identified in human diploid fibroblasts, rabbit lung tissue, and in rat
liver microsomal mediated binding of NP to calf thymus DNA (Beland et al.,
1986; Jackson et al., 1985; Djurić et al., 1986; Patton et al., 1986).  In
these latter studies, the C8-dG-AP adduct was one of the major adducts

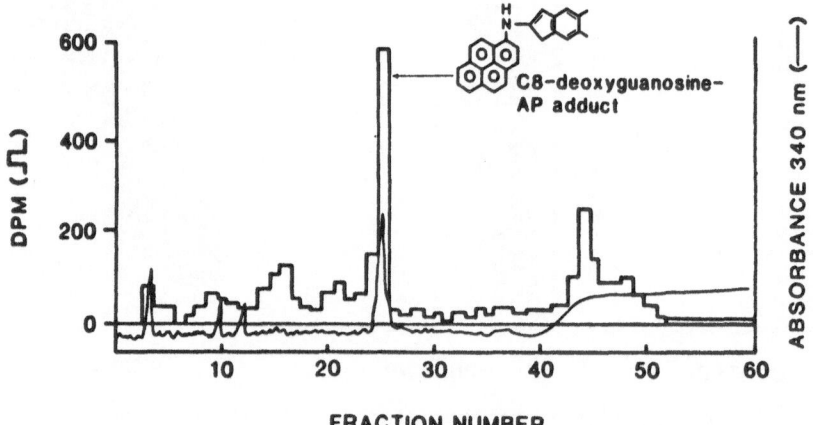

Fig. 5A. Reversed-phase HPLC profile of enzymatically digested lung DNA from mice given 1 mg NP/kg (1.8 Ci/mmol) intratracheally - 1 day post-exposure. The UV absorbance at fractions 24-25 is due to the added marker C8-dG-AP.

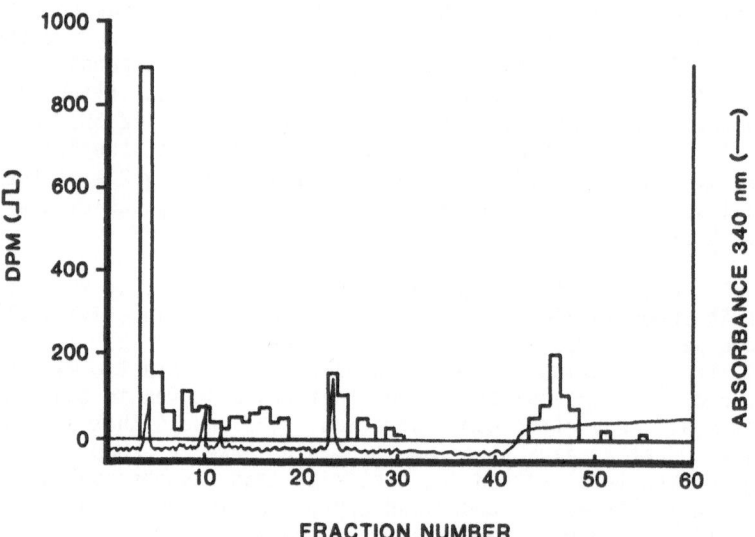

Fig. 5B. Reversed-phase HPLC profile of enzymatically digested lung DNA from mice given 1 mg NP/kg intratracheally - 28 days post-exposure. From Mitchell, 1988; with permission.

formed. Our observation that NP DNA adducts were formed in mouse lung after intratracheal administration of NP is of particular interest, because the lung is the major route of entry for all airborne materials containing NP. In addition, the observation that an adduct that coeluted with C8-dG-AP and other adducts remains in the lung as long as 28 days after administration of NP suggests that the persistence of one or more of these adducts may potentially be associated with the induction of lung tumors (El-Bayoumy et al., 1984).

In summary, both DE and NP induce DNA adduct formation in lung tissue. Our studies with DE, in which DNA adduct levels were independent of exposure concentration and the observation that DNA adducts were elevated at an exposure concentration that did not significantly increase lung tumor incidence, suggest that additional factors probably play a role in DE-induced pulmonary carcinogenicity. Our studies with NP, where the C8-dG-AP DNA adduct was tentatively identified in lung DNA, indicate that lung tissue is capable of metabolically activating NP via nitroreduction to a reactive metabolite that binds to DNA. The pattern of spots produced by the C8-dG-AP adduct in the post-labeling assay (data not shown) was unlike the pattern of spots observed for DE-induced DNA adducts. Therefore, it is unlikely that the C8-dG-AP adduct is one of the DE-induced adducts. It is possible that other adducts of NP (e.g., adducts derived from ring oxidation) may comprise some of the DE adducts.

## ACKNOWLEDGEMENTS

The authors gratefully acknowledge the technical assistance provided by numerous members of the Institute's technical staff throughout the conduct of these studies. The authors also acknowledge the critical review and discussions with a number of our colleagues, particularly Dr. Roger O. McClellan, who provided encouragement for the conduct of this work. The authors wish to also thank Dr. Fred Beland (NCTR) who provided us with the synthetic standard N-(deoxyguanosin-8-yl)-1-aminopyrene. This research was supported by the Office of Health and Environmental Research, U. S. Department of Energy, under Contract No. DE-ACO4-76EVO1013, in facilities fully accredited by the American Association for Accreditation of Laboratory Animal Care.

## REFERENCES

Beland, F. A., Heflich, R. H., Howard, P. C., and Fu, P. P., 1985, The in vitro metabolic activation of nitropolycyclic aromatic hydrocarbons, in: "Polycyclic Hydrocarbons and Carcinogenesis," R. G. Harvey, ed., pp. 371-396, American Chemical Society, Washington, DC.

Beland, F. A., Ribovich, M., Howard, P. C., Heflich, R. H., Kurian, P., and Milo, G. E., 1986, Cytotoxicity, cellular transformation and DNA adducts in normal human diploid fibroblasts exposed to 1-nitrosopyrene, a reduced derivative of the environmental contaminant, 1-nitropyrene, Carcinogenesis, 7:1279.

Bond, J. A., and Mauderly, J. L., 1984, Metabolism and macromolecular covalent binding of $^{14}$C-1-nitropyrene in isolated perfused/ventilated rat lungs, Cancer Res., 44:3924.

Bond, J. A., Sun, J. D., Medinsky, M. A., Jones, R. K., and Yeh, H. C., 1986, Deposition, metabolism and excretion of $^{14}$C-1-nitropyrene coated on diesel exhaust particles as influenced by exposure concentration, Toxicol. Appl. Pharmacol., 85:102.

Bond, J. A., Wolff, R. K., Harkema, J. R., Mauderly, J. L., Henderson, R. F., Griffith, W. C., and McClellan, R. O., 1988, Distribution of DNA adducts in the respiratory tract of rats exposed to diesel exhaust, Toxicol. Appl. Pharmacol., 96:336.

Bond, J. A., Harkema, J. R., Henderson, R. F., Mauderly, J. L., McClellan, R. O., and Wolff, R. K., 1989, Molecular dosimetry of inhaled diesel exhaust, in: "Assessment of Inhalation Hazards," U. Mohr, D. V. Bates, D. L. Dungworth, P. N. Lee, R. O. McClellan, and F. J. C. Roe, eds., Springer-Verlag.

Bond, J. A., Mauderly, J. L., and Wolff, R. K., 1990, Concentration and time-dependent formation of DNA adducts in lungs of rats exposed to diesel exhaust, Toxicology, in press.

Brightwell, J., Fouillet, X., Cassano-Zopi, A. L., Gatz, R., and Duchosal, F., 1986, Neoplastic and functional changes in rodents after chronic inhalation of engine exhaust emissions, in: "Carcinogenicity and Mutagenicity of Diesel Engine Exhaust," N. Ishinishi, A. Koizumi, R. McClellan, and W. Stöber, eds., Elsevier, Amsterdam.

Cheng, Y. S., Yeh, H. C., Mauderly, J. L., and Mokler, B. V., 1984, Characterization of diesel exhaust in a chronic inhalation study, Am. Ind. Hyg. Assoc. J., 45:547.

Djurić, Z., Fifer, E. K., Howard, P. C., and Beland, F. A., 1986, Oxidative microsomal metabolism of 1-nitropyrene and DNA-binding of oxidized metabolites following nitroreduction, Carcinogenesis, 7:1073.

El-Bayoumy, K., Hecht, S. S., Sackl, T., and Stoner, G. D., 1984, Tumori-genicity and metabolism of 1-nitropyrene in A/J mice, Carcinogenesis, 5:1449.

Gallagher, J. E., Jackson, M. A., George, M. H., Lewtas, J., and Robertson, I. G. C., 1989, Differences in detection of DNA adducts in the $^{32}$P-postlabelling assay after either 1-butanol extraction or nuclease $P_1$ treatment, Cancer Lett., 45:7.

Gupta, R. C., Reddy, M. V., and Randerath, K., 1983, $^{32}$P-postlabeling analysis of non-radioactive aromatic carcinogen-DNA adducts, Carcinogenesis, 3:1081.

Gupta, R. C., 1985, Enhanced sensitivity of $^{32}$P-postlabeling analysis of aromatic carcinogen-DNA adducts, Cancer Res., 5:5656.

Gupta, R. C., and Earley, K., 1988, $^{32}$P-adduct assay: Comparative recoveries of structurally diverse DNA adducts in the various enhancement proce-dures, Carcinogenesis, 9:1687.

Hashimoto, Y., and Shudo, K., 1985, Modification of nucleic acids with 1-nitropyrene in the rat: Identification of the modified nucleic acid base, Gann, 76:253.

Hirose, M., Lee, M. S., Wang, C. Y., and King, C. M., 1984, Induction of rat mammary gland tumors by 1-nitropyrene, a recently recognized environ-mental mutagen, Cancer Res., 44:1158.

Ishinishi, N., Koizumi, A., McClellan, R. O., and Stöber, W., 1986, "Carcinogenicity and Mutagenicity of Diesel Engine Exhaust," Elsevier, Amsterdam.

Jackson, M. A., King, L. C., Ball, L. M., Ghayourmanesh, S., Jeffrey, A. M., and Lewtas, J., 1985, Nitropyrene: DNA binding and adduct formation in respiratory tissues, Environ. Health Perspect., 62:203.

Maeda, T., Izumi, K., Otsuka, H., Manabe, Y., Kinouchi, T., and Oshnishi, Y., 1986, Induction of squamous cell carcinoma in the rat lung by 1,6-dinitropyrene, J. Natl. Cancer Inst., 76:693.

Marmur, J. A., 1961, Procedure for the isolation of deoxyribonucleic acid from micro-organisms, J. Mol. Biol., 3:208.

Mauderly, J. L., Jones, R. K., Griffith, W. C., Henderson, R. F., and McClellan, R. O., 1987, Diesel exhaust is a pulmonary carcinogen in rats exposed chronically by inhalation, Fundam. Appl. Toxicol., 9:208.

McClellan, R. O., 1987, Health effects of exposure to diesel exhaust parti-cles, Ann. Rev. Pharmacol. Toxicol., 27:279.

Mitchell, C. E., 1988, Formation of DNA adducts in mouse tissues after intratracheal instillation of 1-nitropyrene, Carcinogenesis, 9:857.

Mokler, B. V., Archibeque, F. A., Beethe, R. L., Kelly, C. P. J., Lopez, J. A., Mauderly, J. L., and Stafford, D. L., 1984, Diesel exhaust exposure system for animal studies, Fundam. Appl. Toxicol., 4:270.

Moon, R. C., 1986, Health Effects Institute Annual Conference, February 16-19, Pacific Grove, CA.

Ohgaki, H., Hasegawa, H., Kato, T., Negishi, C., Sato, S., and Sugimura, T., 1985, Absence of carcinogenicity of 1-nitropyrene, correction of previous results, and new demonstration of carcinogenicity of 1,6-nitropyrene in rats, Cancer Lett., 25:239.

Patton, J. D., Maher, V. M., and McCormick, J. J., 1986, Cytotoxic and mutagenic effects of 1-nitropyrene and 1-nitrosopyrene in diploid human fibroblasts, Carcinogenesis, 7:89.

Raabe, O. G., Bennick, J. E., Light, M. E., Hobbs, C. H., Thomas, R. L., and Tillery, M. I., 1973, An improved apparatus for acute inhalation exposure of rodents to radioactive aerosols, Toxicol. Appl. Pharmacol., 26:264.

Randerath, K., Reddy, M. V., and Gupta, R. C., 1981, $^{32}$P-labeling test for DNA damage, Proc. Natl. Acad. Sci. USA, 78:6126.

Reddy, M. V., Gupta, R. C., Randerath, E., and Randerath, K., 1984, $^{32}$P-postlabeling test for covalent DNA binding of chemical in vivo: Application to a variety of aromatic carcinogens and methylating agents, Carcinogenesis, 5:231.

Reddy, M. V., and Randerath, K., 1986, Nuclease $P_1$-mediated enhancement of sensitivity of $^{32}$P-postlabeling test for structurally diverse DNA adducts, Carcinogenesis, 7:1543.

Schuetzle, D., 1983, Sampling of vehicle emissions for chemical analysis and biological testing, Environ. Health Perspect., 47:65.

Stanton, C. A., Chow, F. L., Phillips, D. H., Grover, P. L., Garner, R. C., and Martin, C. N., 1985, Evidence for N-(deoxyguanosin-8-yl)-1-aminopyrene as a major DNA adduct in female rats treated with 1-nitropyrene, Carcinogenesis, 6:535.

Stöber, W., 1986, Experimental induction of tumors in hamsters, mice and rats after long-term inhalation of filtered and unfiltered diesel engine exhaust, in: "Carcinogenicity and Mutagenicity of Diesel Engine Exhaust," N. Ishinishi, A. Koizumi, R. McClellan, and W. Stöber, eds., pp. 421-439, Elsevier, Amsterdam.

Sun, J. D., Wolff, T. K., Aberman, H. M., and McClellan, R. O., 1983, Inhalation of 1-nitropyrene associated with ultrafine insoluble particles or as a pure aerosol: A comparison of deposition and biological fate, Toxicol. Appl. Pharmacol., 69:185.

Tokiwa, H., Otofuji, T., Horikama, K., Kitamori, S., Otsuka, H., Manabe, Y., Kinouchi, T., and Ohnishi, Y., 1984, 1,6-Dinitropyrene: Mutagenicity in Salmonella and carcinogenicity in BALB/C mice, J. Natl. Cancer Inst., 73:1359.

Wolff, R. K., Sun, J. D., Barr, E. B., Rothenberg, S. J., and Yeh, H. C., 1989a, Lung retention and binding of [$^{14}$C]-1-nitropyrene when inhaled by F344 rats as a pure aerosol or adsorbed to carbon black particles, J. Toxicol. Environ. Health, 26:309.

Wolff, R. K., Bond, J. A., Sun, J. D., Henderson, R. F., Harkema, J. R., Griffith, W. C., Mauderly, J. L., and McClellan, R. O., 1989b, Effect of adsorption of benzo(a)pyrene onto carbon black particles on levels of DNA adduct in lungs of rats exposed by inhalation, Toxicol. Appl. Pharmacol., 97:289.

Wong, D., Mitchell, C. E., Wolff, R. K., Mauderly, J. L., and Jeffrey, A. M., 1986, Identification of DNA damage as a result of exposure of rats to diesel exhaust, Carcinogenesis, 7:1595.

VALIDATION/APPLICATION OF $^{32}$P-POSTLABELING ANALYSIS FOR THE DETECTION OF
DNA ADDUCTS RESULTING FROM COMPLEX AIR POLLUTION SOURCES CONTAINING PAHs
AND NITRATED PAHs

J.E. Gallagher[1], M.J. Kohan[1], M.H. George[2], M.A. Jackson[2],
and J. Lewtas[1]

[1]Genetic Toxicology Division, Health Effects Research
  Laboratory, US EPA, Research Triangle Park, NC
[2]Environmental Health Research and Testing Inc.
  Research Triangle Park, NC.

ABSTRACT

Two recent versions of the $^{32}$P-postlabeling assay (butanol extraction
verses nuclease P1) have been employed to enhance the detection of
polycyclic aromatic hydrocarbon (PAH)-modified DNA. Previously published
studies suggest that the DNA adducts derived from N-substituted aryl
compounds are poorly recovered in the nuclease P1 version of the $^{32}$P-
postlabeling assay. Both versions were employed to ascertain whether the
apparent differences in sensitivity could be used to diagnostically select
for nitroaromatic-derived DNA adducts formed following in vitro (calf
thymus DNA) and in vivo (rodent) exposure to complex air pollution
particle extracts. DNA adduct levels and patterns of radioactivity were
compared to examine putative DNA adducts resulting from treatment with
combustion emission extracts with and without mono- or dinitropyrenes.
Following topical application (50 mg) with nitrated PAH containing
particle extract (diesel) and complex mixture particle extract without
nitrated PAH's (coke oven and coal soot) multiple DNA adducts were
detected along a zone of radioactivity, irrespective of the version of the
assay employed. The patterns of radioactivity, however, were
characteristically different for each of the complex mixture-modified DNA
samples. Nuclease P1-sensitive adducts were not apparent in any of the
mouse skin DNA samples examined. In contrast, calf thymus DNA incubated
with xanthine oxidase (XO) and nitrated-PAH containing particle extracts,
resulted in the formation of DNA adducts detectable only by the butanol
extraction version of the $^{32}$P-postlabeling assay. These nuclease P1-
sensitive adducts were chromatographically similar to mono- and
dinitropyrene-modified DNA adducts formed following in vitro incubation of
rabbit tracheal epithelial cells (RTE) with 1,6- and 1,8 dinitropyrene (10
$\mu$M). Our data suggest that under optimal nitroreducing conditions, the in
vitro calf thymus DNA model described in this study can enhance the
diagnostic potential of the $^{32}$P-postlabeling assay by the identification
of nuclease P1-sensitive, N-substituted aryl-derived DNA adducts from the
array of PAH-adducts formed following in vivo exposure to a variety of air
pollution particle extracts.

INTRODUCTION

Two recent versions of the $^{32}$P-postlabeling assay allow the detection
of DNA adducts at relatively low levels of modification (1 adduct/$10^{9\text{-}10}$

nucleotides) (1,2). The butanol extraction version of the assay selects for adducted nucleotides with the aid of a phase-transfer agent, tetrabutyl ammonium chloride (3). In contrast, the nuclease P1 version of the assay enzymatically and selectively dephosphorylates 3'monophosphates of adducted but not unadducted nucleotides (4). With either version, the adducted mononucleotides are subsequently [32]P-postlabeled at the 5'-hydroxy position by T4 polynucleotide kinase and are resolved on thin-layer chromatography plates using multiple solvents.

The [32]P-postlabeling assay provides an opportunity to detect DNA adducts resulting from exposure of complex mixtures of environmental importance, where prior knowledge of the individual constituents is not a requirement of the assay. The application of the method is limited, however, by the inability to characterize the nature of the DNA adducts of interest. Thus, methodological developments are required to improve the diagnostic capability of the assay.

Nitrosubstituted polycyclic aromatic hydrocarbons have been shown to be associated with the mutagenicity of combustion particles (5,6). We have previously shown that 1-nitropyrene (1-NP) and 3-nitrofluoranthene (3-NF) DNA reactive metabolites bind to rabbit tracheal epithelial cells in vitro (7). These adducts, detected by [32]P-postlabeling analysis, were shown to be nuclease P1-sensitive. In this study, we evaluated whether 1,6-, and 1,8-dinitropyrene derived DNA adducts demonstrated similar nuclease P1-sensitivity. We simultaneously compared DNA adduct profiles and levels to determine whether the differences inherent in the two versions of the assay could be used selectively to identify N-substituted aryl-derived adducts from other PAH-derived DNA adducts resulting from in vitro (calf-thymus) and in vivo (rodent) exposure to various complex mixtures with and without nitrated-PAH constituents.

## MATERIALS AND METHODS

### Materials

[$\gamma$[32]P]ATP (3000 Ci/mmol, 10 mCi/ml aqueous solution containing 5 mM 2 mercaptoethanol) was obtained from Amersham, Arlington Heights, IL. Polyethyleneimine (PEI) cellulose TLC plates were prepared as previously described (1) except that the PEI solution (50% aqueous) was from Aldrich Chemical Co., Milwaukee, WI. Micrococcal nuclease and nuclease P1 were from Sigma Chemical Co., St. Louis, MO.; calf spleen phosphodiesterase was from Boehringer Mannheim, Indianapolis, IN; and T4 polynucleotide kinase was from Pharmacia, Piscataway, NJ. All other chemicals were of analytical grade. 1,6-, and 1,8-dinitropyrenes were provided by Dr. D. Bryant of McMaster University, Ontario, Canada.

### Standard adducts

The N-(deoxyguanosin-8-yl)-1 and 6-aminopyrene were formed by calf thymus reaction by ascorbic acid reduction and provided by Dr. F. Beland, National Center for Toxicological Research, Jefferson, AR. The N-(deoxguanosin-8-yl) 1-aminopyrene adduct was derived with calf-thymus DNA by xanthine oxidase reduction of 1-NP.

### Complex Mixtures

### Diesel and gasoline particle extracts

The diesel and gasoline samples used in this study were obtained from one heavy duty diesel Caterpillar 3304 (1) and two light duty diesel passenger cars: a Nissan (2), a turbocharged Volkswagon Rabbit (3) and a Mercedes(4). The preparation of the extracts has been previously described (8). The concentration of 1-NP in the (2) and (3) was 1587 ppm and 589 ppm, respectively (9). The concentration of 1-NP in Caterpillar (1) and Mercedes (4) was not determined.

**Coke oven main**

The coke oven emissions were collected from a separator between the gas collector and the primary coolers within a coke battery located at Republic Steel in Gadsden, AL as previously described (10). The concentration of 1-NP was <1 ppm (9).

**Coal soot**

Methylene chloride extracts of coal soot were taken and dried under a stream of nitrogen. Dosing solutions were prepared by solvent exchange into dimethyl sulfoxide. Mono-and dinitropyrenes were not detected in this sample at levels exceeding 1 ppm (personal communication). Stock concentrations of all complex mixtures used were 2mg/ml in dimethyl sulfoxide.

## In Vitro treatment: rabbit tracheal epithelial (RTE) Cells

RTE cells were isolated from tracheas of New Zealand White Rabbits (10-12) weeks old as previously described (11). Cells were cultured in the presence of 10 $\mu$M 1,6- and 1,8-dinitropyrene for 2 h in F12 medium.

## In Vitro modification of calf thymus DNA

Calf thymus DNA (1 mg/ml) was modified by incubation with either diesel or gasoline combustion particle extracts (100 $\mu$g/ml), 1-nitro-pyrene (5 $\mu$g/ml) or benzo(a)pyrene (B(a)P) (20 $\mu$g/ml) in the presence of either S9 (0.5 mg/ml) prepared as previously described (12) or xanthine oxidase (0.5 units/ml) pH 7.5 for 1.5 h. The final incubation volume was 2 ml.

## Rodent treatment (diesel, coke oven and coal soot)

The backs of C-57 mice (6-8 weeks old) were shaved prior to 4 separate applications (50 mg total) of various extracts: coke oven, coal soot (without nitrated-PAH's) or those particle extracts containing nitrated-PAH's. 24 h following the last application, the mice were sacrificed, skin removed and stored at -70°C until subsequent DNA isolation.

## DNA extraction and [32]P-postlabeling analysis

DNA was isolated from tissues and cells essentially as described by Gupta (13). 5.0 $\mu$g of test sample DNA was digested to mononucleotides at 37°C for 3.5 h with micrococcal endonuclease and spleen phosphodiester-ase. The digest was then divided into two pools: one extracted by the butanol extraction version of the [32]P-postlabeling assay (3) and the other by the nuclease P1 version essentially as previously described (4). With either version 5.0 $\mu$g of DNA was incubated with approximately 50 $\mu$Ci ATP (Amersham 3000 Ci/mmole) and 3.5 units T4 polynucleotide kinase for approximately 30 min. 5.0 $\mu$g of DNA was spotted onto thin layer plates and adducts resolved using previously described solvent systems (14). To measure the total number of nucleotides, an aliquot (0.5 $\mu$g of mononucleo-tides) was diluted approximately 1200 fold and labeled with an equivalent amount of radiolabeled mix. A further 50 fold dilution was made and 5 $\mu$l spotted on a PEI cellulose sheet previously soaked in 100 mM ammonium formate, pH 3.5. Areas of radioactivity and appropriate control regions were detected by autoradioagraphy and carefully excised. DNA adducts that migrated along a zone of radioactivity were carefully scaped with the aid of a template outlining the boundary of radioactivity. 5 ml of 95% ethanol were added to each vial and Cerenkov counted using a scintillation counter. Relative adduct levels (RAL) were determined from the cpm detected for the excised areas divided by the cpm determined for the total nucleotides after correction for the background radioactivity, dilution factors and volumes spotted.

RESULTS AND DISCUSSION

RTE cells exposed to 1-NP and 3-NF were previously shown to metabolize nitroaromatics to DNA reactive intermediates as measured by DNA adducts detected by $^{32}$P-postlabeling analysis (7). In this study, 1,6- and 1,8-dinitropyrene-derived DNA adducts were detected at levels exceeding those detected for 1-NP-derived RTE cell DNA (ie, $7.2/10^9$ and $8.5/10^9$, respectively) (Table 1 and Figure 1[a,c]). These data are consistent with studies showing a higher mutagenic potentcy of dinitropyrenes compared to mono-substituted nitrated-PAH's (15). The dinitropyrene-derived DNA adducts were also shown to be chromatographically identical to those adducts detected in calf thymus DNA modified by 1,6- and 1,8-dinitropyrene by ascorbic acid-mediated nitroreduction

Table 1. Quantitation of DNA adducts by Nuclease P1 or 1-butanol extraction versions of the $^{32}$P-postlabeling assay

| Adducts[a] | Nuclease P1 | Butanol extraction | Nuclease P1 Recovery(%)[b] |
|---|---|---|---|
| IN VITRO | | | |
| RTE | | | |
| 1,6-dinitropyrene 10 $\mu$M | NDc | 0.73 | - |
| 1,8-dinitropyrene     " | NDc | 0.86 | - |
| | | | |
| Calf Thymus DNA | | | |
| 1,6-dinitropyrene | 0.06± .01 | 22.2± 8.5 | .27 |
| 1,8-dinitropyrene | 2.3 | 278.0 | .82 |
| | | | |
| Calf Thymus DNA + | | | |
| xanthine oxidase + | | | |
| Diesel(1) | 0.23±.01 | 1.8± 0.7 | 13 |
|   "   (2) | 1.4 ±.2 | 7.9± 4.4 | 18 |
|   "   (3) | 0.66±.02 | 3.9± 0.1 | 17 |
|   "   (4) | 0.78±.23(#1)[d] | 10.1±10.0 | 8 |
| | 1.4 ±.5 (#2) | 4.6± 0.8 | 30 |
| 1-nitropyrene | 10.1      (#1) | 1082.2 | 1 |
| | 97.9      (#2) | 158.8 | 62 |
| | | | |
| IN  VIVO e | | | |
| Coke Oven (50 mg) | 248.6±19.1 | 189.2 ± 40.3 | 131 |
| Coal Soot (50 mg) | 124.0± 3.9 | 86.9 ± 35.3 | 143 |
| Diesel (2)(50 mg) | 2.2± 2.0 | 3.2 ±  0.1 | 69 |
| B(a)P     (1.2 mg) | 67.2±16.3 | 72.8 ±  7.2 | 92 |

[a] RAL/$10^8$ See Figure 1-3
[b] (Nuclease P1/1-butanol x 100)
[c] ND (Not detectable after autoradiography for four days at -70°C).
[d] See Figure 2 (e)
[e] skin DNA adducts

**Butanol**

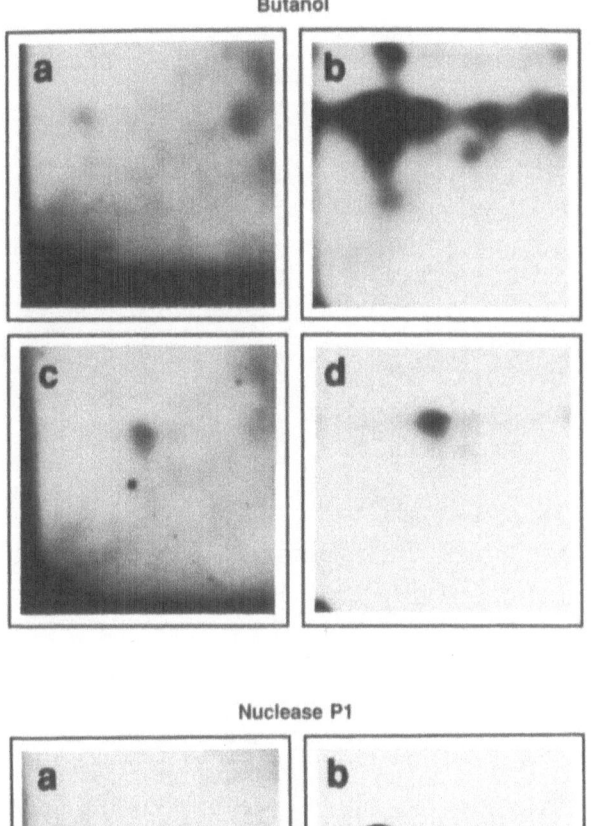

**Nuclease P1**

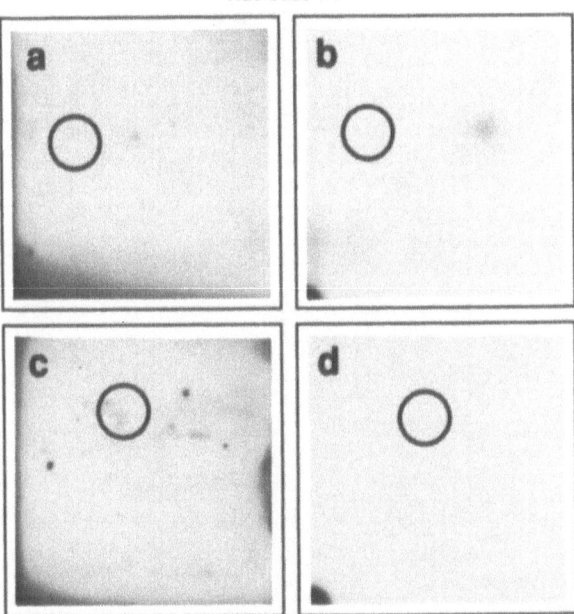

**Figure 1.** Autoradiograms of thin layer chromatography plates showing rabbit tracheal epithelial cell DNA adducts formed following treatment with 10 µM (a) 1,6-dinitropyrene (c) 1,8-dinitropyrene and calf thymus-DNA adducts formed following ascorbic acid mediated nitroreduction of (b) 1,6-dinitropyrene and (d) 1,8-dinitropyrene.

◯ indicates Nuclease P1-sensitive DNA adducts.

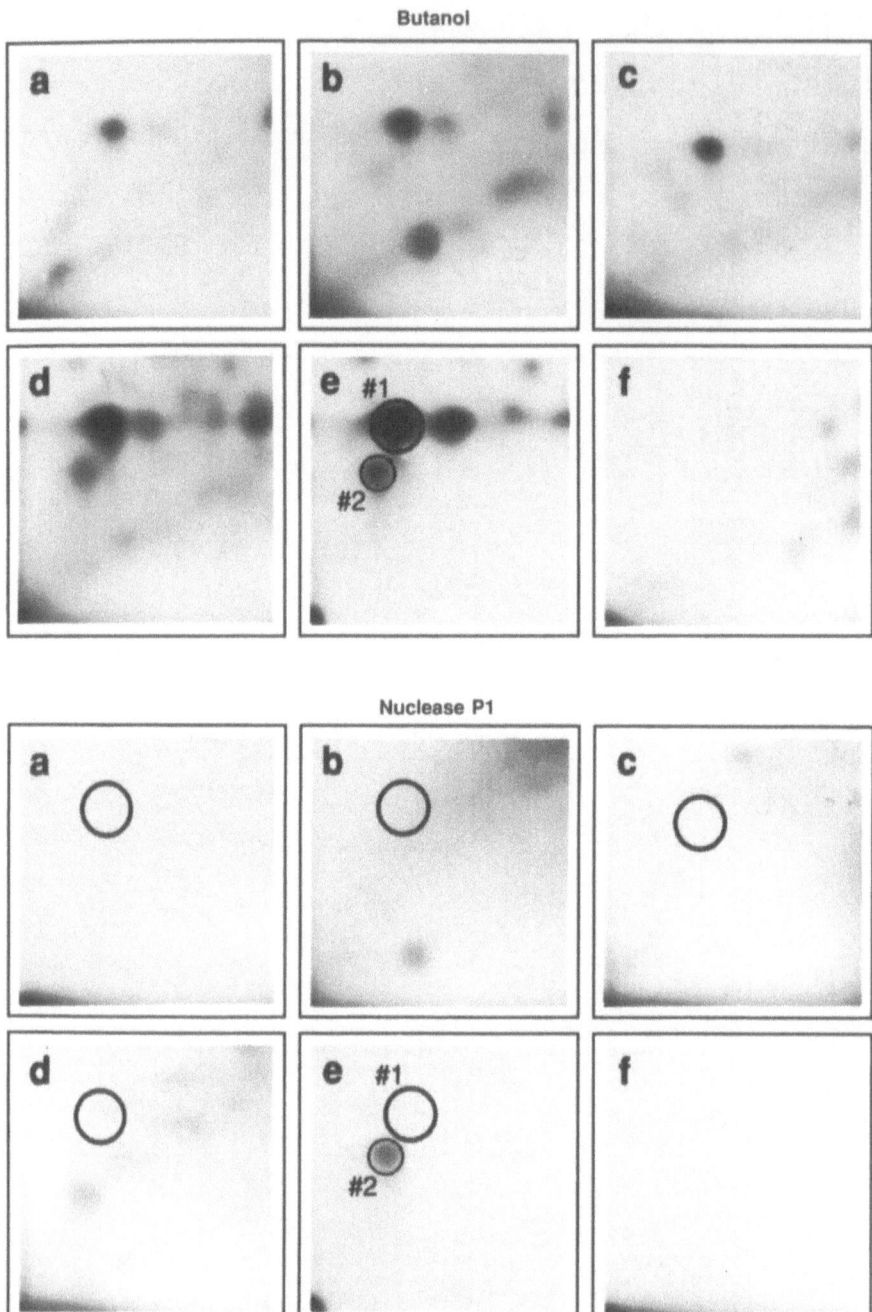

**Figure 2.** DNA adducts detected following *in vitro* xanthine oxidase mediated
nitroreduction of complex mixture particle extracts (a) diesel (1)
(b) diesel (2) (c) diesel (3) (d) diesel (4) (e) calf thymus DNA +
xanthine oxidase and 1-nitropyrene and (f) calf thymus-DNA and
xanthine oxidase.

◯ indicates Nuclease P1-sensitive DNA adducts.

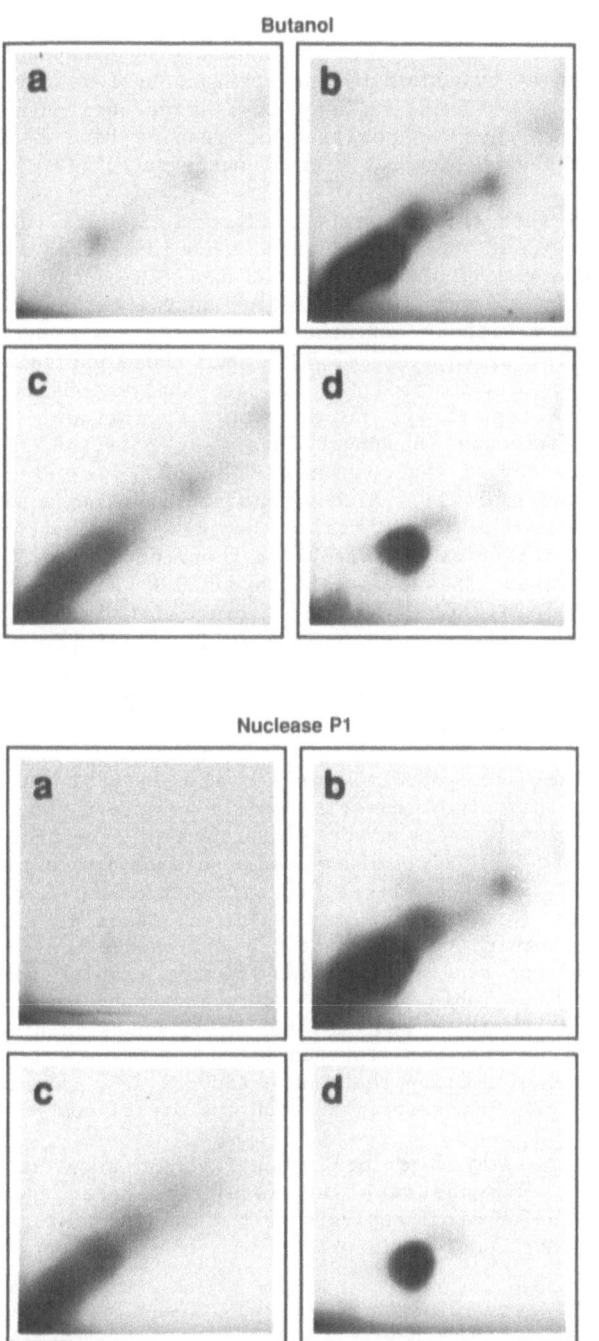

**Figure 3.** *In vivo* modified skin DNA adducts detected following topical treatment with nitrated-PAH containing particle extract (a) diesel (2) and complex mixture extract without nitrated PAH's (b) coke oven (c) coal soot and (d) benzo(a)pyrene modified skin DNA adducts.

(Figure 1 b,d). Consistent with results from a previous study (14), the dinitropyrene-derived calf thymus DNA adducts were nuclease Pl sensitive as measured by a greater than 99% reduction in $^{32}$P-postlabeling efficiency for the nuclease Pl version (Table 1 and Figure 1 [b,d]. Similarly, the nuclease Pl treatment interferes with the $^{32}$P-postlabeling analysis of the guanosine C-8 adduct formed by RTE cells following treatment with 1,6- and 1,8-dinitropyrene as evidenced by the apparent nuclease Pl-sensitivity of this adduct (Figure 1 a,c). Many aryl-amine-derived reactive inter-mediates bound to the C-8 position of guanine have been shown to be sensitive to the phosphatase activity of nuclease Pl (14,16).

To test whether this apparent difference in labeling efficiency of mono and dinitropyrene-derived DNA adducts could be used as a presumptive test for the presence or absence of nitrated PAH-constituents in combus-tion emissions, we selected air pollution source extracts with nitrated-PAH's (diesel) and without nitrated-PAH's (coke oven and coal soot) as determined by chemical analysis (8,9). When these particle extracts were topically applied to C-57 mice (50 mg) and analyzed by both versions of the $^{32}$P-postlabeling assay, no apparent significant qualitative or quantitative differences in adduct levels or patterns of radioactivity were observed for any of the complex mixture or B(a)P-derived DNA adducts (Figure 3[a-d] and Table 1). Although multiple DNA adducts were detected along a diagonal zone of radioactivity, nuclease Pl-sensitive adducts with chromatographic properties comparable to those detected with the mono-and dinitropyrene derived-RTE and/or calf thymus DNA samples (Figure 1, [a-d] were not detected in the DNA adducts formed in vivo (Figure 3 [a-d]). These observations suggest that nuclease Pl-sensitivity alone cannot be used to diagnostically identify di- or mono-nitropyrene-derived DNA adducts from other PAH-DNA adducts formed following in vivo exposure to a variety of complex mixtures.

Two of the four gasoline- or diesel-derived combustion emission extracts used in this study (diesel 2 and 3) have been shown previously to have detectable levels of mono-or dinitropyrene constituents (9). Our data suggest that all four diesel and automobile complex combustion emission particle extracts likely contain nitrated-PAH constituents as evidenced by the formation of one major nuclease Pl-sensitive adduct resulting from xanthine oxidase-mediated nitroreduction of the putative mono and/or dinitropyrene constituents in the complex mixture extracts (Figure 2 [a-d] and Table 1). The chromatographic similarity between these nuclease Pl-sensitive adducts and those generated by ascorbic acid-mediated (Figure 1 [b,d]) nitroreduction of 1,6- and 1,8-dinitropyrene provides further confirmation that these adducts are, infact, derived from nitrated-PAH constituents associated with the diesel combustion emissions.

The in vitro assay described in this study, when coupled with both versions of the $^{32}$P-postlabeling assay, provided DNA adduct data consistent with the chemical analysis of the complex mixtures and may have application to other complex mixtures.

## Acknowledgements

The authors would like to thank Jeff Inmon, Greg O'Brian, Tim Shank and James Scott for excellent technical assistance and Ron Williams and coworkers for the preparation of the particle extracts used in the study.

The research described in this paper has been reviewed by the Health Effects Research Laboratory, U.S. Environmental Protection Agency and approved for publication. Approval does not signify that the contents necessarily reflect the views and policies of the Agency nor does mention of trade names or commercial products constitute endorsement or recommendation for use.

# REFERENCES

1. R.C. Gupta, M.V. Reddy, and K. Randerath, [32]P-postlabeling analysis of nonradioactive carcinogen-DNA adducts, Carcinogenesis 3: 1081-1092 (1982).
2. M.V. Reddy, R.C. Gupta, E. Randerath, and K. Randerath, [32]P-postlabeling for covalent DNA binding of chemicals in vivo:application to a variety of aromatic carcinogens and methylating agents, Carcinogenesis 5: 231-243 (1984).
3. R.C. Gupta, Enhanced sensitivity of [32]P-postlabeling analysis of aromatic carcinogen-DNA adducts, Can. Res. 45: 5656-5662 (1985).
4. M.V. Reddy, and K. Randerath, Nuclease P1-mediated enhancement of sensitivity of [32]P-postlabeling test for structurally diverse DNA adducts, Carcinogenesis 7: 1543-1551 (1986).
5. H.S. Rosenkranz and R. Mermelstein, The genotoxicity, metabolism and carcinogenicity of nitrated polycyclic aromatic hydrocarbons, J. Environ. Sci. Health C3: 221-272 (1985).
6. F.A. Beland, R.H. Heflich, P.C. Howard, and P.P Fu, The in vitro metabolic activation of nitro polycyclic aromatic hydrocarbons, In: "Polycyclic aromatic hydrocarbons and Carcinogensis", Harvey, R.G., (ed), American Chemical Society, Washington, DC. pp 371-396 (1985).
7. J.E. Gallagher, I.C.G. Robertson, M.A. Jackson, A.M. Dietrich, L.M Ball, and J. Lewtas, 32P-postlabeling analysis of DNA adducts of two nitrated polycyclic aromatic hydrocarbons in rabbit tracheal epithelial cells, in: Carcinogenic and Mutagenic Responses to Aromatic Amines and Nitroamines, King, L.J. Ramano and D. Schuetzle, (eds) Elsevier Applied Science Publishers, Barking U.K. 277-281 (1985).
8. J. Lewtas, R.L. Bradlow, R.H. Jungers, B.D. Harris, and R.B. Zweiinger, Mutagenic and carcinogenic potency of extracts of diesel and related environmental emissions: Study design, sample generation,collection and preparation Environ. Intern. 5: 383-387, (1981).
9. R.E. Albert, J. Lewtas, S. Nesnow, T.W. Thorslund, and E. Anderson, Comparative potency method for cancer risk assessment: Application to diesel particle emissions. Risk Analysis, 3: 101-117 (1983).
10. R. Williams, C. Sparacino, B. Petersen, J. Bumgarner, R.H. Jungers, and J. Lewtas, Comparative characterization of organic emissions from diesel particles, coke oven mains, roofing tar vapors and cigarette smoke condensate, Intern. J. Environ. Anal. Chem. 26: 27-49, (1983).
11. L.C. King, M. Jackson, L.M. Ball and J. Lewtas, Metabolism and DNA binding of 1-nitropyrene by isolated rabbit tracheal epithelial cells, Carcinogenesis 8: 675-682 (1987).
12. D.M. Maron, and B.N. Ames, Revised methods for the salmonella mutagenicity test, Mut. Res. 113:173-215 (1983).
13. R.C. Gupta, Nonrandom binding of the carcinogen N-hydroxy-2-acetylaminofluorene to repetitive sequences in rat liver DNA in vivo, Proc. Natl. Acad. Sci. USA (81) 6943-6947 (1984).
14. J.E. Gallagher, M.A. Jackson, M.H. George, J. Lewtas, I.C.G. Robertson, Differences in detection of DNA adducts in the [32]P-postlabeling assay after either 1-butanol or nuclease P1 treatment, Can. Letters (45) 7-12 (1989).
15. J. Lewtas, Genotoxicity of complex mixtures: Strategies for the identification and comparative assessment of airborne mutagens and carcinogens from combustion sources, Fund. and Applied Toxicol. 571-589, (1988).
16. R.C. Gupta, and K. Early, [32]P-postlabeling assay: Comparative recoveries of structurally diverse DNA adducts in the various enhancement procedures, Carcinogenesis, 9: 1687-1693, (1988).

# ANALYSIS OF NO$_2$-PAH DNA ADDUCTS BY MASS SPECTROMETRY

Roger W. Giese and Paul Vouros

Northeastern University
Boston, MA 02118

## INTRODUCTION

Mass spectrometry (MS) is a useful analytical technique for trace organic analysis. It can provide structural information (qualitative analysis) on small amounts of unknown compounds. Also known compounds can be determined (quantitative analysis) with high sensitivity and precision. In part the good precision derives from the addition of an isotopically-labeled internal standard to the sample.

MS has been used to analyze nitropolyaromatic compounds[1,2], aminopolyaromatics[3,4] and DNA adducts derived from or comprising such chemicals.[5-7] In the latter case several micrograms or more of partly-known NO$_2$-PAH or NH$_2$-PAH DNA adducts isolated as deoxynucleosides were analyzed by MS involving sample ionization-volatization by fast atom bombardment (FAB). The DNA adducts were generated from test tube, tissue culture, or animal experiments in which high doses of known NO$_2$-PAH or NH$_2$-PAH compounds or precursors were administered.

We are interested in using MS to analyze NO$_2$-PAH DNA adducts in human samples. For such analysis, ultra-high sensitivity is required not only because the amounts of the samples are limited, but because the samples themselves contain only trace quantities of NO$_2$-PAH DNA adducts. However, the chemical and physical steps needed to achieve the analysis of such adducts with ultra-high sensitivity by MS remain in large measure to be defined. Thus, the first task is to establish these steps. This is the focus of our current work and the subject of this article.

The long-term goal of our work is to provide a more accurate assessment of the risks to health of human exposure to NO$_2$-PAHs. A logical progression of experimental work towards this goal is to first develop appropriate chemical and physical techniques for the analysis of standards of such adducts or model compounds. Then we can proceed to analyze samples from animals followed by human-derived samples. To date our work has been with standards.

For the reasons just given, we are focusing on the most sensitive modes of MS for the analysis of NO$_2$-PAH DNA adducts. In turn, we are pursuing two general strategies of analysis. The first strategy can be called "alternative-sample analysis", which is intended to provide qualitative analysis of human DNA adducts of unknown structure. In this approach, a sensitive technique such as [32]P-post labeling-thin layer chromatography[8,9] is used first to detect a given adduct in both a human and alternative DNA sample. The latter sample would be obtained from exposure of test tube, tissue culture, or animal DNA in vivo to a chemical (or chemical mixture) of interest. The amount of the adduct in the alternative sample would then be scaled up into the range where some structural information could be obtained by MS. This, of course, is the reason for the alternative sample. The amounts of unknown DNA adducts anticipated in routine human samples generally will be too small for much structural information to be obtained directly by MS or, in fact, by any other known analytical technique. As will be discussed, we are particularly interested in using FAB-MS for such qualitative analysis of unknown $NO_2$-PAH DNA adducts derived from alternative samples.

In the second approach, "human-sample analysis," the actual human DNA adduct (but after chemical modification) is measured by GC-MS. This approach will emphasize quantitative determination of known adducts. Such DNA adducts are to be determined by GC-MS or LC-MS as electrophoric derivatives after chemical transformation as explained below.

We will now discuss each of these two general strategies, qualitative analysis of unknown $NO_2$-PAH DNA adducts by FAB-MS, and quantitative analysis of known $NO_2$-PAH adducts by GC-MS, in more detail.

## FAB-MS for Qualitative Analysis

While some DNA adducts have been analyzed in moderate microgram amounts by FAB-MS, as cited above, it is generally difficult to scale adducts up to this level even when the adduct can be obtained by reacting DNA in a test tube with a known chemical. Thus it is important to extend the sensitivity of the FAB-MS technique as much as possible for its application, directly or indirectly, to human samples. As demonstrated in Fig. 1A, it is indeed possible to obtain full-scan mass spectral data at a significant signal/noise on a sub-ug amount of $NO_2$-PAH adduct as a nonderivatized deoxynucleoside. In this example, 6.5 ng of N-(deoxyguanosin-8-yl)-2-(acetylamino)fluorene (dG-8-AAF) was introduced by continuous flow FAB into the MS. Protonation of the parent molecule (M, having a molecular weight of 488) yields a parent ion $(M + H)^+$ at m/z 489. The principal ion fragment derived from this ion occurs at m/z 373, which corresponds to loss of the sugar moiety. Loss of both the sugar and acetyl moieties yields the ion $(M+H-sugar-acetyl)^+$ at m/z 330.

One way to further increase the sensitivity of FAB-MS for such an analyte, and also to obtain additional structural information, is to perform the mass analysis in a $B/E$ linked scan mode. The principle of this technique is shown in Fig. 2, where $B$ and $E$ refer, respectively, to the magnetic and electric regions of the MS. The magnetic and electric field in these regions define the trajectories of the ions. In the $B/E$ mode of such a linked scan, we take advantage of certain ions in the primary ion beam (bulk of the sample ions) that dissociate slowly and thereby tend to continue to fragment after they are repelled out of the source (that is, in the field free region between the source and $E$). This produces so-called metastables. The production of metastable-like daughter ions can also be increased by addition of a collision gas outside of the source. In either case, the metastables or induced daughters, having a fraction of the kinetic energy of the ordinary ions in the primary ion beam, will follow a unique trajectory in the $E$ sector. This is because $E$ focuses ions based on their kinetic energy. While the signal intensity from these special ions relative to that of the ions in the primary ion beam is reduced, the noise is reduced even more. Thus an improvement in signal/noise is obtained, which is equivalent to an

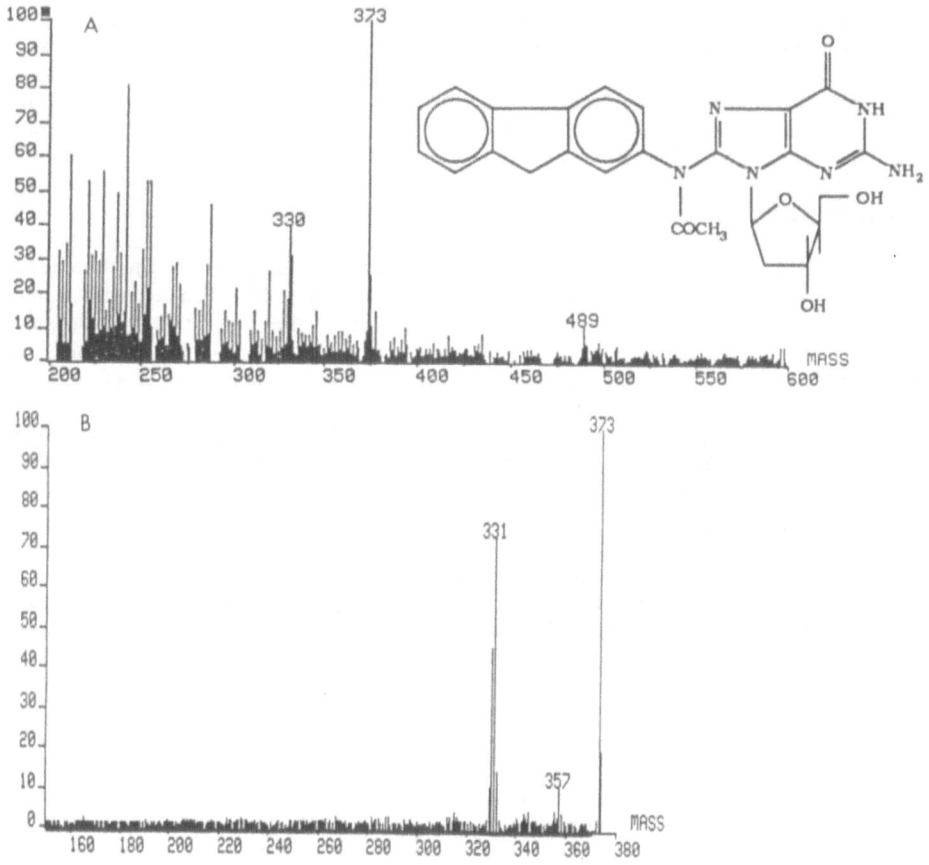

Fig. 1 Continuous flow FAB-MS spectra of 6.5 ng of N-(dG-8-yl)-AAF.[14]
(A) Normal scan. (B) B/E linked scan of daughters from m/z 373;
19 ng of N-(d6-8-yl)AAF was analyzed. This compound was obtained
from Fred Beland.

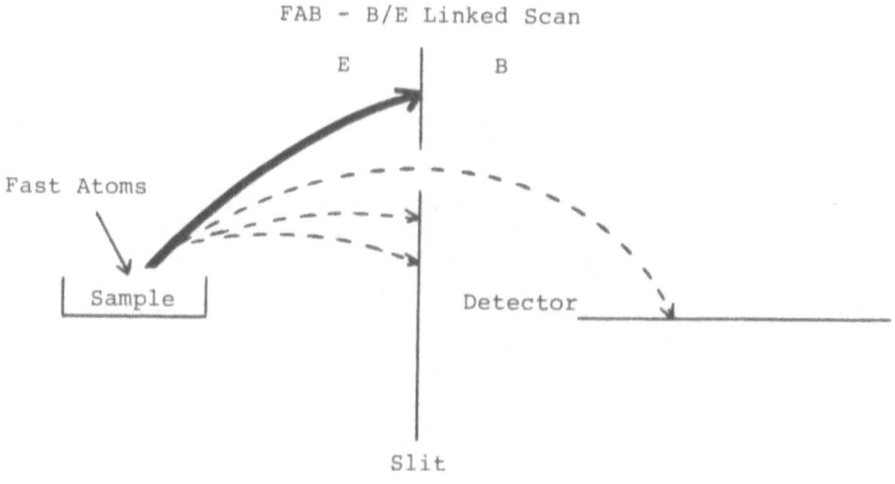

Fig. 2 Concept of B/E linked scan in mass spectrometry.

increase in the overall sensitivity. A given daughter ion in the unique trajectory, after it passes through a slit at the end of $E$, is then subjected to a magnetic field $\underline{B}$ which focuses it on the detector only if it has a certain mass. We thereby can define the parent ion (in terms of its m/z) for this daughter. Further, by scanning $\underline{B}$ and $\underline{E}$ at a constant ratio of their values ($\underline{B}/\underline{E}$ linked scan), we can collect all of the induced daughters and metastables from a given parent ion, which is somewhat equivalent to a family tree. This process can yield significant structural information beyond that provided by the full scan.

As anticipated, an improvement in both sensitivity and structural information is achieved by employing FAB-MS in a $\underline{B}/\underline{E}$ linked scan mode to analyze dG-8-AAF as illustrated in Fig. 1B. The ion at m/z 373 ion was selected as the parent ion for this scan. As seen, a cleaner, more informative spectrum is obtained relative to that of Fig. 1A. The now-visible ion at m/z 357 arises from loss of $NH_2$ from m/z 373, confirming that the $NH_2$ of dG-8-AAF is unsubstituted in the DNA adduct. We also learn that the daughter ion, now predominantly at m/z 331 $(M+2H-sugar-acetyl)^+$, originates in fact from m/z 373.

We anticipate even additional sensitivity by FAB-MS for such adducts, since we have observed (data not shown) that they are about 10 fold more sensitive after trimethylsilylation. Thus, it is becoming increasingly more practical to pursue the structures of unknown human DNA adducts, revealed by [32]P post-labeling or other techniques, that can be matched with equivalent adducts from alternative samples of DNA.

### GC-MS for Quantitative Analysis

The most sensitive mode of mass spectrometry is gas chromatography with electron capture negative ion MS (GC-ECNI-MS), when applied to the detection of volatile strong electrophores. (An electrophore is a compound that readily combines with a low energy electron in the gas phase, which is the basic event in sample ionization by ECNI.) This is largely because of the high efficiency of sample ionization by this technique. For example, we have detected 1.0 fg (3.0 x $10^{-18}$ mole) of an electrophore-derivatized nucleobase, at a signal to noise of 10, by GC-ECNI-MS.[10]

Although a variety of alkyl and related DNA nucleobases can be electrophore-derivatized and detected with high sensitivity by GC-ECNI-MS[11], corresponding $NO_2$-PAH adducts in general can be expected to fall outside the volatility and thermal stability range of GC-ECNI-MS. Thus we have adopted an alternative strategy to detect such adducts with high sensitivity. In this technique the DNA adduct is chemically transformed into a related product which more readily can be derivatized for sensitive detection by GC-ECNI-MS.

As a test case to define the usefulness of chemical transformation as an analytical technique for $NO_2$-PAH DNA adducts, we subjected dG-8-AF to a variety of reactive chemical conditions, and found that hydrazinolysis is the most promising.[12] Reacting this compound with hydrazine at an elevated temperature releases 2-aminofluorene in high yield, as illustrated in Scheme 1. It is attractive that hydrazine is a relatively simple chemical which can be removed by evaporation at the conclusion of the reaction. The current status of this reaction is that we have applied it to amounts of dG-8-AF as small as 5 ug, and are extending it to lower levels prior to its application to samples of DNA containing this adduct.

In parallel, we have optimized the selection of an electrophoric derivative of 2-aminofluorene.[13] Of several derivatives that we

214

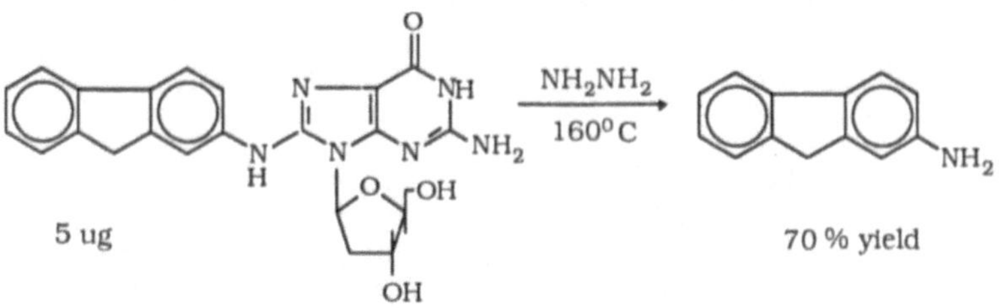

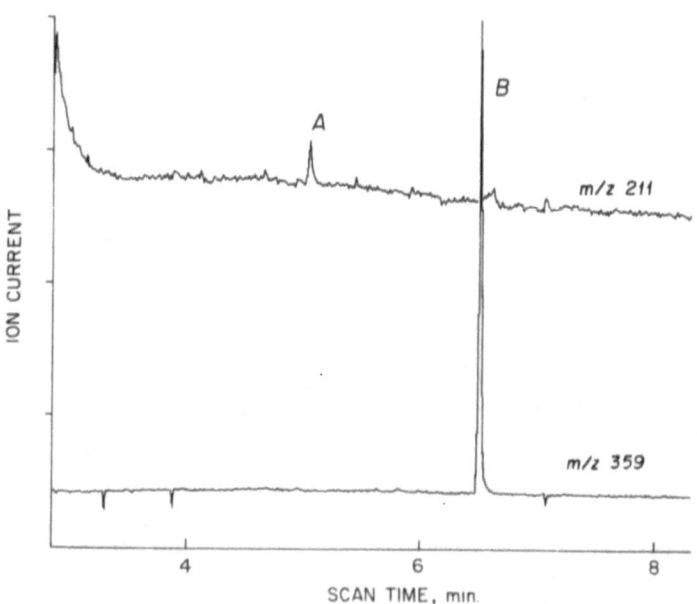

Scheme 1. Hydrazinolysis of N-(deoxyguanosin-8-yl)-2-aminofluorene (dG-8-AF).

examined, two emerged as the best ones. The first is a pentafluorobenzylidene derivative, obtained by reacting 2-aminofluorene with pentafluorobenzaldehyde. The excellent GC-ECNI-MS properties of this derivative are shown in Fig. 3. As seen, the response of this derivative is 20-fold higher than that of a related compound, 2-nitrofluorene. One cannot make a general statement about the relative sensitivity of such compounds by GC-ECNI-MS, however, since we have observed similar responses for pentafluorobenzylidenyl-1-aminopyrene and -1-nitropyrene, apparently reflecting the more extensive conjugation in the 1-nitropyrene vs 2-nitrofluorene derivative. An N-pentafluorobenzyl-N-heptafluorobutyryl derivative of 2-aminofluorene and 1-aminopyrene also has favorable formation and detection properties.

Thus, some promising results have been achieved for both the chemical transformation and the electrophoric derivatization steps of the model adduct dG-8-AF. Hopefully this methodology combined with GC-MS will continue to perform well as we put the overall sequence of steps together

Fig. 3 GC-ECNI-MS of 5.5 pg each of 2-nitrofluorene (A) and N-(pentafluorobenzylidenyl)-2-aminofluorene (B).[13] The mass chromatograms were recorded by monitoring the molecular anions m/z 211 (A) and 359 (B). Reprinted with permission from the Journal of Chromatography.

and extend the method to lower analyte levels of this and related $NO_2$-PAH adducts.

## Isotope Feeding

Although there are anticipated difficulties, the transformation-derivatization-GC-MS method just described potentially can be employed to obtain some clues about the structures of low level, unknown DNA adducts. The strategy (which remains to be tested) is to first isolate an unknown DNA adduct (e.g. as an adduct spot scraped from multiple, overloaded TLC plates in $^{32}$P post-labeling analysis). This sample is then divided and subjected to several transformation-derivatization-GC-MS procedures. Potentially one of these procedures, at least, will yield a fragment ion for the unknown compound. The animal (or even human) from which this adduct is derived can then be administered deuteriated substances (including deuteriated xenobiotics in the case of animals) and the overall analytical procedure repeated. If the fragment ion belonging to the DNA adduct becomes isotopically shifted, then the parent compound for the adduct can be identified. Although metabolic scrambling and isotope effects of the isotopic atoms may complicate the procedure, this may be the most practical way to learn something about the identity of completely unknown trace DNA adducts from human samples having no counterpart in non-human samples.

## CONCLUSION

Mass spectrometry promises to play an increasing role in the future studies of $NO_2$-PAH DNA adducts derived from human and related DNA samples. In part this is likely to be promoted by the strategies and advances in sensitivity that are reported here for FAB and GC modes of MS.

## ACKNOWLEDGMENT

The research described in this article was conducted in part under contract to the Health Effects Institute (HEI), an organization jointly funded by the United States Environmental Protection Agency (EPA) (Assistance Agreement X-812059) and automotive manufacturers. The specific grant was HEI Research Agreement 86-82. The contents of this article do not necessarily reflect the views of HEI, nor do they necessarily reflect the policies of the EPA, or automotive manufacturers. Contribution #390 from the Barnett Institute.

## REFERENCES

1. D. Schuetzle and T.E. Jensen, Analysis of Nitrated Polycyclic Aromatic Hydrocarbons (Nitro-PAH) by Mass Spectrometry, in Nitrated Polycyclic Aromatic Hydrocarbons, C.M. White, ed., Dr. Alfred Huethrg, Verlag, Heidelberg (1985) 121-167.
2. W.A. Korfmacher, J. Djuric, E.K. Fifer and F.A. Beland, Characterization of Nitro-Polycyclic Aromatic Hydrocarbon Metabolites via Methane Enhanced Negative Ion Mass Spectrometry, Spectros. Int. J. 6: 1-16 (1988).
3. U. Sellstrom, B. Jansson, A. Bergman and T. Alsberg, Selective and Sensitive Analysis of Nitro-PAH, Chemosphere. 16: 945-952 (1987).
4. W.A. Korfmacher, L.G. Rushing, R.J. Engelback, J.P. Freeman, Z. Kjuric, E.K. Fifer and F.A. Beland, Analysis of Three Aminonitropyrene Isomers via Fused Silica Gas Chromatography

Combined with Negative Ion Atmospheric Pressure Ionization Mass Spectrometry, J. HRC & CC, 10: 43-45 (1987).

5.  R.K. Mitchum, F.E. Evans, J.P. Freeman and D. Roach, Fast Atom Bombardment Mass Spectrometry of Nucleosides and Nucleotide Adducts of Chemical Carcinogens, Int. J. Mass Spectrum and Ion Physics 46: 383-386 (1983).

6.  D.L. Slowikowski and K.H. Schram, Fast Atom Bombondment Mass Spectrometry of Nucleosides, Nucleotides and Oligonucleotides, Nucleosides and Nucleotides 4: 309-345 (1985).

7.  R.K. Mitchum, J.P. Freeman, F.A. Beland and F.F. Kadlubar, Mass Spectrometric Identification of DNA Adducts Formed by Carcinogenic Aromatic Amines, in "Mass Spectrometry in the Health and Life Sciences," A.L. Burlingame and N. Castagnoli, Jr., eds., Elsevier, Amsterdam (1985), 547-595.

8.  J.G. Liehr, T.A. Avitts, E. Randerath and K. Randerath, Estrogen-Induced Endogenous DNA Adduction: Possible Mechanism of Hormonal Cancer, Proc. Natl. Acad. Sci. 83: 5301-5305 (1986).

9.  W.P. Watson, Post-Labeling for Detecting DNA Damage, Mutagen. 2: 319-331 (1987).

10.  G.B. Mohamed, A. Nazareth, M.J. Hayes, R.W. Giese, Paul Vouros, Gas Chromatography-Mass Spectrometry Characteristics of Methylated Perfluoroacyl Derivatives of Cytosine and 5-Methylcytosine, J. Chromatogr. 314: 211-217 (1984).

11.  M. Saha, R.S. Annan, G.M. Kresbach, P. Vouros and R.W. Giese, Preparation and Mass Spectral Characterization of Pentafluorobenzyl Derivatives of Alkyl and Hydroxyalkyl-Nucleobase DNA Adducts, Biomed. Environ. Mass Spectrom., in press.

12.  J. Bakthavachalam, S. Abdel-Baky and R.W. Giese, unpublished observations.

13.  J. Bakthavachalam, R.S. Annan, F.A. Beland, P. Vouros, R.W. Giese, Selection of Electrophoric Derivatives of 1-Aminopyrene and 2-Aminofluorene for Determination by Gas chromatography with Electron Capture Negative Ion Mass Spectrometry, submitted for publication.

14.  R. Annan, R. Giese, and P. Vouros, in preparation.

# GENERATION OF REACTIVE INTERMEDIATES FROM 2-NITROFLUORENE THAT BIND

# COVALENTLY TO DNA, RNA AND PROTEIN IN VITRO AND IN VIVO IN THE RAT

Gerard J. Mulder, Frank C.J. Wierckx, Rinny Wedzinga
and John H.N. Meerman

Division of Toxicology
Center for Bio-Pharmaceutical Sciences
University of Leiden
Leiden, The Netherlands

## INTRODUCTION

2-Nitrofluorene (2-NF) is an air pollutant that is found in appreciable concentrations in urban air due to the exhaust of especially Diesel engines. Like many other nitroaromatics, it is a direct acting mutagen in bacteria that contain nitroreductase activity: this enzyme converts the nitro-aromatics to reduced products, notably the hydroxylamines, that can interact directly with DNA and can cause mutations. 2-NF is a carcinogen in the rat, especially in the forestomach, where it causes tumors upon oral administration[1-4]. As yet it is unclear which metabolite(s) is (are) responsible for the mutagenic and carcinogenic activity of 2-NF.

Möller et al[5-9] have studied the metabolism of 2-NF in the rat. They showed that nitroreduction by the gut bacteria was not responsible for the presence of mutagenic metabolites in urine: in urine from germ-free animals, in which nitroreduction of 2-NF was severely reduced, the mutagenicity of the urinary metabolites was increased. Mutagenicity was most likely related to mono- and di-hydroxylated 2-NF metabolites. Further, 2-NF was a strong initiator and a weak promotor in a liver model for chemical carcinogenesis in the rat in vivo when it was compared to diethylnitrosamine and 2-acetylaminofluorene (2-AAF) respectively. These results suggest that 2-NF may be converted

---

2-AF, 2-aminofluorene; 2-AAF, 2-acetylaminofluorene; DEM, diethylmaleate; dG-C8-AF, N-(deoxyguanosin-8-yl)-2-aminofluorene; dG-C8-2-AAF, N-(deoxyguanosin-8-yl)-2-acetylaminofluorene; dG-N²-AAF, 3-(deoxyguanosin-N²-yl)-2-acetylaminofluorene; DMSO, dimethylsulfoxide; GSH, glutathione; NAC, N-acetyl-L-cysteine; NAM, N-acetyl-L-methionine; N-OH-2-AAF, N-hydroxy-2-acetylaminofluorene; N-OH-2-AF, N-hydroxy-2-aminofluorene; 2-NF, 2-nitrofluorene; PCP, pentachlorophenol; TFA, trifluoroacetic acid.

to metabolites that interact with DNA to form adducts that subsequently cause mutations. We have done studies with [³H]-labelled 2-NF to detect covalent binding of radioactivity to protein, RNA and DNA. Such binding occurred in several organs, which indicates that 2-NF is activated to (a) reactive intermediate(s) that bind(s) to appropriate groups in these macromolecules.

## MATERIALS AND METHODS

[ring-³H]-2-NF (specific radioactivity 120 mCi/mmol) was obtained by acid-catalyzed ³H exchange of unlabeled 2-NF with tritiated water as described by Breeman et al[10]. In some experiments [9-¹⁴C]-2-NF (specific radioactivity 60 mCi/mmol) was used, obtained from Chemsyn Science Labs, Lenexa, KS, USA. 2-NF was from Aldrich Chem. Comp., Beerse, Belgium. Biochemicals were from Sigma Chemical Comp., St Louis, MO, USA.

Throughout this study male Wistar rats (body weight 180-225 grams) of the Sylvius Laboratories, University of Leiden, were used. They were housed in Macrolon cages on standard hard-wood bedding, and had free access to food and water unless mentioned otherwise. They were maintained on an alternating 12 hours light and dark cycle. All surgery was performed as described elsewhere[11].

## RESULTS

### In vivo metabolism of 2-NF

When [³H]-2-NF is given orally it is readily absorbed from the gut and taken up in blood[6]. In order to study the pharmacokinetics of 2-NF elimination from the blood we administered 2-NF (6 µmol/kg) dissolved in DMSO (0.5 ml/kg) intravenously in pentobarbital-anesthetised rats, with a cannulated bile-duct. Blood samples were drawn from the carotid artery and unchanged 2-NF in blood was extracted with diethyl ether. A typical example of a 2-NF blood disappearence curve is shown in Fig. 1: initially 2-NF was very rapidly removed ($t_{\frac{1}{2}}$ ~ 2.5 minutes), which was followed by a much slower phase with a $t_{\frac{1}{2}}$ of more than 2.5 hours. Radioactivity was rapidly excreted in bile in these rats (Fig. 1); analysis of the biliary excretion rate also revealed the presence of two half-lives with $t_{\frac{1}{2}}$ of 9 minutes and 1 hour respectively. After 2 hours approximately 40 % of the dose had been excreted in bile. The major metabolite was the glucuronide conjugate of 9-hydroxy-2-NF, which was identified by LC-MS as described by Möller et al[5]: the mass spectrum of the β-glucuronidase treated metabolite was identical with the data of Möller et al[6,7]. These experiments were confirmed by some experiments with [¹⁴C]-labelled 2-NF, indicating that the ³H data reflect the metabolism of the whole molecule.

In freely moving rats in metabolism cages 19.7 ± 0.8 % of the dose of orally administered [³H]-2-NF was collected in urine in 24 hours. The main metabolites of 2-NF in urine were identified using the LC-MS system described by Möller et al[5]. In urine hydroxylated 2-NF and 2-acetylaminofluorene (2-AAF) derivatives were found to be the main metabolites.

There is a great difference between the high biliary excretion of 2-NF metbolites on one hand (40 % in 2 hours) and the rather low excretion of the compound and its metabolites in urine (20 % in 24 hours). It is possible that this difference is due to reabsorption of

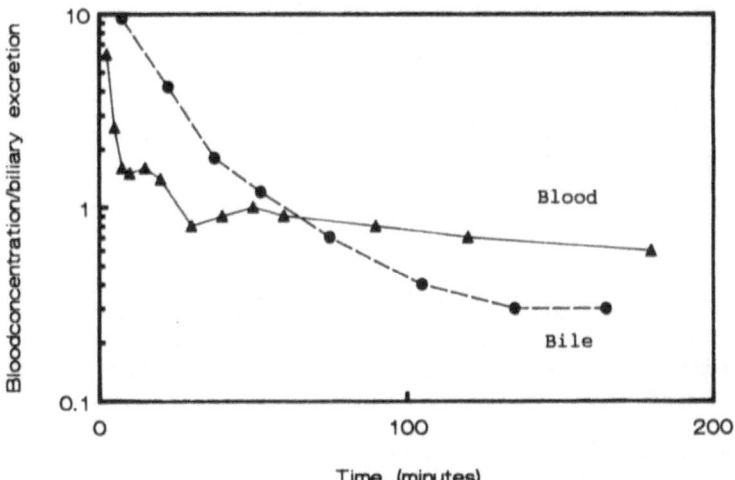

**Fig. 1.** Blood concentration and biliary excretion rate in a bile-duct cannulated rat after intravenous administration of 2-NF (6 μmol/kg).
——▲—— bloodconcentration (μM). ——●—— biliary excretion rate (nmol/min).

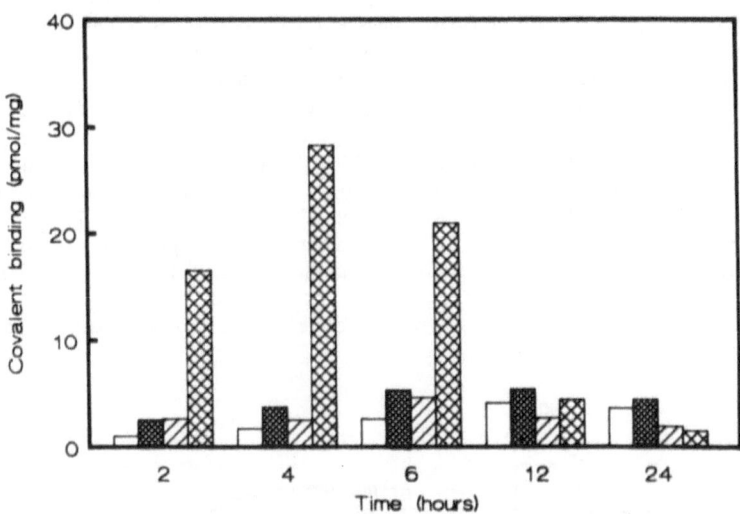

**Fig. 2.** Time course of the covalent binding of 2-NF (6 μmol/kg) in the rat in vivo after oral administration.
☐ liver: protein binding (pmol/mg protein). ▓ liver: total binding to nucleic acids (pmol/mg). ▨ binding to total kidney cellular macromolecules (pmol/mg protein). ▩ binding to total gut mucosa cellular macromolecules (pmol/mg protein).

2-NF or its metabolites from the gut and subsequent enterohepatic recirculation of these compounds, which could also explain the observed very long elimination half-life in blood. Another explanation for this long half-life may be the lipophilic character of 2-NF, which may lead to accumulation of the compound in fat tissues. This storage in fat tissue is also suggested by the apparent volume of distribution, which was approximately 3 l/kg body weight.

## Covalent binding of 2-NF in the rat in vivo

Covalent binding of 2-NF in the rat in vivo was studied after oral administration of[³H]-labelled 2-NF. The time course of covalent binding to various components of a number of organs is shown in Fig. 2. Clearly, the highest covalent binding was reached in the gut: up to 40 pmol/mg protein after 4 hours. This level subsequently decreased again. In the kidneys and the liver maximal binding was reached at 6 hours after administration. The level of binding to protein and RNA in the liver was comparable.

The possibility was considered that covalent binding of 2-NF to macromolecules required nitroreduction in the gut by bacterial nitroreductases. In that case a nitroso- or hydroxylamine-derivative might be generated that could react with thiols as reported previously[12-13]. Such reactive intermediates, if formed in the gut, might be trapped by low molecular weight thiols like glutathione (GSH) and N-acetyl-L-cysteine(NAC) (Fig. 3), which could prevent their macromolecular binding. Therefore, we have given high doses of NAC (0.3 mmol/kg) together with 2-NF (6 µmol/kg), both orally. However, there was no statistically significant effect of NAC on the level of covalent binding, in spite of the fact that a high excess of NAC was given 30 minutes before, at the same time and 30 minutes after 2-NF (Fig. 4). Therefore NAC does not play a protective role under these conditions towards reactive 2-NF intermediates that bind covalently in vivo.

Another possibility is that 2-NF is reduced to 2-aminofluorene (2-AF) and subsequently acetylated to 2-AAF. This compounds then is expected to be metabolically activated through N-hydroxy-2-AAF. In that case the covalent binding in the liver should be very sensitive to inhibition of sulfation by pentachlorophenol (PCP): the major reactive intermediate from N-hydroxy-2-AAF is generated by the formation of a labile sulfate conjugate and binds covalently to guanine residues in

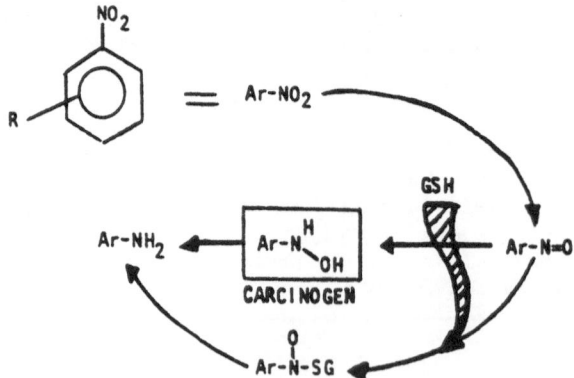

Fig 3. Scheme of the possible activation and inactivation of nitroaromatic compounds by nitroreduction and thiols (such as GSH) respectively.

DNA and RNA[14-15]. However, when the animals were pretreated with PCP at a dose level that effectively reduces covalent binding of N-hydroxy-2-AAF to protein to 20-40 % of control, no effect was observed on 2-NF covalent binding (Fig 5). Therefore, covalent binding of 2-NF is most likely not the result of conversion of 2-NF to N-hydroxy-2-AAF followed by sulfate conjugation. Also N-acetyl-L-methionine (NAM), which efficiently traps reactive intermediates formed from N-hydroxy-2-AAF, which are not trapped by GSH [16-17], was not effective when it was given orally together with 2-NF (Fig. 6).

A rather surprising finding was that depletion of GSH by pretreatment of the rats with diethymaleate (DEM) did not result in increased binding, but, to the contrary tended to reduce covalent binding (Fig. 7). GSH, therefore does not protect the rat against covalent binding of 2-NF to cellular macromolecules.

<u>Covalent binding of 2-NF to protein and DNA after microsomal activation in vitro</u>

When 2-NF ([³H] or [¹⁴C] labelled) was incubated with microsomes isolated from livers of Aroclor 1254 pretreated rats in the presence of NADPH, radioactivity became covalently bound to protein (Table 1). 2-NF also became covalently bound to DNA and RNA. The level of covalent binding to DNA was approx. 40 pmol/mg DNA in the presence of NADPH; if NADPH was omitted it was approx. 6 pmol/mg DNA. For RNA the level of covalent binding was 30 pmol/mg RNA in the presence of NADPH and 1 pmol/mg RNA if NADPH was omitted. If the microsomal preparation was boiled for 1 minute before it was added to incubations with 2-NF, the level of covalent binding was equal to that found in incubations without NADPH; we therefore assume that the low level of covalent binding found in the incubations without NADPH represents non-enzymatic

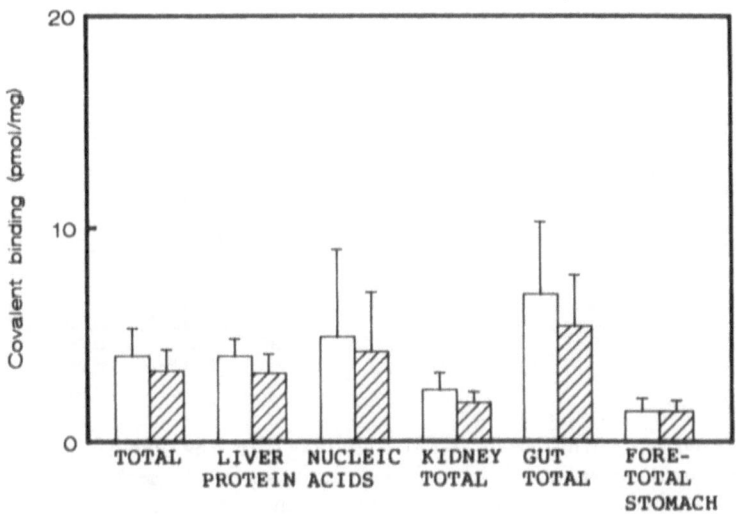

Fig. 4. Covalent binding of 2-NF (6 μmol/kg) in the rat in vivo 6 hours after an oral dose of 2-NF in NAC treated (0.3 mmol/kg orally, 30 minutes before, at the same time and 30 minutes after the administration of 2-NF) and control rats.

⬜ Control rats.    ▨ NAC treated rats.

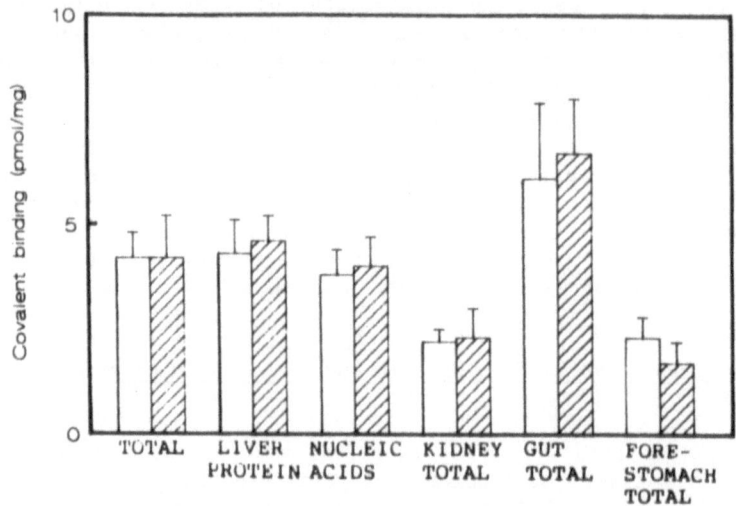

**Fig. 5.** Covalent binding of 2-NF (6 μmol/kg) in the rat in vivo 6 hours after an oral dose of 2-NF in PCP treated (40 μmol/kg, intraperitoneally, 45 minutes before the administration of 2-NF) and control rats.
☐ Control rats. ▨ PCP treated rats.

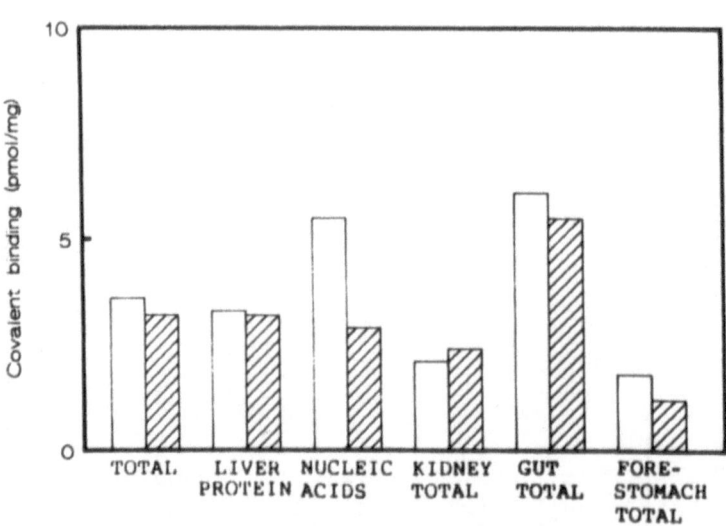

**Fig. 6.** Covalent binding of 2-NF(6 μmol/kg) in the rat in vivo 6 hours after an oral dose of 2-NF in NAM treated (0.3 mmol/kg, orally 30 minutes before, at the same time and 30 minutes after the administration of 2-NF) and control rats.
☐ Control rats. ▨ NAM treated rats.

**TABLE 1.**

Effect of various compounds on the covalent binding of 2-NF in Aroclor 1254 induced rat liver microsomes.

| Compound | Covalent binding ( % of complete) |
|---|---|
| Complete system | 100 [a] |
| Without NADPH | 12 |
| N-acetyl-L-cysteine [b] | 94 |
| Glutathione [b] | 85 |
| N-acetyl-L-methionine [b] | 100 |
| β-mercaptoethanol [b] | 93 |
| Disulfiram [b] | 48 |
| Diethyldithiocarbamate [b] | 65 |
| SKF 525A [b] | 12 |

[a] 100 % equaled a covalent binding of 700 pmol/mg protein
[b] final concentration in the incubation mixture was 5 mM
All values are a mean of 3 determinations.

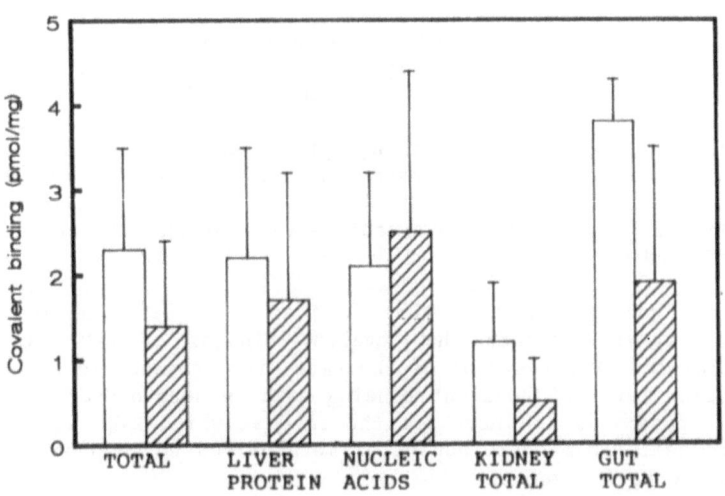

Fig. 7. Covalent binding of 2-NF (6 μmol/kg) in the rat in vivo 6 hours after an oral dose of 2-NF in DEM treated (6.2 mmol/kg, intraperitoneally 30 minutes before the administration of 2-NF and 3.1 mmol/kg 3 hours after the first dose of DEM) and control rats.
☐ Control rats.  ▨ DEM treated rats.

binding of radioactivity to protein. The level of covalent binding of 2-NF to macromolecules increased with time.

Addition of SKF 525A the to incubation mixture prevented covalent binding (Table 1). These data, therefore indicate that covalent binding of 2-NF is most likely catalyzed by cytochrome $P_{450}$. In addition covalent binding can also partially be prevented by disulfiram and diethyldithiocarbamate, which are also inhibitors of the cytochrome $P_{450}$ enzyme system[18-19] (Table 1).

Thiols, like GSH and NAC could not prevent the covalent binding of 2-NF (Table 1). This is in agreement with our observations in the rat in vivo and in freshly isolated rat hepatocytes in which GSH depletion by DEM did not give rise to a significantly higher level of covalent binding (see below).

## Covalent binding of 2-NF in isolated hepatocytes: effect of Aroclor pretreatment

When 2-NF was incubated with isolated rat hepatocytes obtained by collagenase perfusion, it was rapidly converted to ring hydroxylated metabolites of 2-NF, which were subsequently glucuronidated (Fig. 8). These metabolites were identified by LC-MS as described by Möller et al[5-7]. In addition, radioactivity became covalently bound to macromolecules (Fig. 9). When the cells were depleted of their GSH by preincubation with DEM this covalent binding did not change appreciably. In isolated hepatocytes from Aroclor treated rats, the covalent binding occurred much faster than in those from normal rats (data not shown). This correlated with a very fast metabolism of 2-NF in the Aroclor-treated cells: within a few minutes all 2-NF had disappeared from the incubation, while in the control cells the unchanged 2-NF slowly disappeared.

When rats were pretreated with Aroclor and covalent binding of 2-NF was determined in the rat in vivo, covalent binding was lower in Aroclor treated animals than in controls, if expressed per mg of protein. However, when the increased liver weight is considered, the total covalent binding as percentage of the dose is the same in control and treated animals. That the percentage of covalent binding is similar may be due to the limited amount of substrate (2-NF) available for the formation of the reactive intermediate of 2-NF and therefore, for the covalent binding.

These results indicate that in freshly isolated hepatocytes GSH did not protect against the covalent binding of 2-NF to cellular macromolecules. Pretreatment of the rats with Aroclor did not give rise to a higher level of covalent binding both in hepatocytes and in vivo, but in hepatocytes it increased the rate at which 2-NF was covalently bound to cellular macromolecules or metabolised to compounds that were not able to bind covalently.

## DNA adduct formation from 2-NF in vivo

DNA was isolated from livers of rats treated orally with [³H]-2-NF of high specific radioactivity (110 mCi/mmol, 60 μmol/kg bodyweight) and hydrolysed with trifluoroacetic acid (TFA) as described by van de Poll et al[20]. The radioactive adducts were analyzed by HPLC; the amount of adducts were quantitated by the amount of radioactivity eluting under each peak (Fig. 10). Only 20 % of the recovered DNA adducts

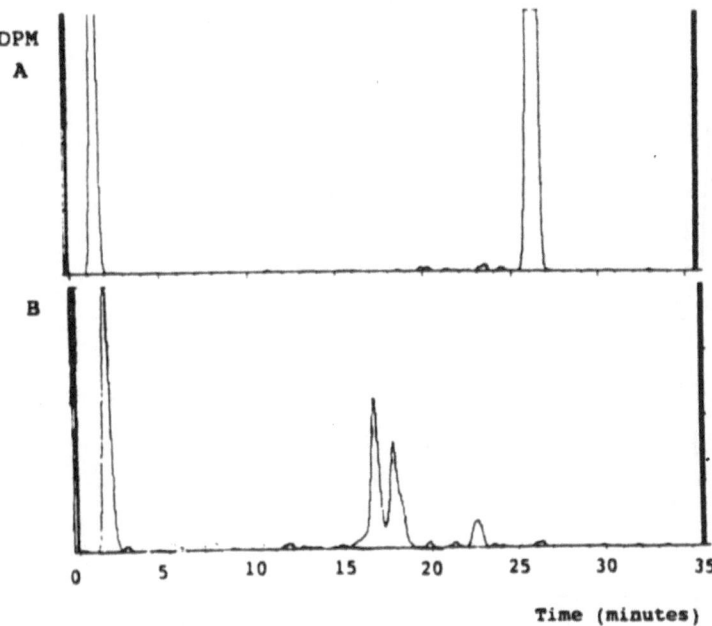

Fig. 8. HPLC chromatograms of 2-NF metabolites formed in incubations with isolated rat hepatocytes. One volume of methanol to cell suspension in order to percipitate protein. Cells were incubated with 50 μM [³H]-2-NF.
Panel A represents a sample obtained immediately after the addition of 2-NF to the incubation mixture. In this supernatant only 2-NF is present.
Panel B represents a supernatant at t = 30 minutes after addition of 2-NF. In this supernatant no 2-NF is left over and only conjugated hydroxylated 2-NF metabolites are present.

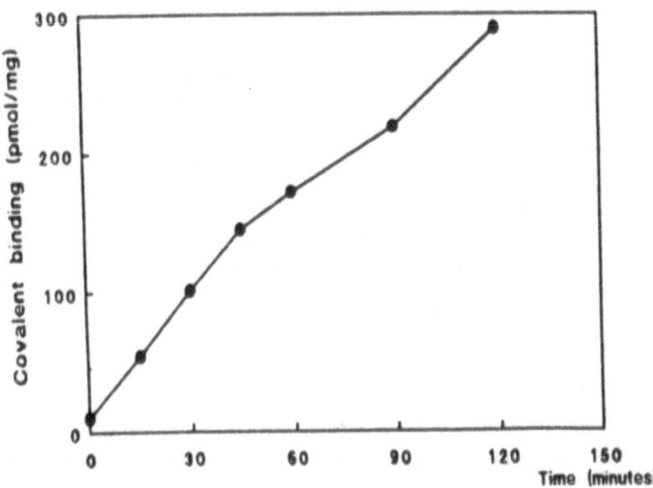

Fig. 9. Time course of the covalent binding of 2-NF to cellular macromolecules in hepatocytes. Cells were incubated with 50 μM [³H]-2-NF.

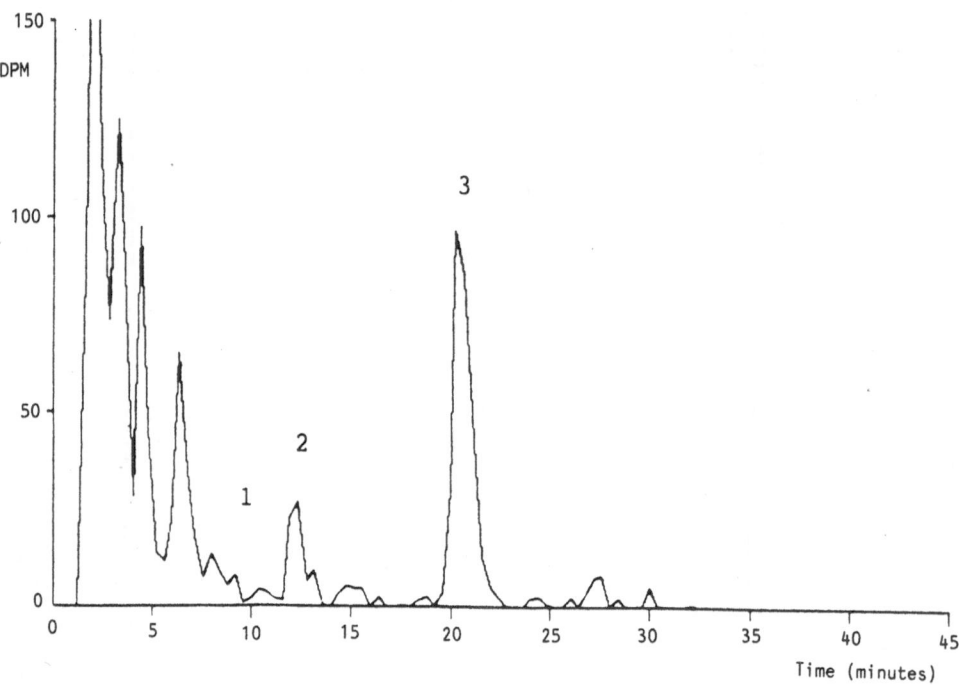

**Fig. 10.** Chromatogram of DNA adducts (radioactivity) eluting from a HPLC column after injection of a sample of TFA hydrolysed DNA that was isolated from the liver of a rat treated orally with [$^3$H]-2-NF (60 μmol/kg). Peak 1 indicates the position of dG-N$^2$-AAF, peak 2 represents dG-C8-AAF and peak 3 represents dG-C8-AF.

could be identified as the known adducts N-(deoxyguanosin-8-yl)-2-aminofluorene (dG-C8-AF) (15 %), N-(deoxyguanosin-8-yl)-2-ace-tylaminofluorene (dG-C8-AAF) (4 %) and 3-(deoxyguanosin-N$_2$-yl)-2-acetylaminofluorene (dG-N$_2$-AAF) (2 %). These adducts are most likely formed from the 2-NF metabolites N-hydroxy-2-aminofluorene (N-OH-AF) and N-hydroxy-2-acetylamino-fluorene (N-OH-AAF). The other 80 % eluted before these known adducts. Upon hydrolysis of hepatic DNA of rats that had received [ring $^3$H]-N-OH-2-AAF, more than 63 % of the DNA adducts could be accounted for as dG-C8-AF, dG-C8-AAF and dG-N$^2$-AAF [20]. These data indicate that other types of adducts are formed from 2-NF in vivo then from N-OH-2-AAF. Thus the formation of covalently bound DNA adducts from 2-NF is not mainly mediated through the hydroxamic acid N-OH-2-AAF. Administration of the hydroxylamine, N-OH-AF, to rats leads to the formation of dG-C8-AF adduct[21]. Whether other types of adducts (and how much) are formed from N-OH-AF, which might explain the amount of early eluting adducts found after 2-NF administration, is not known because of a different method of determining DNA-adducts used in this study. Presently we are investigating the nature of these unknown DNA adducts formed from 2-NF in vivo.

## Conclusions

2-NF can be converted to (a) reactive intermediate(s) that bind(s) covalently to protein and DNA, both in vitro and in the rat in vivo. GSH and other thiols do not protect against this covalent binding. Covalent binding of 2-NF can not be inhibited by PCP; therefore, sulfation does not play a role in this covalent binding. In vitro covalent binding is NADPH dependent and can be prevented by SKF 525A;

in freshly isolated hepatocytes the rate of covalent binding can be increased by pretreatment of the animals with Aroclor 1254. In vivo Aroclor 1254 pretreatment does not give rise to changes in the percentage of the dose that is bound covalently to macromolecules. Therefore cytochrome $P_{450}$ seems to play a role in the bioactivation of 2-NF to reactive intermediates. Analysis of the DNA adducts formed in the rat in vivo after oral administration of [$^3$H]-2-NF gave rise to the formation of DNA adducts of which only 20 % could be identified as adducts derived from N-OH-AF and N-OH-AAF.

## ACKNOWLEDGEMENTS

We wish to thank Professor Dr. J. van der Greef, Dr. W.M.A. Niessen and Mr. E.R. Verheij (Center for Bio-Pharmaceutical Sciences, Division of Analytical Chemistry, University of Leiden) for their liquid chromatographic-mass spectrometric analysis of our samples. This work was supported by grant IKW 87-1 of the Dutch Cancer Foundation (Koningin Wilhemina Fonds).

## REFERENCES

1.  Beije, B. and Möller, L. (1988) 2-Nitrofluorene and related compounds, prevalence and biological effects. Mutat. Res. 196, 177-209.
2.  McCoy, E.C, Rosenkranz, E.J, Rosenkranz, H.S. and Mermelstein, R. (1981) Nitrated fluorene derivatives are potent frameshift mutagens. Mutat. Res. 90, 11-20.
3.  Arey, J, Zielinska, B, Harger, W.P, Atkinson, R. and Winer, A.M. (1988) The contribution of nitrofluoranthenes and nitropyrenes to the mutagenincity of ambient particulate organic matter collected in Southern California. Mutat. Res. 207, 45-51.
4.  Miller, J.A, Sandin, R.B, Miller, E.C. and Rusch, H.P. (1955) The carcinogenicity of compounds related to 2-acetylaminofluorene. II. Variations in the bridges and 2-substituent. Cancer Res. 15, 188-199.
5.  Möller, L. and Gustafsson, J-Å. (1986) Liquid Chromatographic-mass spectrometric analysis of 2-nitrofluorene and its derivatives. Biomed. Mass Spectrom. 13, 681-688.
6.  Möller, L, Rafter, J. and Gustafsson J-Å. (1987) Metabolism of the carcinogenic air pollutant 2-nitrofluorene in the rat. Carcinogenesis, 8, 637-645.
7.  Möller, L, Törnquist, S, Beije, B, Rafter, J, Toftgärd, R. and Gustafsson J-Å. (1987) Metabolism of the carcinogenic air pollutant 2-nitrofluorene in the isolated perfused rat lung and liver. Carcinogenesis, 8, 1847-1852.
8.  Möller, L, Corrie, M, Midtvedt, T, Rafter, J. and Gustafsson J-Å. (1988) The role of the intestinal microflora in the formation of mutagenic metabolites from the carcinogenic air pollutant 2-nitrofluorene. Carcinogenesis, 9, 823-830.
9.  Möller, L, Torndal, U.-B, Eriksson, L.C. and Gustafsson, J-Å. (1989) The air pollutant 2-nitrofluorene as initiator and promotor in a liver model for chemical carcinogenesis. Carcinogenesis, 10, 435-440.
10. Breeman, G.E.M, Kaspersen, F.M. and Westra, G.J. (1978) Non specific tritiation of some carcinogenic aromatic amines. J. Labeled Compounds, 14, 741-750.
11. Mulder, G.J, Scholtens, E. and Meijer, D.K.F. (1981) Collection of urine and bile. In Jakoby, W.B. (ed) Methods in enzymology 77, 21-30, Academic Press, New York.

12. Mulder, G.J, Unruh, L.E, Evans, F.E, Ketterer, B. and Kadlubar, F.F. (1982) Formation and identification of glutathione conjugates from 2-nitrosofluorene and N-hydroxy-2-aminofluorene. Chem. Biol. Int. 39, 111-127.

13. Mulder, G.J, Kadlubar, F.F, Mays, J.B. and Hinson, J.A. (1984) Reaction of mutagenic phenacetin metabolites with glutathione and DNA. Possible implications for toxicity. Molec. Pharmacol. 26, 342-347.

14. Meerman, J.H.N, Beland, F.A. and Mulder, G.J. (1981) Role of sulfation in the formation of DNA adducts from N-hydroxy-2-acetylaminofluorene in rat liver in vivo. Inhibition of N-acetylated aminofluorene adduct formation by pentachlorophenol. Carcinogenesis, 2, 413-416.

15. Meerman, J.H.N. and Mulder, G.J. (1981) Prevention of hepatotxic action of N-hydroxy-2-acetylaminofluorene in the rat by inhibition of N-O-sulfation by pentachlorophenol. Life Sci. 28, 2361-2365.

16. Mulder, G.J, Hinson, J.A. and Gilette, J.R. (1978) Conversion of the N-O-glucuronide and N-O-sulfate of N-hydroxy-phenacetin to reactive intermediates. Biochem. Pharmacol. 27, 1641-1649.

17. Meerman, J.H.N. and Tijdens, R.B. (1985) Effect of glutathione depletion on the hepatotoxicity and covalent binding to rat liver macromolecules of N-hydroxy-2-acetylaminofluorene. Cancer Res. 45, 1132-1139.

18. Hunter, A.L. and Neal, R.A. (1975) Inhibition of hepatic mixed function oxidase by various thione-sulfer containing compounds. Biochem. Pharmacol. 24, 2199-2205.

19. Zemaitis, M.A. and Greene, F.E. (1976) Impairment of hepatic microsomal drug metabolism in the rat during daily disulfiram administration. Biochem. Pharmacol. 25, 1355-1360.

20. Van de Poll, M.L.M, van der Hulst, D.A.M, Tates, A.D, Mulder, G.J. and Meerman, J.H.N. (1989) The role of specific DNA adducts in the induction of micronuclei by hydroxy-2-acetyaminofluorene in rat liver in vivo. Carcinogenesis, 10, 717-722.

21. Lai, C.C, Miller, E.C, Miller, J.A. and Liem, A. (1988) The essential role of microsomal deacetylase activity, DNA-(deoxyguanosin-8-yl)-2-aminofluorene adduct formation and initiation of liver tumors by N-hydroxy-2-acetylaminofluorene in the liver of infant B6C3F1 mice. Carcinogenesis, 9, 1295-1302.

# ACTIVATION OF CARCINOGENIC N-ARYLHYDROXAMIC ACIDS BY

## PEROXIDASE/H₂O₂/HALIDE SYSTEMS:  ROUTE TO C-NITROSO AROMATICS

Danuta Malejka-Giganti, Clare L. Ritter,
and Lauri J. Sammartano

University of Minnesota and
Veterans Administration Medical Center
Minneapolis, MN 55417

ABSTRACT

Evidence that locally carcinogenic N-arylhydroxamic acids may be activated by oxidants generated by leukocytes has been obtained from model systems including tissue peroxidases, myeloperoxidase (MPO)/$H_2O_2$/ halide ($X^-$) or hypohalous acids (HOX).  Peroxidative oxidations of N-hydroxy-N-2-fluorenylacetamide (N-OH-2-FAA) $\underline{via}$ one electron ($1e^-$) to equimolar 2-nitrosofluorene (2-NOF) and N-acetoxy-2-FAA, and $\underline{via}$ oxidative cleavage to 2-NOF were determined.  The latter oxidation predominated with peroxidase-rich extracts of rat uterus and mammary gland and with eosinophils from intraperitoneal fluid in the presence of $H_2O_2$, $Br^-$ and cationic detergent.  Contribution to 2-NOF by $1e^-$ oxidation was up to 50% at 0.1 mM $Br^-$ and pH 7.4, but negligible at 10 mM $Br^-$ and pH 5.5.  Oxidation of N-OH-2-FAA by MPO of human neutrophils in the presence of physiologic concentrations of $Cl^-$ (0.1 M) or $Br^-$ (0.1 mM) or their mixture was examined in the pH range of 4 to 6.5.  At the respective pH optima (4 for $Br^-$ and 5 for $Cl^-$ or $Cl^- + Br^-$), oxidation of N-OH-2-FAA to 2-NOF by MPO/ $H_2O_2$ was much more rapid with $Br^-$ and $Br^- + Cl^-$ than with $Cl^-$.  Since HOBr oxidized N-OH-2-FAA to 2-NOF much more rapidly than did HOCl, MPO/ $H_2O_2$-catalyzed oxidation in the presence of $Cl^- + Br^-$ was possibly due in part to HOBr.  $1e^-$ oxidation of N-OH-2-FAA occurred to a lesser extent in chemical (HOX) than enzymatic systems (MPO/$H_2O_2$/$X^-$), in which it appeared to be stimulated by $Br^-$ at pH $\geq$ 5.5.  In the presence of taurine, a scavenger of hypohalous acids $\underline{in\ vivo}$, oxidation of N-OH-2-FAA to 2-NOF by MPO/$H_2O_2$ was unaffected with $Br^-$, inhibited with $Cl^-$, and partially inhibited with $Cl^- + Br^-$.  These results were linked to N-halotaurine formation since it was found that N-bromotaurine, but not N-chlorotaurine, oxidized N-OH-2-FAA chiefly to 2-NOF.  The possibility that 2-NOF through its interactions with unsaturated lipids may initiate lipid peroxidation was investigated by measurements of malondialdehyde (MDA), a lipid degradation product, following aerobic incubations of 2-NOF with arachidonic, linolenic or linoleic acid at a molar ratio of 1:1.  The amounts of MDA were optimal after 4 hr and were increased by 50% when 2-NOF was preincubated with the fatty acids for 24 hr under anaerobic conditions.  Our studies indicate potential significance of:  1) $Br^-$-derived oxidants in activation of carcinogenic N-arylhydroxamic acids to C-nitroso aromatics, and 2) interactions of C-nitroso aromatics with unsaturated lipids leading to lipid peroxidation and thus, genotoxicity.

*Nitroarenes,* Edited by P. C. Howard *et al.*
Plenum Press, New York, 1990

Metabolic or chemical N-hydroxylation of N-arylamides yielded N-aryl-hydroxamic acids which showed greater carcinogenic potencies than the parent amides [1-4]. Local carcinogenicity was also acquired through N-hydroxylation [Rev. in 5]. Thus, N-hydroxy-N-2-fluorenylacetamide (N-OH-2-FAA[*]) and related hydroxamic acids induced mammary gland tumors (adenocarcinomas, sarcomas, fibroadenomas) upon application to the rat mammary gland and sarcomas upon ip, im or sc injections. The tumorigenicity and histology data implied that N-arylhydroxamic acids were activated in their local targets. These sites might be infiltrated by leukocytes in response to various chemical or physical stimuli. PMNL stimulated to undergo a respiratory burst were shown to mediate metabolic activation of N-arylamines [6,7] and polycyclic aromatic hydrocarbons [8] resulting in their covalent binding to DNA. These activations were linked to MPO/$H_2O_2$-catalyzed oxidations. Leukocytic peroxidases: MPO (neutrophils and monocytes) and EPO (eosinophils) use $H_2O_2$ generated in the respiratory burst of leukocytes to oxidize halides to hypohalous acids, which are powerful oxidants [9,10] (Fig. 1). Both MPO and EPO oxidize $Cl^-$, $Br^-$ and $I^-$. $Cl^-$ is reportedly more efficiently oxidized by MPO than by EPO, which preferentially oxidizes $Br^-$ [9-15]. We therefore investigated oxidations of carcinogenic N-2-fluorenylhydroxamic acids by enzymatically or chemically generated oxidants of $Cl^-$ and/or $Br^-$.

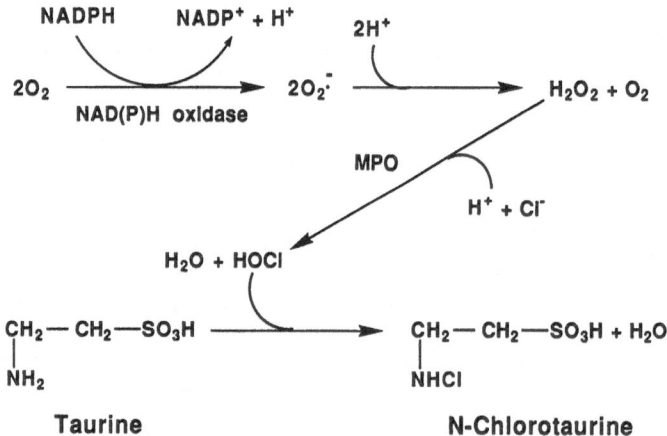

Fig. 1.    Generation of $X^-$ oxidants by leukocytic peroxidase.

---

[*]Abbreviations used: N-OH- or N-AcO-2-FAA, N-hydroxy- or N-acetoxy-N-2-fluorenylacetamide; N-OH-2-FBA, N-hydroxy-N-2-fluorenylbenzamide; 2-NOF, 2-nitrosofluorene; NOB, nitrosobenzene; MPO, myeloperoxidase; EPO, eosinophil peroxidase; LPO, lactoperoxidase; GU, guaiacol unit; $X^-$, halide ($Cl^-$ or $Br^-$); HOX, hypohalous acid (HOCl or HOBr); PMNL, polymorphonuclear leukocytes; le[-], one electron; ip, intraperitoneal; im, intramuscular; sc, subcutaneous; Cetab or Cetac, cetyltrimethylammonium $Br^-$ or $Cl^-$; MDA, malondialdehyde; PUFA, polyunsaturated fatty acids.

Peroxidative oxidations of N-arylhydroxamic acids by haloperoxidase/ H₂O₂/X⁻ or HOX systems: route to C-nitroso aromatics. Peroxidative oxidation of N-OH-2-FAA and related compounds by bovine milk LPO and human neutrophil MPO in the absence of halides was first shown by Bartsch et al. [16]. The reaction involves 1e⁻ oxidation of the hydroxamic acid such as N-OH-2-FAA by peroxidase/H₂O₂ (compound I) to nitroxyl free radical (N-Ȯ-2-FAA) which dismutates to equimolar N-AcO-2-FAA and 2-NOF [17] (Fig. 2). 2-NOF is a common product for all N-2-substituted fluorenyl hydroxamic acids, whereas the acyl groups of the ester are those of the hydroxamic acid used as the hydrogen donor. This reaction sequence was confirmed with a variety of enzymic and non-enzymic oxidants [18-23]. Subsequently, we discovered that LPO or rat uterine peroxidase (presumably EPO) in the presence of Br⁻ and H₂O₂ oxidized N-OH-2-FAA to nearly stoichiometric amounts of 2-NOF [24] (Fig. 2). The Br⁻-dependency of this reaction, its simulation with chemically generated HOBr, and loss of oxidizing potential in the presence of OBr⁻-trapping reagents or H₂O₂ led us to postulate that the enzyme-catalyzed reaction yields HOBr which oxidatively cleaves N-OH-2-FAA to 2-NOF, possibly via reactivity of Br⁺ with oxygen of the N-OH group [25]. Thus, peroxidative metabolism of N-OH-2-FAA yielded N-AcO-2-FAA from 1e⁻ oxidation and 2-NOF from both 1e⁻ oxidation and oxidative cleavage. Both 2-NOF and the ester are considered active metabolites of N-OH-2-FAA in mutagenesis and carcinogenesis Rev. in [5]. 2-NOF was a particularly potent frameshift mutagen for S. typhimurium and a direct acting teratogen in the rat embryo culture system [26].

In our earlier studies, we showed that Br⁻-dependent peroxidative oxidation of N-OH-2-FAA to 2-NOF was enhanced in the presence of a cationic detergent (Cetab or Cetac) [24,27]. Although the mechanism of the detergent action is not understood at present, it is possible that cationic protein(s) in PMNL similarly stimulate peroxidative metabolism in vivo. We used Cetab to extract peroxidase from rat uterus, mammary gland and eosinophil-rich cell fractions from ip fluid [5,24,27]. In the incubation media, we retained Cetab concentrations of 0.004% or 0.4% to simulate either physiologic (0.1 mM) or elevated (10 mM) Br⁻ levels, respectively. At the lower Br⁻ concentration, uterine peroxidase/H₂O₂-catalyzed 1e⁻ oxidation of N-OH-2-FAA, as indicated by N-AcO-2-FAA formation, contributed ~40% 2-NOF at pH 7.4 and only ~10% 2-NOF at pH 5.5 (Table 1). At the higher Br⁻ concentration oxidation of N-OH-2-FAA was chiefly via cleavage to 2-NOF [27]. Likewise, peroxidative metabolism of N-OH-2-FAA or N-OH-2-FBA by eosinophil-rich cells from ip fluid in the presence of Cetab (0.4%) and H₂O₂ (0.1 mM) yielded large amounts of 2-NOF and only small (or trace) amounts of the respective ester [5]. By contrast, no N-AcO-2-FAA was formed by extracts of mammary gland peroxidase/ H₂O₂ which yielded less 2-NOF per unit of peroxidative activity (measured with guaiacol) than uterine extracts (Table 1). The low specific activities of mammary gland extracts necessitated using greater amounts of protein in the incubations which might have interfered with peroxidative metabolism of N-OH-2-FAA. LPO was found in mammary glands of pregnant and lactating, but not virgin rats [28]. Although the peroxidase activity of virgin rat mammary gland has not yet been characterized, the presence of various hemeproteins [29], infiltration of the glands by eosinophils in response to estrogen treatment [30] and of mammary gland tumors by mast cells [31] (reportedly containing EPO [10]) have been described. Thus, it would appear that peroxidative activity of virgin mammary gland may partly be that of EPO.

Induction of local sarcomas following ip, sc or im injections of N-OH-2-FAA or related hydroxamic acids (e.g., N-OH-2-FBA) as suspensions or solutions suggested that the deposits of these compounds as well as the mechanical injury to the tissue might be attracting PMNL (primarily

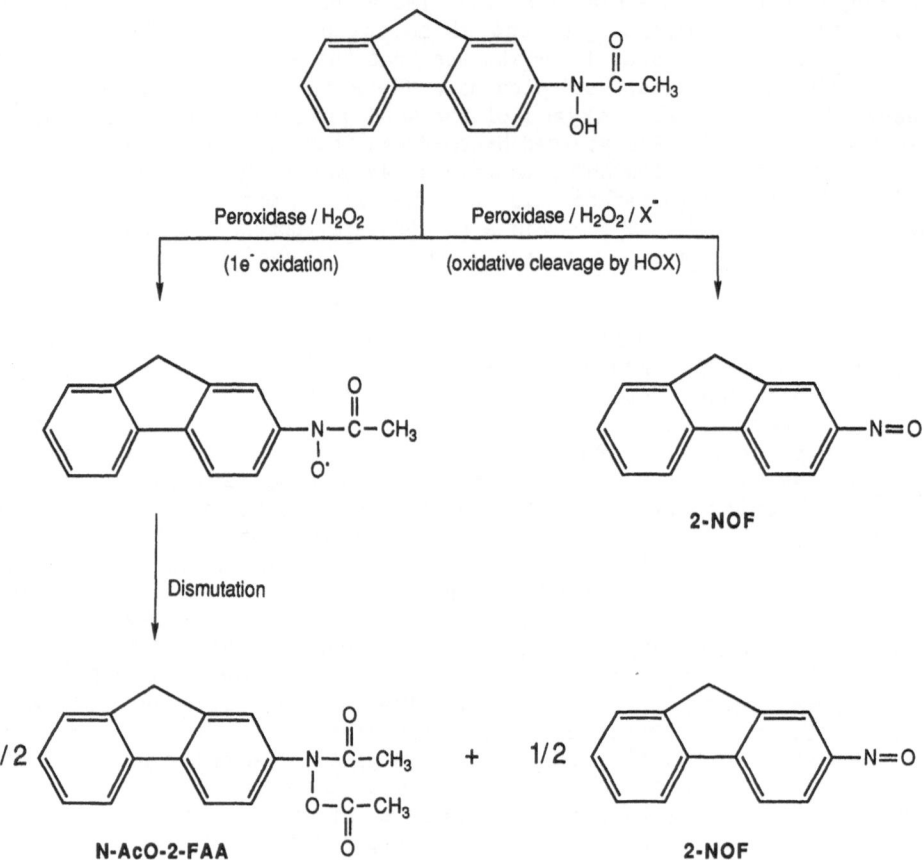

Fig. 2. Peroxidative oxidations of N-OH-2-FAA.

Table 1. Oxidations of N-OH-2-FAA by Peroxidase/$H_2O_2$ ± X⁻ or HOX ± X⁻

| Oxidant[a] | GU[b] $10^3$ | pH | Initial rate, 2-NOF $10^3$(mM) min⁻¹ | Amounts, $10^3$(mM) 2-NOF | N-AcO-2-FAA | Termination time (min) |
|---|---|---|---|---|---|---|
| UT/$H_2O_2$/Br⁻[c] | 8.0 | 5.5 | N.D.[d] | 12.1 | 1.32 | 8 |
| | | 7.4 | N.D. | 15.7 | 5.90 | 8 |
| MG/$H_2O_2$/Br⁻[c] | 8.0 | 5.5 | N.D. | 8.80 | 0 | 8 |
| | | 7.4 | N.D. | 2.59 | 0 | 8 |
| MPO/$H_2O_2$/Br⁻ | 3.6 | 4 | 2.11 | 2.66 | 0.17 | 3 |
| MPO/$H_2O_2$/Br⁻ | 7.2 | 4 | N.D. | 5.66 | 0.67 | 3 |
| MPO/$H_2O_2$/Cl⁻ | 3.6 | 5 | 0.90 | 4.00 | 0.17 | 3 |
| MPO/$H_2O_2$/Cl⁻ | 7.2 | 5 | N.D. | 6.55 | 0.27 | 3 |
| HOBr/Br⁻ | - | 5 | -[e] | 37.0 | 0.73 | 20 |
| HOCl | - | 5 | 3.40 | 23.2 | N.D. | 20 |
| HOCl/Cl⁻ | - | 5 | 84.0 | 30.6 | 0.80 | 20 |
| N-BrT[f]/Br⁻ | - | 5 | 5.60 | 27.4 | 0 | 20 |
| N-ClT[f]/Cl⁻ | - | 5 | 0 | 0 | 0 | 20 |

[a]HOX (0.04 mM), Br⁻ (0.1 mM) or Cl⁻ (0.1 M), $H_2O_2$ (0.0025 to 0.1 mM) were used where noted.
[b]Guaiacol unit.
[c]Uterine (UT) or mammary gland (MG) peroxidase. Br⁻ added as 0.004% Cetab.
[d]Not determined.
[e]Too fast to determine.
[f]N-bromo- or N-chlorotaurine (0.04 mM).

neutrophils) to the injection site and leading to their respiratory burst with release of MPO. Since MPO/$H_2O_2$ preferentially oxidizes Cl⁻ [9,12,14] and in vivo Cl⁻ is present at 1000 times greater concentration than Br⁻ [32], it was of interest to examine metabolism of N-OH-2-FAA in model systems containing MPO/$H_2O_2$/Cl⁻ and/or Br⁻ at the physiologic levels of halides. These oxidations were determined in the pH range of respiratory burst (4 to 6.5 [9,33]). The initial rates of 2-NOF formation were optimal at pH 4 in the presence of 0.1 mM Br⁻ and at pH 5 in the presence of 0.1 M Cl⁻ or the X⁻ mixture [25] (Table 1). The formation of 2-NOF was essentially completed by 20 min and the greatest amounts of 2-NOF in the presence of Br⁻, Cl⁻ or the X⁻ mixture were determined at the respective pH optima (Fig. 3A and 4). When Br⁻ was increased to 1 mM, all of the $H_2O_2$ was consumed at pH 4 and stoichiometric amounts of 2-NOF were formed [25]. The contribution of 1e⁻ oxidation was determined from the amounts of N-AcO-2-FAA (Fig. 3B). With relatively high $H_2O_2$ (0.05 mM): peroxidase (0.0036 GU) ratios, very small amounts of N-AcO-2-FAA ranging from ~0.2 µM at pH 4 to 0.4 µM at pH 6 were measured in the absence of halides, and were approximately equimolar to 2-NOF. In the presence of Cl⁻ or Br⁻ the contribution from 1e⁻ oxidation to 2-NOF ranged from 3 to 8% or 1.4 to 30%, respectively, at pH 4 to 6. The apparent stimulation of 1e⁻ oxidation at higher pH, especially by Br⁻, is not understood at present.

In addition to the two products of oxidation shown in Fig. 3, small amounts of 2-FAA (<1% of N-OH-2-FAA) and occasional 2-NO$_2$F were present in the extracts of the incubations. The amide was probably the reduction product of the hydroxamic acid itself [22] and/or N-AcO-2-FAA [34]. 2-NO$_2$F was most likely formed from 2-NOF [27]. Since the unreacted N-OH-2-FAA was quantitatively recovered, we concluded that no other products were formed during MPO/H$_2$O$_2$/X$^-$-catalyzed oxidations. The latter most likely involved hypohalous acids. Thus, oxidations of N-OH-2-FAA by chemically generated HOCl or HOBr in amounts (~0.04 mM) approximating the maximum amounts generated by the enzymatic systems were examined. While formation of 2-NOF was almost instantaneous upon addition of HOBr at pH 4 to 6.5, it was considerably slower upon addition of HOCl (shown at pH 5 in Table 1) and decreased with increasing pH. Addition of 0.1 M Cl$^-$ to the buffer increased the rate of oxidation of N-OH-2-FAA by HOCl considerably, although not to rates from HOBr. Under the experimental conditions, oxidation of N-OH-2-FAA to 2-NOF ceased by 20 min. At this time, the extent of oxidation by HOBr in the presence of Br$^-$ was independent of pH, was nearly stoichiometric (i.e. ~1 mole of 2-NOF was formed per mol of HOBr) (Fig. 3A), and was unaffected by Cl$^-$. The extent of oxidation of N-OH-2-FAA by HOCl at pH 4 to 6 was less than that by HOBr, but it increased at pH 6.5 to yield amounts of 2-NOF similar to those formed by

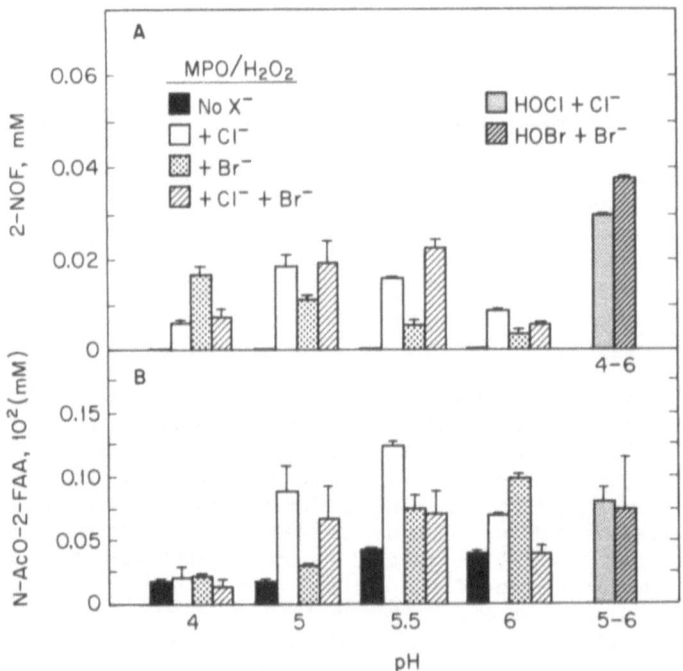

Fig. 3.  Effect of halides (0.1 M Cl$^-$ and/or 0.1 mM Br$^-$) on yields of 2-NOF (A) and N-AcO-2-FAA (B) formed from N-OH-2-FAA by MPO/ H$_2$O$_2$ or HOX. MPO (0.0036 GU) was added to cuvettes containing sodium phosphate/citric acid buffer ± X$^-$ and 0.08 mM N-OH-2-FAA. Reaction was started by the addition of H$_2$O$_2$ (0.05 mM). After 20 min of incubation at room temperature, 2-NOF was determined spectrophotometrically in the reaction mixtures and also in the extracts of the incubations by HPLC [25]. Both methods gave closely corresponding amounts. N-AcO-2-FAA was determined by HPLC. Values are the means of 2 determinations ± SE.

HOBr [25]. The presence of 0.1 M Cl⁻ in the buffers significantly increased the extent of oxidation of N-OH-2-FAA by HOCl (Fig. 3A). HOCl added
to the buffers containing both Cl⁻ and Br⁻ yielded 2-NOF in the amounts
similar to those produced by HOBr indicating oxidation by HOCl of Br⁻ to
HOBr. The amounts of N-AcO-2-FAA determined in these experiments (Fig.
3B) indicated that 1e⁻ oxidation occurred to a lesser extent in chemical
than enzymatic systems.

Hypohalous acids are most likely short-lived *in vivo* but may form
more stable oxidants such as N-haloamines. Since taurine is an abundant
free amino acid [35,36], its halogenation by HOCl or HOBr to N-chloro- or
N-bromotaurine may provide long-lived oxidants (Fig. 1). Hence, we
investigated oxidation of N-OH-2-FAA by these N-haloamines (Table 1).
N-Bromotaurine (~0.04 mM) oxidized N-OH-2-FAA to 2-NOF at a rate slower

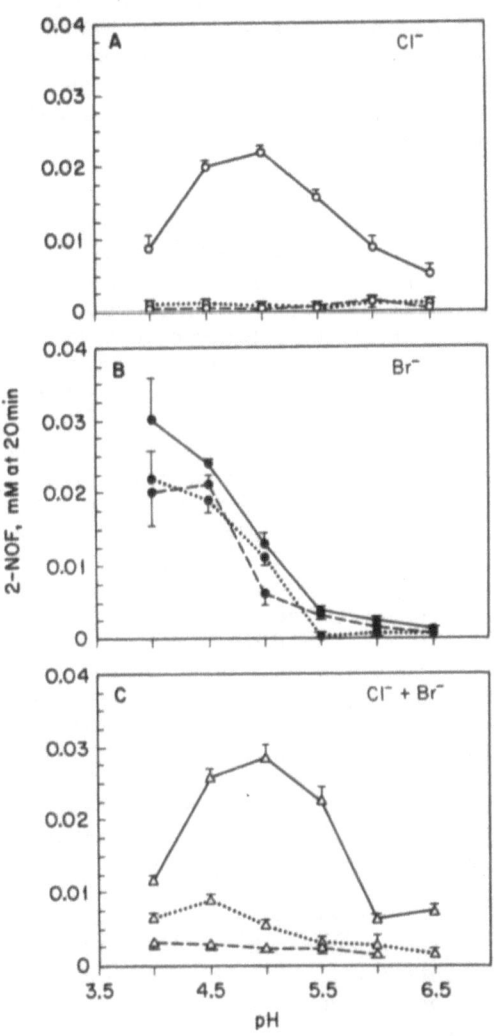

Fig. 4. Effect of 1 mM (····) or 10 mM (----) taurine on yields of
2-NOF from oxidation of N-OH-2-FAA by MPO/$H_2O_2$ (———) in the
presence of 0.1 M Cl⁻ (A), 0.1 mM Br⁻ (B) and 0.1 M Cl⁻ + 0.1
mM Br⁻ (C). Incubations and assays of 2-NOF are described in
the legend of Fig. 3.

than HOBr but faster than HOCl. By contrast, N-chlorotaurine alone or in the presence of 0.1 M Cl⁻ did not oxidize N-OH-2-FAA. Since N-chlorotaurine might oxidize N-OH-2-FAA slowly, the incubations were monitored for up to 20 hr for the presence of 2-NOF [25]. At no time was any 2-NOF detectable in the incubations containing N-chlorotaurine + Cl⁻ . Whereas the presence of Cl⁻ had no effect on the amount of 2-NOF generated from N-OH-2-FAA by N-bromotaurine + Br⁻, addition of Br⁻ to incubations with N-chlorotaurine + Cl⁻ resulted in 2-NOF formation which decreased with increasing pH. This was most likely due to X⁻ exchange with formation of N-bromotaurine [25].

Our results with chemically generated oxidants of Cl⁻ and Br⁻ indicated that taurine and possibly other HOX scavengers in vivo might affect oxidations by MPO/H₂O₂/X⁻-derived oxidants. We thus determined the effect of 1 or 10 mM taurine on the oxidation of N-OH-2-FAA by MPO/H₂O₂ in the presence of physiologic Cl⁻, Br⁻ or their mixture. Taurine inhibited the oxidation of N-OH-2-FAA by MPO/H₂O₂/Cl⁻ (Fig. 4A), had essentially no effect on the oxidation by MPO/H₂O₂/Br⁻ (Fig. 4B) and partially inhibited the oxidation in the presence of the X⁻ mixture (Fig. 4C). Thus, the data were in agreement with our findings from chemical systems, in that the inhibition of the oxidation was most likely due to N-chlorotaurine formation, and lack of inhibition or partial inhibition was due to oxidation by N-bromotaurine.

Potential significance of Br⁻-derived oxidants in vivo. At physiologic concentrations of halides and in the presence of taurine (up to 10 mM), which reportedly protects against cellular damage due to MPO-generated oxidants of Cl⁻ [37], formation of N-bromotaurine (as shown in our studies) may actually retain or enhance oxidative damage. This may occur when Br⁻ concentrations become elevated as a result of environmental exposure, diet or medication. In an excellent recent review on toxicology of Br⁻, the authors [38] described the sources of human exposure resulting in increased Br⁻ levels. The major source of exposure is the presence of Br⁻ residues in food due to the use of bromine-containing fumigants for treatments of soil and harvest. Ethylene dibromide has been largely replaced by methyl bromide for the above purposes, but the dibromide is used as a lead scavenger in gasoline and thus, is present in automobile exhaust. Certain sedatives and hypnotics available as non-prescription drugs in Europe contain bromine. Br⁻ toxicity is strongly dependent on Cl⁻ intake and may be manifested in individuals with reduced salt (NaCl) intake since the half-life of Br⁻ would then be prolonged. Under such circumstances, Br⁻-derived oxidants may participate in metabolism and activation of xenobiotics, including carcinogens. Our data showed that peroxidative oxidations of N-2-fluorenylhydroxamic acids by peroxidase/H₂O₂/X⁻-generated oxidants, particularly those formed at higher Br⁻ concentrations, would yield relatively large amounts of 2-NOF chiefly from oxidative cleavage and small amounts of the respective esters from 1e⁻ oxidation [5,25,27].

Interaction of C-nitroso aromatics with PUFA: route to lipid peroxidation. 2-NOF and other C-nitroso aromatics lack electrophilicity and hence are unreactive with DNA. Electrophilic species may be derived from enzymatic and/or non-enzymatic reduction of the nitroso to the hydroxylamine derivative [39-41], which yields arylamine-type adduct with DNA [42,43]. Alternatively, damage to DNA may be effected by products resulting from reactivity of C-nitroso compounds with other macromolecules such as unsaturated lipids. Methyl oleate, cholesterol and other lipids interacted with 2-NOF or NOB according to a pseudo Diels-Alder mechanism [44-47]. The resulting hydroxylamine adduct might be oxidized by another molecule of the nitroso compound (and/or oxygen) to nitroxyl free radical. This mechanism was originally proposed by Sullivan [48] for the

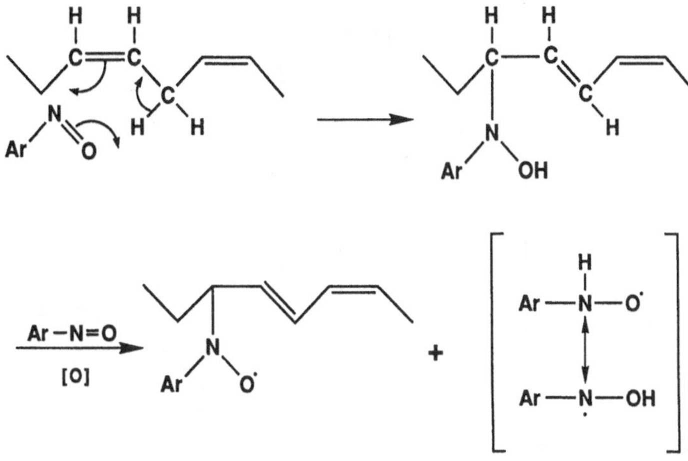

Fig. 5. Proposed scheme for interaction of C-nitroso aromatics with PUFA (based on ref. [45,48]).


interaction of NOB with the double bond of 2,3-dimethyl-2-butene. We adapted this mechanism for interaction of C-nitroso aromatics with PUFA (Fig. 5) and propose that the oxidation of the hydroxylamine derivative by the nitroso compound would also yield the radicals of the latter (shown in brackets). These radicals would be preferentially formed under anaerobic conditions and be capable of initiating lipid peroxidation. Thus, we examined whether the formation of a lipid peroxidation product MDA [49] was increased following 24 hr anaerobic preincubation of 2-NOF or NOB with PUFA at a molar ratio of 1:1. We found that anaerobic preincubation of 2-NOF with linoleic or linolenic acid significantly increased the amounts of MDA formed between 2 and 6 hr under aerobic conditions with the maximum effect after 4 hr (Fig. 6). Similar increases in MDA

Table 2. Generation of MDA after interaction of
C-nitroso aromatics with arachidonic acid

| Time of incubation (hr) | | MDA (nmol/mg PUFA)[a] | |
|---|---|---|---|
| Aerobic | Anaerobic | 2-NOF | NOB |
| 2 | 0 | 1.50 | 2.73 |
| | 24 | 4.64[b] | 4.87[b] |
| 3 | 0 | 2.63 | 5.13 |
| | 24 | 5.24[c] | 8.71[c] |
| 4 | 0 | 3.38 | 5.61 |
| | 24 | 6.46[d] | 9.94[d] |
| 6 | 0 | 4.91 | 5.34 |
| | 24 | 6.68 | 9.42[c] |

[a]Incubations and assays of MDA are described in the legend of Fig. 6.
The effect of anaerobic incubation was significant at:
[b]$P<0.050$; [c]$P<0.005$; [d]$P<0.025$.

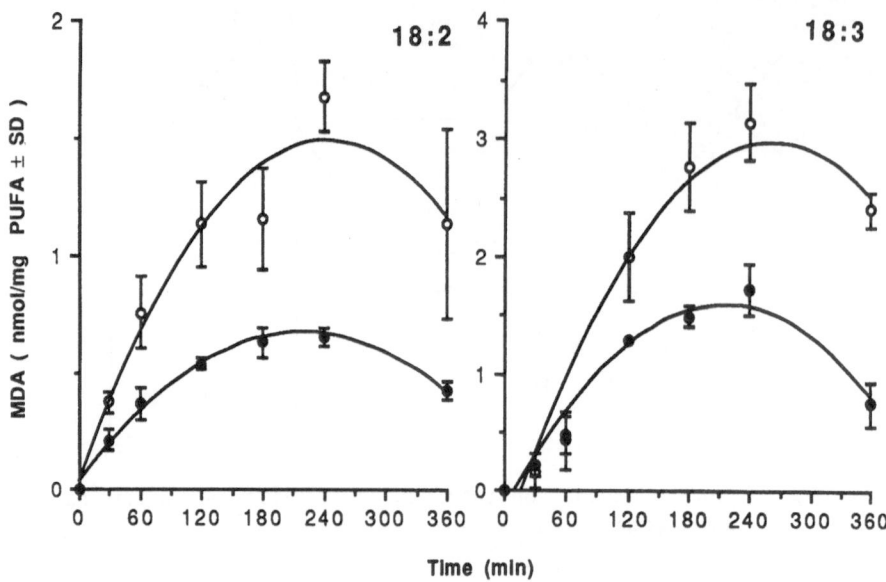

Fig. 6. Generation of MDA after interaction of 2-NOF with PUFA:linoleic
(18:2) or linolenic (18:3) acid. Following 24 hr anaerobic
preincubation of PUFA or PUFA:2-NOF (1:1) in $CHCl_3$, the solvent
was evaporated under $N_2$ and the residues were treated with
Lubrol PX (0.125%) in 50 mM Tris-maleate buffer, pH 7.4, for
0.5 hr under anaerobic conditions. 2-NOF in methanol or
methanol alone was added to the incubations containing PUFA (•)
or PUFA:2-NOF (o), respectively. MDA assays [49] were carried
out under aerobic conditions at the indicated time intervals.
The values are the means ± SD from 3 determinations and are
corrected for the amounts of MDA present in PUFA alone.

formation were also determined following 24 hr anaerobic preincubation of
2-NOF or NOB with arachidonic acid (Table 2).

Interactions of C-nitroso aromatics with unsaturated lipids may be
of far reaching consequences in vivo. Firstly, "adduction" of C-nitroso
aryl compounds to double bonds of lipids [46] or "spin-trapping" of
nitroxyl free radicals by membrane components [50] may result in physico-
chemical changes in membranes affecting their permeability and conceiv-
ably transport of carcinogens or metabolites thereof. Secondly,
induction or enhancement of lipid peroxidation by C-nitroso aromatics, as
shown herein, may lead to accumulation of cytotoxic lipid degradation
products such as MDA and unsaturated aldehydes. These compounds have
been reported to modify nucleic acid bases in systems in vitro including
cell cultures [51] and intact rat liver nuclei [52]. Several novel adduct
structures resulting from interactions of the aldehydes with guanine
nucleosides have been identified [53-56]. Moreover, the unsaturated
aldehydes rapidly reacted with -SH groups which might impair cellular
defense mechanisms such as DNA repair via inhibition of DNA repair enzyme
$O^6$-methylguanine-DNA methyltransferase [57].

C-Nitroso aromatics reacted readily with -SH groups of proteins and
GSH [39,41,58-61]. The interactions with GSH yielded the reduced derivatives of the nitroso compounds (hydroxylamines (electrophilic), amines
and azoxy compounds) and GSH adducts. Among the latter, labile [41,59,60]
or radical [61] intermediates have been reported. Hence, conjugations of
C-nitroso aromatics with GSH may result in their activation. Because

multiple metabolic routes lead directly or indirectly to C-nitroso aromatics, elucidation of the mechanism of action of these highly reactive intermediates, which are potent mutagens, carcinogens and teratogens, is of particular importance for future investigations.

ACKNOWLEGEMENTS

The authors thank Mr. Richard Decker for assistance with HPLC analyses, and Miss Cindy Oliva for typing the manuscript. These studies were supported by grants from USPHS (CA-28000) and US Department of Veterans Affairs.

REFERENCES

1. E. C. Miller, J. A. Miller, and H. A. Hartmann, N-Hydroxy-2-acetyl-aminofluorene: a metabolite of 2-acetylaminofluorene with increased carcinogenic activity in the rat, Cancer Res. 21: 815-824 (1961).
2. J. A. Miller, C. S. Wyatt, E. C. Miller, and H. A. Hartmann, The N-hydroxylation of 4-acetylaminobiphenyl by the rat and dog and the strong carcinogenicity of N-hydroxy-4-acetylaminobiphenyl in the rat, Cancer Res. 21:1465-1473 (1961).
3. H. R. Gutmann, S. B. Galitski, and W. A. Foley, The conversion of noncarcinogenic aromatic amides to carcinogenic arylhydroxamic acids by synthetic N-hydroxylation, Cancer Res. 27:1443-1455 (1967).
4. H. R. Gutmann, D. S. Leaf, Y. Yost, R. E. Rydell, and C. C. Chen, Structure-activity relationships of N-acylhydroxylamines in the rat, Cancer Res. 30:1485-1498 (1970).
5. D. Malejka-Giganti, and C. L. Ritter, Peroxidase-mediated metabolism of N-arylhydroxamic acids in relation to tumorigenesis, in: "Carcinogenic and Mutagenic Responses to Aromatic Amines and Nitroarenes", C. M. King, L. J. Romano, and D. Schuetzle, eds., Elsevier Science Publishing Co., Inc., New York, pp. 109-209 (1988).
6. Y. Tsuruta, V. V. Subrahmanyam, W. Marshall, and P. J. O'Brien, Peroxidase-mediated irreversible binding of arylamine carcinogens to DNA in intact polymorphonuclear leukocytes activated by a tumor promoter, Chem. -Biol. Interact. 53:25-35 (1985).
7. M. D. Corbett, B. R. Corbett, M. -H. Hannothiaux, and S. J. Quintana, Metabolic activation and nucleic acid binding of acetaminophen and related arylamine substrates by the respiratory burst of human granulocytes, Chem. Res. Toxicol. 2:260-266 (1989).
8. M. A. Trush, J. L. Seed, and T. W. Kensler, Oxidant-dependent meta-bolic activation of polycyclic aromatic hydrocarbons by phorbol ester-stimulated human polymorphonuclear leukocytes: Possible link between inflammation and cancer, Proc. Natl. Acad. Sci. USA 82:5194-5198 (1985).
9. S. J. Klebanoff, Oxygen metabolism and the toxic properties of phagocytes, Ann. Int. Med. 93:480-489 (1980).
10. S. J. Weiss, and A. F. LoBuglio, Biology of disease. Phagocyte-generated oxygen metabolites and cellular injury, Lab. Invest. 47:5-18 (1982).
11. E. C. Jong, W. R. Henderson, and S. J. Klebanoff, Bactericidal activity of eosinophil peroxidase. J. Immunol. 124:1378-1382 (1980).
12. A. R. J. Bakkenist, J. E. G. De Boer, H. Plat, and R. Wever, The halide complexes of myeloperoxidase and the mechanism of the halogenation reactions, Biochim. Biophys. Acta 613:337-348 (1980).

13. R. Cramer, M. R. Soranzo, and P. Patriarca, Evidence that eosino-
    phils catalyze the bromide-dependent decarboxylation of amino
    acids, Blood 58:1112-1118 (1981).
14. S. J. Weiss, S. T. Test, C. M. Eckmann, D. Roos, and S. Regiani,
    Brominating oxidants generated by human eosinophils, Science 234:
    200-203 (1986).
15. A. N. Mayeno, A. J. Curran, R. L. Roberts, and C. S. Foote,
    Eosinophils preferentially use bromide to generate halogenating
    agents, J. Biol. Chem. 264:5660-5668 (1989).
16. H. Bartsch, J. A. Miller, and E. C. Miller, N-Acetoxy-N-acetylamino-
    arenes and nitrosoarenes. One-electron non-enzymatic and
    enzymatic oxidation products of various carcinogenic aromatic
    acethydroxamic acids, Biochim. Biophys. Acta 273:40-51 (1972).
17. H. Bartsch, M. Traut, and E. Hecker, On the metabolic activation of
    N-hydroxy-N-2-acetylaminofluorene. II. Simultaneous formation of
    2-nitrosofluorene and N-acetoxy-N-2-acetylaminofluorene from
    N-hydroxy-N-2-acetylaminofluorene via a free radical inter-
    mediate, Biochim. Bophys. Acta 237:556-566 (1971).
18. H. Bartsch, and E. Hecker, On the metabolic activation of the
    carcinogen N-hydroxy-N-2-acetylaminofluorene. III. Oxidation
    with horseradish peroxidase to yield 2-nitrosofluorene and
    N-acetoxy-N-2-acetylaminofluorene. Biochim. Biophys. Acta 237:
    567-578 (1971).
19. R. A. Floyd, L. M. Soong, and P. L. Culver, Horseradish peroxidase/
    hydrogen peroxide-catalyzed oxidation of the carcinogen
    N-hydroxy-N-acetyl-2-aminofluorene as effected by cyanide and
    ascorbate, Cancer Res. 36:1510-1519 (1976).
20. R. A. Floyd, L. M. Soong, R. N. Walker, and M. Stuart, Lipid hydro-
    peroxide activation of N-hydroxy-N-acetylaminofluorene via a free
    radical route, Cancer Res. 36:2761-2767 (1976).
21. R. A. Floyd, and L. M. Soong, Obligatory free radical intermediate
    in the oxidative activation of the carcinogen N-hydroxy-2-acetyl-
    aminofluorene, Biochim. Biophys. Acta 498:244-249 (1977).
22. C. L. Ritter, D. Malejka-Giganti, and C. F. Polnaszek, Cytochrome
    $c/H_2O_2$-mediated one electron oxidation of carcinogenic N-
    fluorenylacetohydroxamic acids to nitroxyl free radicals, Chem.
    -Biol. Interact. 46:317-334 (1983).
23. M. D. Corbett, and B. R. Corbett, HRP-catalyzed bioactivation of
    carcinogenic hydroxamic acids. The greater reactivity of
    glycolyl- versus acetyl-derived hydroxamic acids, Chem. -Biol.
    Interact. 63:249-264 (1987).
24. C. L. Ritter, and D. Malejka-Giganti, A novel oxidation of the
    carcinogen N-hydroxy-N-2-fluorenylacetamide catalyzed by
    peroxidase/$H_2O_2$/$Br^-$, Biochem. Biophys. Res. Commun. 131:174-181
    (1985).
25. C. L. Ritter, and D. Malejka-Giganti, Oxidations of the carcinogen
    N-hydroxy-N-2-fluorenylacetamide by enzymatically or chemically
    generated oxidants of chloride and bromide, Chem. Res. Toxicol.
    2:325-333 (1989).
26. E. M. Faustman-Watts, J. C. Greenaway, M. J. Namkung, A. G. Fantel,
    and M. R. Juchau, Teratogenicity in vitro of two deacetylated
    metabolites of N-hydroxy-2-acetylaminofluorene, Toxicol. Appl.
    Pharmacol. 76:161-171 (1984).
27. D. Malejka-Giganti, C. L. Ritter, R. W. Decker, and J. M. Suilman,
    Peroxidative metabolism of a carcinogen, N-hydroxy-N-2-fluorenyl-
    acetamide, by rat uterus and mammary gland in vitro, Cancer Res.
    46:6200-6206 (1986).
28. J. M. Strum, Hormonal activation of mammary gland peroxidase, Tissue
    & Cell 10:505-514 (1978).
29. D. Malejka-Giganti, R. W. Decker, C. L. Ritter, and M. R. Polovina,
    Microsomal metabolism of the carcinogen, N-2-fluorenylacetamide,

by the mammary gland and liver of female rats. I. Ring- and
N-hydroxylations of N-2-fluorenylacetamide, Carcinogenesis
6:95-103 (1985).

30. R. Arriagada, C. Unda, E. Sentis, H. Kong, and A. N. Tchernitchin,
    Estrogen-induced tissue eosinophilia in the rat mammary gland,
    Med. Sci. Res. 15:1505-1506 (1987).

31. M. Kh. Dabbous, R. Walker, L. Haney, L. M. Carter, G. L. Nicolson,
    and D. E. Woolley, Mast cells and matrix degradation at sites of
    tumor invasion in rat mammary adenocarcinoma, Br. J. Cancer
    54:459-465 (1986).

32. K. Diem, and C. Lentner, eds., "Scientific Tables", Documenta Geigy,
    7th Ed., Geigy Pharmaceuticals, Ardsley, NY, pp. 563, 644, 663
    (1970).

33. M. S. Jensen, and D. F. Bainton, Temporal changes in pH within the
    phagocytic vacuole of the polymorphonuclear neutrophilic leuko-
    cyte, J. Cell. Biol. 56:379-388 (1973).

34. M. Pelecanou, and M. Novak, Oxidation-reduction reactions of
    N-sulfonoxyacetanilides: mechanisms of the halide-induced
    reduction of models for carcinogenic metabolites of aromatic
    amides. J. Am. Chem. Soc. 107:4499-4503 (1985).

35. J. A. Sturman, and K. C. Hayes, The biology of taurine in nutrition
    and development, in: "Advances in Nutritional Research", Vol. 3,
    Plenum Press, New York, pp. 231-299 (1980).

36. M. B. Grisham, M. M. Jefferson, D. F. Melton, and E. L. Thomas,
    Chlorination of endogenous amines by isolated neutrophils.
    Ammonia-dependent bactericidal, cytotoxic, and cytolytic
    activities of the chloramines, J. Biol. Chem. 259:10404-10413
    (1984).

37. C. E. Wright, T. T. Lin, Y. Y. Lin, J. A. Sturman, and G. E. Gauli,
    Taurine scavenges oxidized chlorine in biological systems, in:
    "Taurine: Biological Actions and Clinical Perspectives", Alan R.
    Liss, Inc., New York, pp. 137-146 (1985).

38. F. X. R. Van Leeuwen, and B. Sangster, The toxicology of bromide
    ion, CRC Critical Reviews in Toxicology 18:189-213 (1987).

39. P. D. Lotlikar, E. C. Miller, J. A. Miller, and A. Margreth, The
    enzymatic reduction of the N-hydroxy-derivatives of 2-acetyl-
    aminofluorene and related carcinogens by tissue preparations,
    Cancer Res. 25:1743-1752 (1965).

40. L. A. Sternson, Detection of arylhydroxylamines as intermediates in
    the metabolic reduction of nitro compounds, Experientia 31:
    268-269 (1975).

41. G. J. Mulder, L. E. Unruh, F. E. Evans, B. Ketterer, and F. F.
    Kadlubar, Formation and identification of glutathione conjugates
    from 2-nitrosofluorene and N-hydroxy-2-aminofluorene, Chem.
    -Biol. Interact. 39:111-127 (1982).

42. E. Kriek, On the interaction of N-2-fluorenylhydroxylamine with
    nucleic acids in vitro, Biochem. Biophys. Res. Commun. 20:793-799
    (1965).

43. D. T. Beranek, G. L. White, R. H. Heflich, and F. A. Beland, Amino-
    fluorene-DNA adduct formation in Salmonella typhimurium exposed
    to the carcinogen N-hydroxy-2-acetylaminofluorene, Proc. Natl.
    Acad. Sci. USA 79:5175-5178 (1982).

44. R. A. Floyd, L. M. Soong, M. A. Stuart, and D. L. Reigh, Free
    radicals and carcinogenesis. Some properties of the nitroxyl
    free radicals produced by covalent binding of 2-nitrosofluorene
    to unsaturated lipids of membranes, Arch. Biochem. Biophys. 185:
    450-457 (1978).

45. R. Sridhar, M. J. Hampton, J. E. Steward, and R. A. Floyd, Studies
    on the mutagenicity and electron spin resonance spectra of
    nitrosofluorene-lipid adducts, Applied Spectroscopy 34:289-293
    (1980).

46. R. A. Floyd, Free radicals in arylamine carcinogenesis, in: "Free Radicals in Biology", Vol. IV, W. A. Pryor, ed., Academic Press, Inc., New York, pp. 187-208 (1980).

47. R. Sridhar, and R. A. Floyd, An electron paramagnetic resonance study of the reaction of nitrosobenzene with cholesterol, Can. J. Chem. 60:1574-1576 (1982).

48. A. B. Sullivan, Electron spin resonance studies of a stable aryl-nitroso-olefin adduct free radical, J. Org. Chem. 31:2811-2817 (1966).

49. D. J. Kornburst, and R. D. Mavis, Relative susceptibility of microsomes from lung, heart, liver, kidney, brain and testes to lipid peroxidation. Correlation with vitamin E content, Lipids 15:315-322 (1979).

50. A. Stier, R. Clauss, A. Lücke, and I. Reitz, Redox cycle of stable mixed nitroxides formed from carcinogenic aromatic amines, Xenobiotica 10:661-673 (1980).

51. B. R. Brooks, and O. L. Klamerth, Interaction of DNA with bifunctional aldehydes, Eur. J. Biochem. 5:178-182 (1968).

52. C. E. Vaca, and M. Harms-Ringdahl, Nuclear membrane lipid peroxida-tion products bind to nuclear macromolecules, Arch. Biochem. Biophys. 269:548-554 (1989).

53. V. Nair, G. A. Turner, and R. J. Offerman, Novel adducts from the modification of nucleic acid bases by malondialdehyde, J. Am. Chem. Soc. 106:3370-3371 (1984).

54. L. J. Marnett, A. K. Basu, S. M. O'Hara, P. E. Weller, A. F. M. M. Rahman, and J. P. Oliver, Reaction of malondialdehyde with guanine nucleosides: formation of adducts containing oxadiazbi-cyclononene residues in the base-pairing region, J. Am. Chem. Soc. 108:1348-1350 (1986).

55. C. K. Winter, H. J. Segall, and W. F. Haddon, Formation of cyclic adducts of deoxyguanosine with the aldehydes trans-4-hydroxy-2-hexenal and trans-4-hydroxy-2-nonenal in vitro, Cancer Res. 46: 5682-5686 (1986).

56. A. K. Basu, S. M. O'Hara, P. Valladier, K. Stone, O. Mols, and L. J. Marnett, Identification of adducts formed by reaction of guanine nucleosides with malondialdehyde and structurally related alde-hydes, Chem. Res. Toxicol. 1:53-59 (1988).

57. H. Krokan, R. C. Grafstrom, K. Sundqvist, H. Esterbauer, and C. C. Harris, Cytotoxicity, thiol depletion and inhibition of $O^6$-methylguanine-DNA methyltransferase by various aldehydes in cultured human bronchial fibroblasts, Carcinogenesis 6:1755-1759 (1985).

58. E. J. Barry, D. Malejka-Giganti, and H. R. Gutmann, Interaction of aromatic amines with rat-liver proteins in vivo. III. On the mechanism of binding of the carcinogens, N-2-fluorenylacetamide and N-hydroxy-2-fluorenylacetamide, to the soluble proteins, Chem. -Biol. Interact. 1:139-155 (1969/70).

59. B. Dölle, W. Töpner, and H.-G. Neumann, Reaction of arylnitroso compounds with mercaptans, Xenobiotica 10:527-536 (1980).

60. K. Saito, and R. Kato, Glutathione conjugation of arylnitroso compound: detection and monitoring labile intermediates in situ inside a fast atom bombardment mass spectrometer, Biochem. Biophys. Res. Commun. 124:1-5 (1984).

61. N. Takahashi, V. Fischer, J. Schreiber, and R. P. Mason, An ESR study of nonenzymatic reactions of nitroso compounds with biological reducing agents, Free Rad. Res. Commun. 4:351-358 (1988).

BIOCHEMICAL STUDIES ON THE PUTATIVE NITROSO METABOLITE OF CHLORAMPHENICOL:

A NEW MODEL FOR THE CAUSE OF APLASTIC ANEMIA

Michael D. Corbett and Bernadette R. Corbett

Department of Pharmacology and Therapeutics and
Department of Food Science and Human Nutrition
University of Florida
Gainesville, FL 32611-0163

Chloramphenicol (CAP) is one of the oldest and more potent broad spectrum antibiotics available for the treatment of certain severe infections. CAP is an unusual natural product since it possesses both the nitro functional group and C-Cl bonds, neither of which are particularly common among terrestrial natural products. The necessity of the nitro functional group to the antimicrobial action of CAP has been thoroughly studied (reviewed in ref. 1). Unfortunately, the clinical use of CAP often causes toxic reactions, most notably to the hemopoietic system (2). The most serious of these actions is the rare, but generally fatal condition of aplastic anemia (hypoplastic marrow failure). In 1980 Yunis made the hypothesis that reduced intermediates (nitroso, hydroxylamine) of the nitro group of CAP were responsible for the aplastic anemia associated with CAP (2); however, this hypothesis had actually been proposed by the Weisburgers in 1967 (3). It was the Weisburgers' hypothesis that prompted us to synthesize CAP-NO, CAP-NHOH and related analogs in 1978 (1), and to investigate the potential toxicity of such putative metabolites to hemopoietic precursor cells (4,5).

Numerous studies with CAP-NO have shown that this putative metabolite is much more toxic than CAP (2,6). Irreversible inhibition of DNA synthesis, single strand breaks in DNA, and irreversible binding to bone marrow cells have been well documented. Covalent binding of CAP-NO to sulfhydryl groups of protein has been observed (7), and is typical of aromatic C-nitroso compounds in general. Single strand breaks in DNA could be due to the generation of reactive oxygen species such as superoxide and eventually hydroxyl radical, which might result from a redox cycling (8) of CAP-NO and its $1e^-$ reduction product, or of the latter with CAP-NHOH. The reduction of CAP-NO to CAP-NHOH is a facile reaction (1), which is also readily accomplished with ascorbic acid as described below. There have been no reports on the ability of CAP-NO or CAP-NHOH to bind covalently to nucleic acids.

Although there is little doubt as to the potent toxicity of CAP-NO to hemopoietic cells, a problem in explaining aplastic anemia as the result of CAP-NO arises in accounting for how this putative metabolite is transported to, or produced within, the target organ. It has been proposed that any CAP-NO produced in the liver is much too reactive to survive transport to the hemopoietic system (9); however, the possible transport of CAP-NHOH or a latent derivative (10) of CAP-NO or CAP-NHOH has not been excluded. It

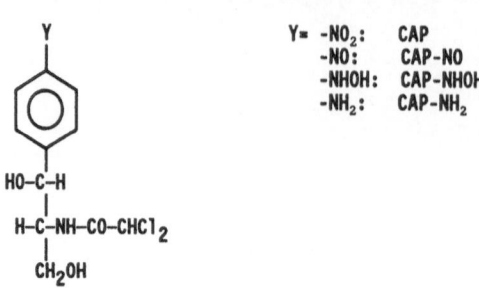

Y= -NO₂: CAP
-NO: CAP-NO
-NHOH: CAP-NHOH
-NH₂: CAP-NH₂

appears unlikely that reduction of CAP to CAP-NO or CAP-NHOH occurs in bone marrow (6,11), although it could be argued that a critical idiosyncrasy of patients predisposed to aplastic anemia is the unique ability of their bone marrow to effect this nitroreduction. An interesting recent development is the report that the highly-toxic bacterial metabolite, dehydro-CAP, is susceptible to nitroreduction by bone marrow, although neither the putative nitroso or hydroxylamine metabolites of this compound were identified (11). It is likely that the nitro group of dehydro-CAP is more susceptible to bioreduction than is the nitro group of CAP, based upon the fact that the ketone group exerts a more electronegative effect on aromatic rings than does the hydroxyalkyl group ($\sigma_p^o$=0.47 vs $\sigma_p^o$=0.01, from ref. 12).

Recent research results in our laboratory prompted us to consider a totally different mechanism by which toxic metabolites of CAP might be generated within the hemopoietic system. We found that arylamines, in general, undergo metabolic activation, and subsequently bind covalently to nucleic acids and protein as the result of the respiratory burst of human granulocytes (13,14). The relative amount of binding to granulocyte DNA and RNA was found to depend strongly on the nature of aromatic ring substituents. More DNA binding was observed for those arylamines which possess higher electron densities in the aromatic ring, which we suspect is a reflection of the production of more stable 1e⁻ oxidation products. The more stable activated metabolites can reasonably be expected to have a greater opportunity to penetrate the nuclear membrane and react with DNA.

This general observation of arylamine bioactivation by granulocytic cells led us to investigate a similar bioactivation process for CAP-NH₂, which is a known metabolite of CAP in humans. We now report that the respiratory burst of human granulocytes resulted in the binding of CAP-NH₂ to both DNA and RNA of the cells. Similar studies with CAP showed a much lower level of nucleic acid binding. In addition, we have summarized important aspects concerning the synthesis, purification and analytical aspects of CAP-NO, CAP-NHOH and CAP-NH₂.

EXPERIMENTAL

Synthesis of CAP-NO. CAP (20 mmole, 6.5 g, Sigma Chemical Co.) was dissolved in 25 mL of 95% EtOH in a 250 mL Erlenmeyer flask equipped with a magnetic stirrer. With stirring, a solution of NH₄Cl (40 mmole, 2.1 g) in 100 mL of Ar-gassed H₂O was added to give a fine suspension of CAP in the aq EtOH. Zn-dust (50 mmole, 3.2 g) was added in numerous small portions in the course of 5 min, while maintaining vigorous stirring. A water bath at room temp was employed to keep the reaction temp below 30°C. The mixture was stirred for an additional 10 min after the Zn had been added and a stream of Ar gas was directed over the vigorously stirred mixture throughout the reaction. The mixture was filtered in vacuo, and the filter cake was washed with 25 mL of Ar-gassed $H_2O$. The filtrate was quickly extracted twice with 70 mL of $Et_2O$ to remove unreacted CAP, the aq portion quickly chilled to < 5°C, then treated with a pre-cooled (< 5°C) solution of $FeCl_3 \cdot 6H_2O$ (60 mmole, 16.2 g) in 100 mL of $H_2O$. After 30 sec the green solution was extracted with 150 mL of EtOAc, then the organic layer washed twice with 20 mL of $H_2O$. The organic phase was dried (anhyd $Na_2SO_4$), then evaporated in vacuo to yield a green oil (~4 g), which was dissolved in 100 mL of 5% $MeOH/CH_2Cl_2$. The emerald-green solution was rapidly chromatographed on a silica gel 60 column (E. Merck, 70-230 mesh, 2.2 X 18 cm bed vol) with 5% $MeOH/CH_2Cl_2$, and the entire blue band that eluted was collected in a single fraction and evaporated in vacuo to give 3 g of a blue-green solid. The residue was dissolved in minimal EtOAc and chromatographed on a silica gel 60 column (2.2 X 40 cm) with EtOAc as eluant. The first blue band to elute was monitored by HPLC on a $\mu$Bondapak $C_{18}$ column (3.9 mm X 30 cm) with 50% aq. MeOH (7), and pure fractions were combined and evaporated in vacuo to give 1.9 g (31% overall yield) of pure CAP-NO as a pale blue crystalline solid. The second pale blue band was also monitored by HPLC and the appropriate fractions were combined and evaporated to give 0.4 g (6% overall yield) of deschloro-CAP-NO as a green, hygroscopic solid.

Synthesis of $^3H$-CAP-NH$_2$. [Ring-3,5-$^3H$]CAP (New England Nuclear, spec act = 58 Ci/mmole, 250 $\mu$Ci total) was combined with 4.0 mg (12.4 $\mu$moles) of unlabelled CAP in a 5 mL R.B. flask equipped with a magnetic stirrer and the ethanol solvent and transfer solvent removed with a gentle stream of Ar. The residue was dissolved in 100 $\mu$L of 95% EtOH, then treated with $NH_4Cl$ (10.8 mg) in 400 $\mu$L of $H_2O$ (all solvents were previously purged with Ar). The solution was stirred vigorously, then treated in a single portion with 2.5 mg (39 $\mu$moles) of Zn-dust, and stirring was continued at R.T. for 30 min while maintaining an Ar atmosphere. The reaction mixture was then transferred to a centrifuge tube and centrifuged to pellet ZnO and unreacted Zn dust. The residue was washed with 200 $\mu$L of 95% EtOH, recentrifuged and the supernatant combined with the first in a 15 mL centrifuge tube, which was then cooled to near 0°C. The solution was treated with cold $FeCl_3 \cdot 6H_2O$ (10.1 mg, 37 $\mu$moles) in 100 $\mu$L of $H_2O$ and kept at 0°C for 5 min. The solution was extracted twice with 4 mL of EtOAc, then the combined extract was washed with 100 $\mu$L of $H_2O$ and centrifuged. The organic layer was dried (anhyd $Na_2SO_4$) and reduced in volume to ~1 mL with an Ar stream. The solution was chromatographed on silica gel 60 (1.1 X 18 cm bed vol) with EtOAc, while monitoring fractions by HPLC with exclusion of any fractions containing deschloro-CAP-NO (estimated to comprise ~2% of crude product). The combined desired fractions contained CAP-NO contaminated with ~30% CAP, but no deschloro-CAP-NO contaminant. The CAP-NO and CAP product was evaporated with an Ar stream, then dissolved in 0.7 mL of 0.2 M $Na_2HPO_4$ buffer, pH 6.6, in a 5 mL R.B. flask under Ar. A solution of reduced glutathione (3.5 mg, 11.4 $\mu$mole) in 0.50 mL of the same phosphate buffer was added to the vigorously stirred solution of CAP-NO during the course of 11 min at R.T. After stirring for an additional 3 min, the solution was treated with ~350 mg of solid NaCl, then extracted three times with 3.5 mL of EtOAc in a 15 mL centrifuge tube. The combined EtOAc extracts were saved for the eventual recovery of unreacted CAP. The remaining aqueous portion was cooled to 0°C in an ice bath, then treated with 50 $\mu$L of conc HCl for 40 min at 0°C. About 63 mg of solid $NaHCO_3$ was added in portions, which sufficed to adjust the pH to ~6, then the solution was extracted three times with 3.5 mL of EtOAc. The combined EtOAc extract was evaporated with an Ar stream, treated with 1 mL of MeOH and further evaporated, finally under high vacuum, to yield high purity CAP-NH$_2$ (1.65 mg, 5.6 $\mu$moles, 45% overall yield) with a spec act = 17.4 mCi/mmole. The CAP-NH$_2$ was dissolved in Ar-gassed 95% EtOH to give a 10 mM solution which was stored at -20°C under Ar, and used entirely within 4 days.

<u>Studies on the covalent binding of $^3$H-CAP and $^3$H-CAP-NH$_2$ to cellular nucleic acids</u>.  Freshly collected and heparinized whole human blood was obtained from healthy volunteers in the early morning, and granulocyte preparations were immediately prepared as described previously (13).  Incubations were conducted on the same day with the substrates at 10 $\mu$M concentrations with and without the use of phorbol myristate acetate (PMA) to induce the respiratory burst.  Following induction of the respiratory burst, which was monitored by the production of $O_2^-$, the incubations were carried out for 30 min, then DNA, RNA and protein were isolated as previously described (13) and modified (14).

<u>Attempted detection of respiratory burst generated oxidation products of CAP-NH$_2$</u>.  A 90 mL granulocyte suspension of 2 X 10$^6$ cells/mL was freshly prepared as previously described, then treated with 180 $\mu$L of a 10 mM ethanolic solution of the CAP-NH$_2$ substrate to give a final concentration of 20 $\mu$M.  The suspension was divided into two equal portions contained within 125 mL polycarbonate flasks.  One flask was treated with ascorbic acid to make it 1 mM, then both flasks were treated with 45 $\mu$L of PMA in DMSO (100 $\mu$g/mL) to induce the respiratory burst.  The incubates were agitated at 190 rpm on a New Brunswick Gyrotory shaker at 37°C, and 12 mL aliquots were sampled from each of the two flasks at T = 15, 30 and 45 min following PMA treatment.  The aliquots were immediately filtered through Acrodisc filters (0.45 $\mu$ pore diameter, Gelman Scientific) and 10 mL of each filtrate were transferred to lyophylization flasks, which were immediately frozen in a dry ice/isopropanol bath and then lyophillized.  Each residue was thoroughly mixed with 1.0 mL of 1-propanol to dissolve CAP-related organics and filtered to remove inorganic salts.  Aliquots of 20 $\mu$L were injected onto an HPLC system consisting of a normal phase partitioning column (LiChrosorb Diol, 4 mm X 25 cm, E. Merck) and eluted with the solvent 1-propanol/CH$_2$Cl$_2$/EtOAc (1:10:89) at 1.5 mL/min, with UV detection of eluted peaks conducted simultaneously at $\lambda_{280}$ and $\lambda_{313}$.  Standard CAP-NHOH to use as an HPLC reference was freshly prepared daily by treating 1.0 mL of a 50 $\mu$M solution of CAP-NO in 1-propanol with 10 $\mu$L of 0.1 M aq ascorbic acid to make it 100 $\mu$M in ascorbic acid.  The in situ production of CAP-NHOH was complete within 15 min, and remained stable for several hours at 0°C in the dark.  The elution times for the peaks of interest were:  CAP and CAP-NO at 3.15 min; CAP-NH$_2$ at 4.7 min and CAP-NHOH at 5.5 min.  Differentiation between CAP and CAP-NO as metabolites was accomplished through the use of another HPLC system, which consisted of a $\mu$Bondapak C$_{18}$ column (3.9 mm X 30 cm, Waters) with a solvent system made of 1 mM Na$_2$EDTA, 15 mM Na-acetate and 8% 1-propanol in water.  CAP eluted at 9.8 min and CAP-NO eluted at 11 min.


RESULTS AND DISCUSSION

<u>Synthesis and purification of CAP-NO</u>.

We first described CAP-NO as a bright blue, highly crystalline material.  This form of CAP-NO could be obtained by a somewhat difficult crystallization process.  We also observed a second form of the material that was more amorphous in nature and lacked the bright blue color of the pure crystalline CAP-NO.  Several other laboratories were evidently unable to obtain pure CAP-NO by our crystallization process, and thus had to employ additional chromatographic methods in order to obtain CAP-NO in a pure form (7,15). Heys reported in 1981 that the mono-dechlorinated analog of CAP-NO was a major by-product of our synthetic method, but that column chromatography readily separated the two nitroso compounds (15). The synthetic method, which is described in the <u>Experimental</u> section, has incorporated the chromatographic purification method described by Heys (15) along with several additional minor changes into what we now consider to be an optimal method for obtaining several grams of CAP-NO in a highly pure form. The necessity for inducing CAP-NO to crystallize in the obviously monomeric form of blue crystals has been eliminated, since the pale blue solid form of CAP-NO obtained by this improved method is highly pure.

Synthesis of [Ring-3,5-$^3$H]-CAP-NH$_2$.

Since the chemical reduction of CAP to CAP-NH$_2$ would likely result in some de-chlorination (7), and thus a product from which it could be hard to obtain pure CAP-NH$_2$, an alternate route to this compound was developed. CAP-NO was previously converted to CAP-NH$_2$ by reaction with reduced gluta-thione, followed by acid-catalyzed hydrolysis (7). This reaction proceeds in high yield with no loss of C-Cl bonds. Since CAP-NO could be obtained in a highly pure form, this pathway proved suitable for obtaining labelled CAP-NH$_2$ in high purity. The scheme also allowed for the recovery of unreacted CAP, which is desirable since the synthesis starts with expensive radiolabelled CAP. To rid the recovered CAP from traces of CAP-NO, we subjected the impure CAP to the nitroso-glyoxylate reaction (16). This reaction converted the CAP-NO impurity to the corresponding formyl-based hydroxamic acid (Scheme I), which was then readily removed from CAP by chromatography.

Nucleic acid binding of $^3$H-CAP and $^3$H-CAP-NH$_2$ effected by the respiratory burst of human granulocyte suspensions.

In view of our recent observations that arylamines are susceptible to metabolic activation by respiratory burst products (13,14), we expected that CAP-NH$_2$ would bind to cellular nucleic acids under such conditions. Such binding was indeed observed in studies conducted with blood from two human volunteers. In a single experiment conducted with granulocytes prepared from the blood from Subject 1, the binding to DNA and RNA was determined, along with protein binding, in the presence (PMA-treated) and absence (Control) of the respiratory burst (Table I). In this study, the extent of metabolism of CAP-NH$_2$ by the induced cells was estimated (HPLC methods) to be about 11% of the original substrate present, which is significantly less than that observed for most other arylamines, which typically ranged from 40-100% metabolism (a notable exception is p-nitroaniline, which is also metabolized in small amounts of 10-15% by the respiratory burst, as reported in ref. 13). Of the CAP-NH$_2$ that was metabolized, about 40% could be ac-counted for as CAP. No other possible metabolites were evident in our HPLC chromatograms. In agreement with our prior findings, the covalent binding of CAP-NH$_2$ to nucleic acids required the occurrence of the respiratory burst. Furthermore, binding to protein was extensive only for those granu-locyte suspensions that were induced to the respiratory burst. This is the first report of nucleic acid binding by any actual or potential metabolite of CAP.

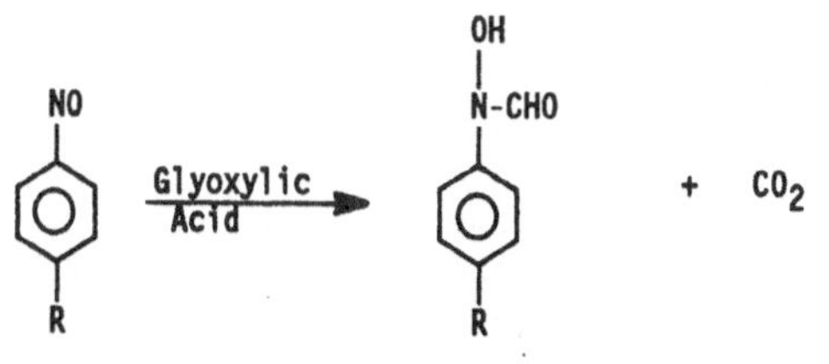

Scheme I. The nitroso-glyoxylate reaction on CAP-NO.

Table I. Nucleic Acid and Protein Binding of $^3$H-CAP-NH$_2$ in Granulocytes.

| Treatment[a] | nmol of CAP-NH$_2$/mg of DNA[b] | nmol of CAP-NH$_2$/mg of RNA[b] | nmol of CAP-NH$_2$/mg of Protein[c] |
|---|---|---|---|
| None | 0 | 0 | 0.02 |
| PMA | 0.10 | 0.49 | 0.60 |

[a]The respiratory burst was induced by the addition of phorbol myristate acetate (PMA) to granulocyte suspensions containing 10 $\mu$M $^3$H-CAP-NH$_2$

[b]Covalent binding to DNA and RNA was determined on the same granulocyte suspensions after 30 min incubation time.

[c]Covalent binding to protein was determined from separate, but otherwise identical, granulocytes suspensions than those used to determine nucleic acid binding.

CAP-NH$_2$ is a relatively more complex arylamine than the other arylamines that have been studied in activated granulocytes (13,14). CAP-NH$_2$ contains other functional groups that might be affected by respiratory burst products, including the dichloroacetamido group, which might be oxidized to the oxamic acid group via an oxamyl chloride. Such metabolic oxidation of this group has been well documented to occur via Cyt P-450-dependent microsomal oxidations, and its occurrence can result in the suicide inactivation of such microsomal oxidases (17,18). Because of the possibility that macromolecule binding via the respiratory burst could occur through oxidative bioactivation of functional groups other than the arylamine group, we designed an experiment to determine if CAP itself might result in such binding as the result of the respiratory burst. Such binding by CAP should be a good indicator of the contribution to macromolecule binding by these other (non-arylamine) functional groups present in both CAP and CAP-NH$_2$. The studies summarized in Table II were conducted with granulocyte preparations obtained from the blood of Subject 2. CAP-NH$_2$ gave about 10 times as much binding to DNA than did CAP when subjected to the respiratory burst, and about 5-6 times as much binding to RNA. In the absence of the burst, neither CAP-NH$_2$ nor CAP gave any binding to any nucleic acid. We concluded that most covalent binding of CAP-NH$_2$ to nucleic acids is due to metabolic activation of the arylamine group rather than bioactivation of one of the side chain functional groups.

Studies on the ability of the myeloperoxidase (MPO) inhibitor, azide ion, to prevent nucleic acid binding by CAP-NH$_2$ are also summarized in Table II. Azide totally blocked CAP-NH$_2$ binding to the DNA of activated granulocytes, but still allowed significant binding (18%) to RNA relative to that seen in uninhibited granulocytes. This difference in inhibition could be indicative of more than one pathway for the bioactivation of the amino group of CAP-NH$_2$ or arylamines in general. On the other hand, the residual binding of CAP-NH$_2$ to RNA in the presence of azide could also be due to bioactivation processes occurring on the side chain, and which are not sensitive to azide inhibition. Like the situation with Cyt P-450 inactivation, it is likely that the bioactivated dichloroacetamido group is much too reactive to ever reach DNA, but might react with the more accessible RNA. Unfortunately, studies on azide inhibition of CAP binding have not yet been conducted, and the results of such studies could shed light on this possiblity.

Table II.  Nucleic Acid Binding of [3]H-CAP-NH$_2$ and [3]H-CAP in Granulocytes.

| Treatment[a] | nmol of substrate/ mg of DNA | nmol of substrate/ mg of RNA |
|---|---|---|
| CAP-NH$_2$, PMA (n=4) | 0.07 ± 0.02 | 0.71 ± 0.02 |
| CAP-NH$_2$, PMA, Azide, (n=3) | -0- | 0.13 ± 0.02 |
| CAP, PMA | 0.007 | 0.13 |
| CAP, (Control) | -0- | -0- |

[a]Substrates were employed at 10 $\mu$M concentrations, with or without 1 mM azide (CAP-NH$_2$ only), and with or without PMA stimulation for CAP.  All data were obtained from a total of two experiments conducted with granulocytes obtained from the same human subject.  The data for CAP studies were generated in paired experiments such that studies on CAP-NH$_2$ were conducted simultaneously.

It is noteworthy that the binding of CAP-NH$_2$ to granulocyte DNA is totally blocked by azide.  Previously we found that azide caused only partial inhibition (40-50%) of DNA binding by arylamine substrates such as 2-aminofluorene and 4-chloroaniline (13,14).  Acetaminophen binding to DNA was subsequently found to be more sensitive to azide inhibition (~85% inhibition) than these arylamines; however, that is probably a reflection of the fact that it is a phenolic functional group that is responsible for the biactivation of acetaminophen (14).  Further studies on the effect of MPO inhibitors on arylamine binding to both DNA and RNA as a function of arylamine structure are planned, since such studies might reveal multiple bioactivation pathways for these substrates.  At this time, we suspect that only the MPO pathway is azide sensitive.

The extent of DNA binding by CAP-NH$_2$ is similar to that previously observed for 4-chloroaniline; however, CAP-NH$_2$ appears to be considerably more resistant to respiratory burst-dependent metabolism than is 4-chloroaniline.  Given a similar degree of metabolism, we would expect CAP-NH$_2$ to result in 2-3 times more DNA binding than does 4-chloroaniline.  The ratio of RNA to DNA binding for CAP-NH$_2$ was about 5-10, which is low compared to most other arylamines.  This suggests that the bioactivated metabolite(s) of CAP-NH$_2$ are relatively stable (13).

Attempted detection of CAP-NHOH and CAP-NO during the oxidative bioactivation of CAP-NH$_2$ by the respiratory burst.

We originally proposed that the bioactivation of arylamines by the respiratory burst probably proceeded via 1e$^-$ (free radical) oxidative reactions (13).  This was based on the inability to readily detect metabolites typical of 2e$^-$ peroxidative reactions on arylamines, particularly the nitroso metabolite (19), either by the action of purified MPO or activated granulocytes.  On the other hand, Uetrecht et al. (20) has proposed that procainamide is oxidized via the respiratory burst to hydroxylamine and nitroso metabolites, although the latter is too reactive to allow for its detection.  In the presence of exogenous ascorbic acid, the hydroxylamine metabolite was reported to be readily detected by HPLC methods (20), presumably since oxidation of the hydroxylamine to nitroso is either blocked by ascorbate or the nitroso is reduced back to the hydroxylamine by this reductant. In the absence of added ascorbic acid, only the nitro metabolite of procainamide was produced by activated granulocytes (20). A more recent report on procainamide oxidation products supported some of Uetrecht's conclusions (21), but there seems to be some confusion concerning the absolute identification of several of the possibile metabolites of procainamide.

Our attempts to detect CAP-NHOH and CAP-NO as respiratory burst metabolites of CAP-NH$_2$ were convincingly negative. Even with an ability to easily detect a 1% conversion of CAP-NH$_2$ (20 $\mu$M) to CAP-NHOH by granulocytes, we were unable to detect this metabolite either in the presence or absence of exogenous ascorbic acid. In the absence of ascorbic acid, CAP was readily detected, and was the only obvious metabolite. We have concluded that metabolites which arise from 2e$^-$ oxidation reactions on arylamines are not major products, if indeed they are produced at all by the respiratory burst. Certainly for the case of CAP-NH$_2$, the hydroxylamine and nitroso metabolites did not accumulate in detectable amounts. However, extensive macromolecule binding does occur with CAP-NH$_2$ during the respiratory burst, and such binding could not be correlated with the production of either CAP-NHOH or CAP-NO. A study of the effect of ascorbic acid on CAP-NH$_2$ binding to granulocyte macromolecules has not yet been conducted.

A reconciliation of the data of Uetrecht with our observations might arise from a consideration of the potential ability for arylhydroxylamines to undergo further oxidations to nitroso, nitro or other oxidation products. It is reasonable to propose that arylhydroxylamines with a higher electron density in the aromatic ring (and hence at the hydroxylamine-N atom) might be much more readily oxidized by the various oxidants of the respiratory burst. The $\sigma_p^0$ values are 0.01 for the hydroxyalkyl substituent of CAP-NHOH and 0.31 for the carboxamido substituent of procainamide (12), thus CAP-NHOH has a higher electron density than does procainamide-NHOH and might be so rapidly oxidized by the respiratory burst that it cannot accumulate as can procainamide hydroxylamine. Perhaps this provides a sufficient basis for the differences in the reactivities of these hydroxylamines. On the other hand, in several peroxidative enzyme systems, we have found that oxidation of arylhydroxylamines to the nitroso state proceeds much more rapidly than does the subsequent oxidation of the nitroso compounds to the nitro level (19,22). This explains why chloroperoxidase is so efficient in converting many arylamines selectively to the nitroso oxidation state. Finally, our experiences with arylamine oxidations by MPO have always indicated that the preponderance of metabolites arose from free radical processes (unpublished results), in marked contrast to the hydroxylamine and nitroso metabolites produced by chloroperoxidase and pea seed peroxygenase (19,22). In agreement with our general observations concerning MPO oxidations, Bakkenist et al. reported that 4-chloroaniline is oxidized by this enzyme to a complex mixture of products indicative of free radical oxidation reactions (23); however, that group made no attempt to detect reactive intermediates. Unfortunately, it will be very difficult to distinguish between the hydroxylamine pathway and the highly likely possibility that the true toxic metabolites of arylamines produced by the respiratory burst are radical species produced independently of any pathway that gives rise to hydroxylamine metabolites. This major mechanistic difference certainly justifies further investigation. An examination of the nature of respiratory burst induced arylamine binding to protein might be particularly informative. If protein binding arises mostly from the nitroso metabolite as proposed (20,21), then the predominant form of arylamine binding should be of the sulfinanilide type (7). Free radical mediated protein binding would most likely yield a much more complex type of binding.

CONCLUSION

Although CAP-NO and CAP-NHOH have never been detected as in vivo metabolites of CAP, their extreme toxicity has led biochemical toxicologists to implicate one or both of them as possible causes of aplastic anemia. On the other hand, such cytotoxicity alone is certainly no proof that one of these possible metabolites is, in fact, the ultimate cause of this toxic action. Essentially nothing is known concerning the actual lesion that results in the condition of aplastic anemia. The rare incidence of this side effect creates a dilemma in attempting to determine a simple correlation between the production of a certain toxic species and marrow failure. If indeed there is a biochemical mechanism that allows for the interaction of CAP-NO with critical hemopoietic cells, it is difficult to imagine why such a pathway can occur in only one out of approximately 40,000 cases. Irregardless of what the toxic metabolite might be, it seems more reasonable to expect its production to be a relatively common event, and that aplastic anemia remains an idiosyncratic reaction to something that the metabolite does to the cell.

Our original hypothesis was that toxic metabolites such as CAP-NHOH and CAP-NO might be produced within hemopoietic cells via oxidation of CAP-$NH_2$, rather than by partial nitroreduction of CAP (Scheme II). This proposal has also been made by other groups (8). Given the possibility that some cells within the hemopoietic system can generate active oxygen species in a manner similar to the respiratory burst of granulocytes (24), we have used these differentiated cells of the peripheral blood to mimic possible bioactivation reactions within the bone marrow. Although we were unable to detect the low-level production of any CAP-NHOH or CAP-NO, the action of the respiratory burst still resulted in the binding of CAP-$NH_2$ to cellular macromolecules including nucleic acids. Thus, we propose that oxidative metabolism of CAP-$NH_2$ in hemopoietic cells (Scheme II) provides a pathway for macromolecule binding by CAP metabolites, and that such binding might cause the lesion that ultimately results in the development of aplastic anemia. The necessity for the intermediary production of nitroso or hydroxylamino metabolites of CAP in order to explain the pathogenesis of aplastic anemia is highly doubtful.

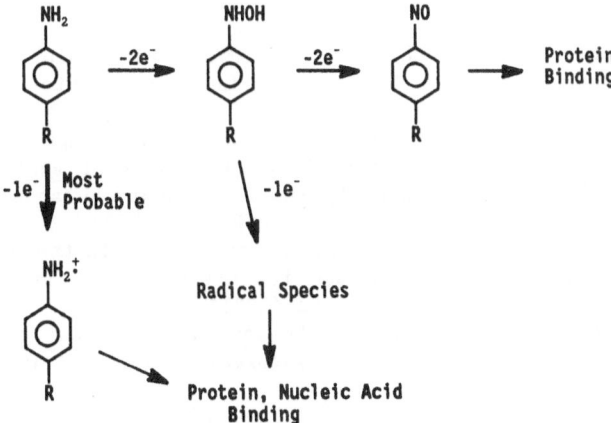

Scheme II. Oxidative Bioactivation of CAP-$NH_2$ in Bone Marrow.

ACKNOWLEDGEMENT

This work was supported by Grant No. ES03631 from the National Institute of Environmental Health Sciences, DHHS.

REFERENCES

1. M.D. Corbett and B.R. Chipko, "Synthesis and antibiotic properties of chloramphenicol reduction products", Antimicro. Agents Chemother., 13: 193-198 (1978).
2. A.A. Yunis, "Chloramphenicol: relation of structure to activity and toxicity", Ann. Rev. Pharmacol. Toxicol., 28: 83-100 (1988).
3. J.H. Weisburger, Y. Shirasu, P.H. Granthan and E.K. Weisburger, "Chloramphenicol, protein synthesis, and the metabolism of the carcinogen N-2-fluorenyldiacetamide in rats". J. Biol. Chem., 242: 372-378 (1967).
4. T.L. Pazdernik and M.D. Corbett, "Effects of chloramphenicol reduction products on hemopoietic precursor cells in vitro", Pharmacology, 19: 191-195 (1979).
5. T.L. Pazdernik and M.D. Corbett, "Role of chloramphenicol reduction products in aplastic anemia", Pharmacology, 20: 87-94 (1980).
6. B.J. Gross, R.V. Branchflower, T.R. Burke, D.E. Lees and L.R. Pohl, "Bone marrow toxicity in vitro of chloramphenicol and its metabolites", Toxicol. Appl. Pharmacol., 64: 557-565 (1982).
7. P. Eyer and M. Schneller, "Reactions of the nitroso analogue of chloramphenicol with reduced glutatione", Biochem. Pharmacol., 32: 1029-1036 (1983).
8. S. Teo, L. Pohl and J. Halpert, "Production of superoxide anion radicals during the oxidative metabolism of aminochloramphenicol", Biochem. Pharmacol., 35: 4584-4586 (1986).
9. M. Ascherl, P. Eyer and H. Kampffmeyer, "Formation and disposition of nitrosochloramphenicol in rat liver", Biochem. Pharmacol., 34: 3755-3763 (1985).
10. M.D. Corbett, L.O. Lim, B.R. Corbett, J.J. Johnston and P. Wiebkin, "Covalent binding of N-hydroxy-N-acetyl-2-aminofluorene and N-hydroxy-N-glycolyl-2-aminofluorene to rat hepatocyte DNA: in vitro and cell-suspension studies", Chem. Res. Toxicol., 1: 41-46 (1988).
11. M.Isildar, W.H. Abou-Khalil, J.J. Jimenez, S. Abou-Khalil and A.A. Yunis, "Aerobic nitroreduction of dehydrochloramphenicol by bone marrow", Toxicol. Appl. Pharmacol., 94: 305-310 (1988).
12. N.S. Isaacs, 1987, "Physical Organic Chemistry", pp. 133-135, J. Wiley & Sons, New York.
13. M.D. Corbett and B.R. Corbett, "Nucleic acid binding of arylamines during the respiratory burst of human granulocytes", Chem. Res. Toxicol., 1: 356-363 (1988).
14. M.D. Corbett, B.R. Corbett, M.-H. Hannothiaux and S.J. Quintana, "Metabolic activation and nucleic acid binding of acetaminophen and related arylamine substrates by the respiratory burst of human granulocytes", Chem. Res. Toxicol. 2: In Press (1989).
15. J.R. Heys, "The use of the acetonide derivative in the preparation of D-threo-chloramphenicol-1-$^3$H and its nitroso derivative", J. Label. Cpds. Radiopharm., 18: 1743-1753 (1981).
16. M.D. Corbett and B.R. Corbett, "Reaction of nitroso aromatics with glyoxylic acid. A new path to hydroxamic acids", J. Org. Chem., 45: 2834-2839 (1980).
17. N.E. Miller and J. Halpert, "Analogues of chloramphenicol as mechanism-based inactivators of rat liver cytochrome P-450: modifications of the propanediol side chain, the p-nitro group and the dichloromethyl moiety", Molec. Pharmacol., 29: 391-398 (1986).

18. L.R. Pohl, S.D. Nelson and G. Krishna, "Investigation of the mechanism of the metabolic activation of chloramphenicol by rat liver microsomes", Biochem. Pharmacol., 27: 491-496 (1978).

19. M.D. Corbett, B.R. Chipko and A.O. Batchelor, "The action of chloride peroxidase on 4-chloroaniline", Biochem. J., 187: 893-903 (1980).

20. J. Uetrecht, N. Zahid and R. Rubin, "Metabolism of procainamide to a hydroxylamine by human neutrophils and mononuclear leukocytes", Chem. Res. Toxicol., 1: 74-78 (1988).

21. R.L. Rubin and J.T. Curnutte, "Metabolism of procainamide to the cytotoxic hydroxylamine by neutrophils activated in vitro", J. Clin. Invest., 83: 1336-1343 (1989).

22. M.D. Corbett and B.R. Corbett, "Arylamine N-oxidation by the microsomal fraction of germinating pea seedlings", J. Agric. Food Chem., 31: 1276-1282 (1983).

23. A.R.J. Bakkenist, H. Plat and R. Wever, "Oxidation of 4-chloroaniline catalyzed by human myeloperoxidase", Bioorg. Chem., 10: 324-328 (1981)

24. L.E. Twerdok and M.A. Trush, "Neutrophil-derived oxidants as mediators of chemical activation in bone marrow", Chem.-Biol. Interact., 65: 261-273 (1988).

# MECHANISMS OF FANFT/ANFT INDUCED BLADDER CANCER

Terry V. Zenser, Michael B. Mattammal, Vijaya M. Lakshmi,
Ruthellen Dawley, Suhas V. Sohani, Leslie A. Spry, and
Bernard B. Davis

Geriatric Research, Education, and Clinical Center
Veterans Administration Medical Center and
Departments of Biochemistry and Internal Medicine
St. Louis University School of Medicine
St. Louis, MO 63104

## INTRODUCTION

Bladder cancer has long been associated with exposure to chemicals. The high incidence of bladder cancer among smokers and workers in dye, rubber and chemical industries is associated with their exposure to aromatic amines (1-3). Progress in fully understanding bladder cancer is hampered by the lack of appropriate animal models.

N-[4-(5-nitro-2-furyl)-2-thiazolyl]formamide (FANFT) provides an excellent model for studying the pathogenesis of experimental bladder cancer (4). This 5-nitrofuran substituted 2-aminothiazole represents a diverse group of compounds that are used as food additives and human and veterinary drugs. FANFT induces bladder cancer in rat, mouse, hamster, and dog with similar results observed in males and females (5). Neither FANFT nor ANFT is carcinogenic in guinea pig. FANFT induced bladder neoplasia in susceptible species has histologic characteristics similar to those reported for humans, with most tumors being transitional cell carcinomas. Studies have demonstrated a relationship between FANFT induced bladder cancer and (i) duration of exposure to a given dose, (ii) dose administered, and (iii) two-stage process of carcinogenesis. Because these characteristics permit a great deal of flexibility in experimental design, the FANFT model was assessed.

ANFT, deformylated FANFT, is considered the proximate carcinogen in FANFT induced bladder cancer (6). This is because the mutagenicity of urine from FANFT-fed animals and the susceptibility of different species to FANFT induced urothelial cancer correlate with urinary levels of ANFT rather than FANFT. Paradoxically, dietary FANFT is more carcinogenic than dietary ANFT (7). Hence, metabolism and disposition of both FANFT/ANFT were evaluated.

A holistic model illustrated in Figure 1 has served as the basis for our studies assessing aromatic and heterocyclic amine induced bladder cancer. The four hypotheses that form the basis of this model are: (i) Unique endogenous patterns of metabolism and disposition of aromatic and heterocyclic amines determine biological characteristics such as species and tissue specificity. (ii) Urothelial cells (target cells) activate aromatic and heterocyclic amines to toxins and carcinogens. (iii) Peroxidases, including the hydroperoxidase activity of prostaglandin H synthase (PHS), are important activating enzymes in urothelial cells. (iv) Arachidonic acid cascade, as exemplified by

*Nitroarenes,* Edited by P. C. Howard *et al.*
Plenum Press, New York, 1990

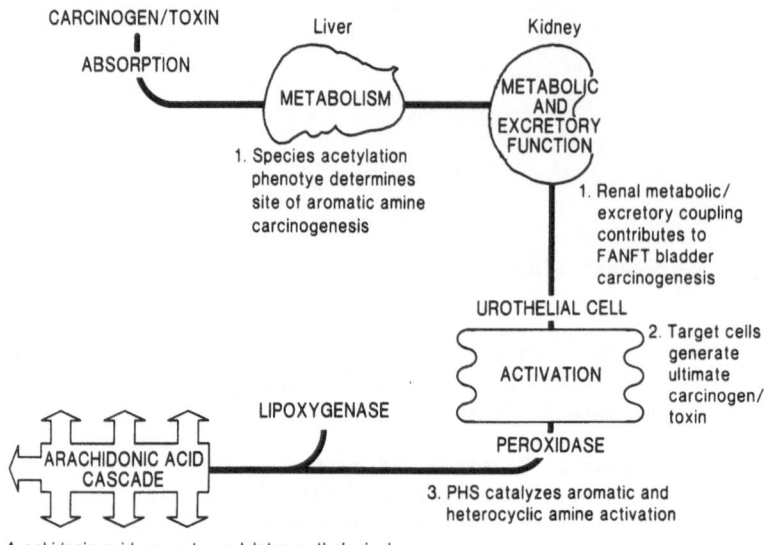

Figure 1.  Holistic model describing aromatic and heterocyclic
amine induced bladder cancer.

transmembrane signaling and second messenger systems, plays an important role
in toxic or carcinogenic processes of the bladder.

A working model based on results from our laboratory and general
concepts of xenobiotic metabolism and disposition has been developed for
initiation of FANFT induced bladder cancer (Figure 2).  Hepatic metabolism of
FANFT is envisioned to result in ANFT and other metabolites (Met.) in addition
to unmetabolized FANFT.  Biliary metabolites may be returned to
liver by the enterohepatic circulation.  Renal metabolism results in
deformylation (Deform.) of FANFT and concomitant excretion of ANFT (metabolic/
excretory coupling).  Unmetabolized FANFT is not observed in urine.
Significantly more ANFT is excreted in urine as a result of FANFT
deformylation than by filtration of plasma ANFT.  Urine is the major site of
excretion.  Detoxified metabolites (Detox. Met.) are observed in urine.  The
proximate carcinogen, ANFT, and perhaps other as yet unidentified metabolites
are activated by enzyme systems within urothelial cells. Activated carcinogen
binds to macromolecules such as DNA and initiates bladder cancer.  Both
oxidative and reductive pathways for activation of FANFT/ANFT have been
demonstrated.  Experiments demonstrating each aspect of this model will be
described.

RENAL METABOLIC/EXCRETORY COUPLING

Metabolism and disposition of FANFT and ANFT have been assessed using
the intact rat and the isolated perfused rat kidney.  In rats prepared for
clearance experiments, equimolar concentrations of either FANFT or ANFT were
administered (8).  After 2 hours, the urinary excretion rate of ANFT with i.v.
ANFT administration was $0.9 \pm 0.1$ nmol/min.  Following i.v. FANFT
administration, the urinary excretion rate for ANFT was $49.7 \pm 8.6$ nmol/min.
The plasma elimination half-lives ($t_{1/2}$) were $23 \pm 3$ and $< 5$ min for ANFT and
FANFT, respectively.  Approximately 27% of the radioactivity from ANFT dosed
rats was excreted in urine compared with 44% with FANFT.  In bile, 20% of
radiolabel from either FANFT or ANFT dosed rats was excreted. The differences
in renal handling of ANFT and FANFT could not be accounted for by differences
in plasma protein binding.  Results consistent with this have been reported in
the isolated perfused kidney (9).  These results demonstrate deformylation

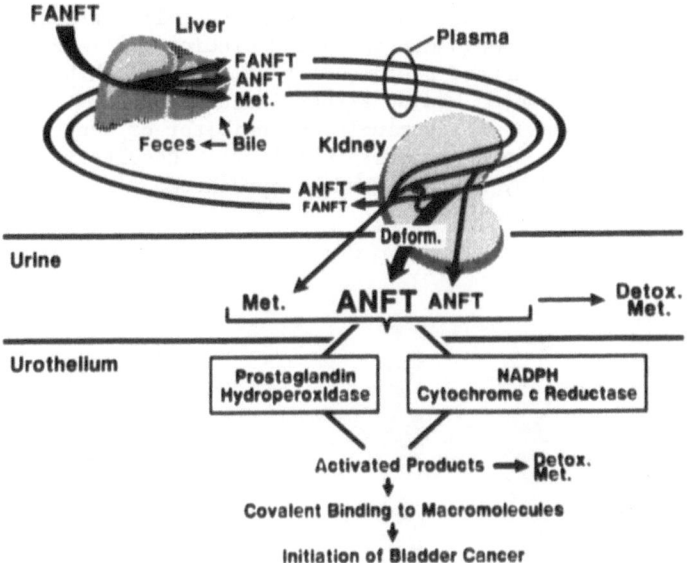

Figure 2.  Illustration of FANFT induced bladder cancer.

dependent excretion for FANFT.  FANFT administration therefore results in much
higher concentrations of ANFT in the urinary tract than with ANFT
administration.  This phenomenon is called renal metabolic/excretory coupling.

Purified proximal tubules were used to further assess this phenomenon
(10).  FANFT and ANFT uptake into tubules achieved equilibrium within 60 sec.
Greater uptake of FANFT relative to ANFT was demonstrated in viable as well as
nonviable tubules.  FANFT partitioned into oil better than ANFT.  Similar
effects were observed with or without albumin.  The mechanism at the cellular
level can be explained by the combined energy-independent uptake of FANFT with
metabolism to ANFT (deformylation), resulting in enhanced urinary ANFT
excretion.

Renal FANFT deformylase activity was assessed (10).  Renal deformylase
was predominantly cytosolic and yielded apparent $K_m$ and $V_{max}$ of 6.7 uM and 6.1
nmol ANFT min$^{-1}$ mg protein$^{-1}$, respectively.  Deformylase activity was specific
for N-formylated compound and was not inhibited by Paraoxon.  N-formylation of
arylamine carcinogens has recently been attributable to soluble formamidase
activity requiring N-formyl-L-kynurenine (11).  The relationship between the
latter activity and FANFT deformylase activity is not yet known.


NITROREDUCTION OF FANFT/ANFT

Anaerobic reductive metabolism of several 5-nitrofuran substituted 2-
aminothiazoles was assessed.  Microsomal reductase activity from rat and
rabbit was observed in liver, bladder epithelium, and renal cortex, outer and
inner medulla.  Rabbit bladder transitional epithelial microsomes contained
the highest specific activity of cytochrome $c$ reductase observed in kidney and
bladder (12).  FANFT reductase was highest in epithelium with similar amounts
observed in each area of the kidney.  Nonepithelial microsomes exhibited very
low amounts of reductase activity.  Renal hydronephrosis abolishes cortical
cytochrome P-450 content and mixed-function oxidase activities (13).  However,
neither the level of NADPH-cytochrome $c$ reductase nor 5-nitrofuran reductase
was affected.

Reductive metabolism was further characterized by the use of inhibitors
(13).  This is illustrated for the acetylated analogue of ANFT in Figure 3.

NADPH was a better cofactor than NADH. Aerobic conditions and heat-treated microsomes (not shown) completely inhibited binding. The antioxidants glutathione and vitamin E completely inhibited binding. 2'-AMP, an inhibitor of the NADPH-dependent reductase system, also completely inhibited nucleic acid binding. Compounds which had no effect included SKF-525A (an inhibitor of mixed-function oxidases), allopurinol (an inhibitor of xanthine oxidase), and aspirin (a prostaglandin H synthase inhibitor). Results suggest metabolism to be mediated by NADPH-dependent cytochrome $c$ reductase.

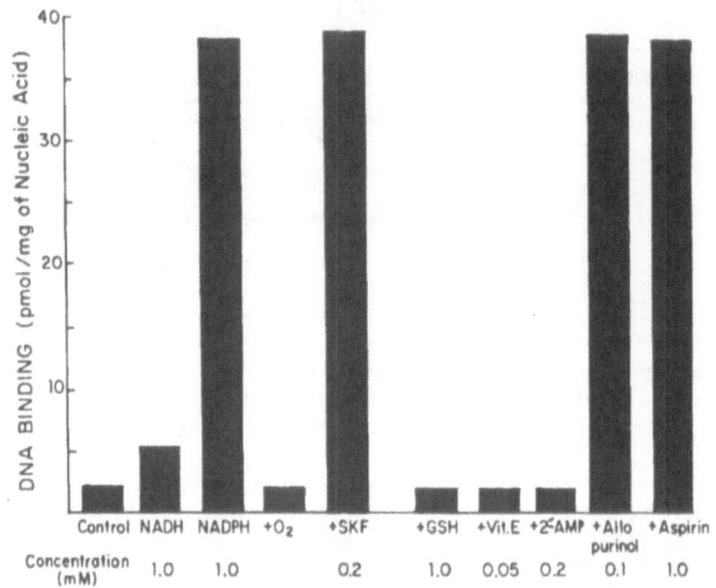

Figure 3.  Reductive metabolism of the ANFT analogue NFTA.

Nitroreduction products that have been identified include 5-nitroso and 5-N-hydroxyfurylthiazoles (13,14). The amide resulting from ring cleavage of the 5-amino-furylthiazole metabolite was also identified (15). N-Hydroxy but not keto-nitrile reduction products react with glutathione and deoxyguanosine to form conjugates. Inhibition by antioxidants is consistent with formation of a nitro-anion radical. Comparative reductive studies between nitroaromatic and nitrofuran 4-substituted-thiazoles indicated the increased potency of the former was due to the formation of a more stable amine intermediate that could then be activated by oxidation (16). Thus, the nitroreduction pathway may play an important role in initiation of FANFT induced bladder cancer.

OXIDATION OF FANFT/ANFT

Peroxidatic metabolism of FANFT/ANFT was assessed. ANFT is a much better substrate than FANFT for metabolism by the hydroperoxidase activity of PHS (17). Furthermore, the ability of PHS to metabolize ANFT was much greater than horseradish peroxidase, chloroperoxidase or lactoperoxidase (18). Metabolism of ANFT resulted in binding to nucleic acids, proteins and glutathione. Peroxidatic metabolism of aromatic amines has been demonstrated with dog and human bladder urothelial cells (19,20).

To assess peroxidatic activation of ANFT in vivo, a mercapturic acid product of activated ANFT (ANFT-MA) was used as a marker (21). Peroxidase inhibitors, 6-n-propyl-2-thiouracil and methimazole, significantly reduced ANFT binding to protein and glutathione conjugate formation in vitro. Isolated perfused rat kidneys rapidly converted the glutathione conjugate to its corresponding mercapturic acid. ANFT-MA was identified in urine of rats administered FANFT. ANFT was the major urinary metabolite with FANFT not detected. A 30-min pretreatment with peroxidase inhibitors significantly reduced urinary excretion of ANFT-MA in rats administered FANFT (150 mg/kg) from $14.8 \pm 2.1$ to $7.9 \pm 0.8$ and $6.2 \pm 1.1$ nmoles/18 h, respectively. Inhibitor pretreatment did not alter excretion of ANFT or $PGE_2$. These experiments provide the first demonstration of the presence of a peroxidatic metabolite of activated ANFT in vivo and the first in vivo screen to select compounds for their ability to alter FANFT/ANFT metabolism and disposition.

Evidence for peroxidatic metabolism in man was obtained. Peroxidatic metabolism of nicotine, a major constituent of cigarette smoke, was determined and found to produce a unique metabolite, 3-(2,3-dihydro-1-methyl-2-pyrrolyl)pyridine (22). This new product was found to represent a significant amount of nicotine metabolism in rabbits administered nicotine and a smoker. As indicated below, PHS was demonstrated in human urothelial cells. The latter is consistent with the potential for peroxidatic metabolism by these cells.

The mechanism of ANFT peroxidative metabolism was evaluated. ANFT was shown to be a reducing co-substrate for PHS. Peroxidatic oxidation of the 2-aminothiazole ring was demonstrated. This includes electrochemical measurements and identification of 2-amino-4-(5-nitro-2-furyl)-5-(N-acetylcystein-S-yl)thiazole (ANFT-MA) as a thioether conjugate of PHS activated ANFT (23,24). In addition, the configurations of $^{13}C$-and $^{15}N$-labeled FANFT and ANFT were analyzed in DMSO solutions (25). Data suggest the transfer of the C-5 proton of ANFT to the endocyclic nitrogen N-3. In contrast, FANFT showed a proton-transfer reaction involving the amide proton. ANFT is envisioned as existing in a "zwitterion" configuration. PHS may facilitate the "zwitterion" formation and subsequent withdrawal of electrons from ANFT. Thus, differences in the chemistry of the thiazole ring caused by deformylation may partially explain the differential peroxidatic activation of ANFT compared to FANFT.

UV/VIS spectra changes for FANFT/ANFT were recorded at different pHs to provide further evidence for the "zwitterion" configuration. For ANFT, a hypsochromic shift of 31 nm (404 to 373 nm) was observed in the nitrofuran region as the pH decreased from 3.54 to 1.75 (Figure 4). Changes in the thiazole region (260 to 280 nm) were also noted. The spectra observed at pH 3.54 was similar to that observed at all higher pHs. The spectrum of FANFT was not altered by pH. These results are consistent with protonation at the N-3 position of ANFT at lower pH. If this was the case, part of the electron density on sulfur is expected to be attracted towards the endocyclic nitrogen (charge stabilization by resonance), causing the observed hypsochromic shift. Thus, these experimental findings are consistent with the protonation of N-3 in the proposed "zwitterion" configuration of ANFT.

## UROTHELIAL CELL SYNTHESIS OF EICOSANOIDS

Any scientific approach to understanding bladder cancer at the target cells must consider that carcinogenesis is a multistage process. Peroxidatic activation by PHS can utilize the arachidonic acid cascade for initiation. However, eicosanoid products of arachidonic acid metabolism can modulate promotion, differentiation and metastatic stages of carcinogenesis (26). In addition, several oncogenes are homologous to various components in signal transduction pathways (27) and ectopic expression of receptors can cause alignant transformation (28). This emphasizes the importance of transmembrane signaling and second messenger systems in the multistage carcinogenic process. For this reason, a detailed study of the regulation of the arachidonic acid cascade has begun in human and dog urothelial cells.

The first characterization of signal transduction events by dog urothelial cells was conducted (29). Primary and subcultured cells were morphologically identified as epithelial in nature. A decreased hormonal responsiveness was observed with serum deprivation, excess serum and subculturing (Figure 5). Results were similar to those reported for human cells with respect to the kinds of agonists, mechanism of agonist stimulation (release of arachidonic

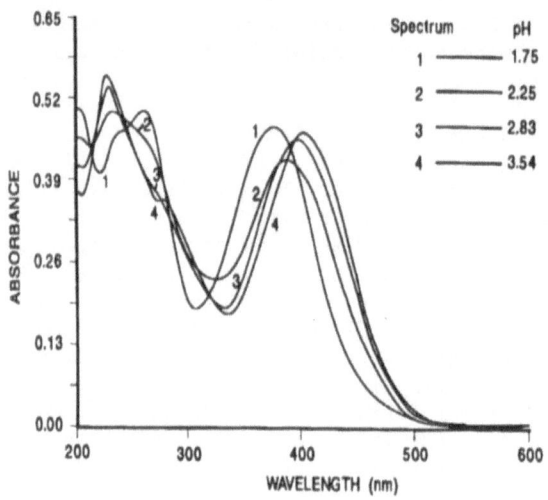

Figure 4.   UV/VIS spectral changes of ANFT obtained at different pHs.

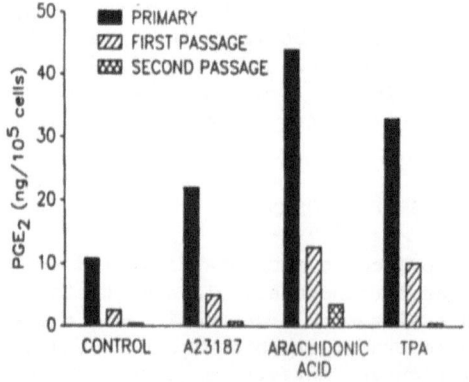

Figure 5.   Hormonal responsiveness of primary and subcultured dog urothelial cells monitored by $PGE_2$ synthesis.

acid), delayed time course of TPA response, and inhibition of TPA response by cycloheximide (30,31). Both TPA and bradykinin increase prostaglandin synthesis by increasing availability of arachidonic acid. However, these agonists appear to function by distinct receptors. Human and dog urothelial cells exhibit very similar mechanisms of signal transduction involving the arachidonic acid cascade.

Signal transduction pathways affect a multitude of cellular processes. Altered responsiveness by these pathways can lead to abnormal cellular responses. These studies demonstrate that common variables such as the presence of serum and subculturing can alter the responsiveness of signal transduction pathways. Thus, results with cells in which the responsiveness of these pathways is not known should be evaluated with caution.

## ACKNOWLEDGMENTS

This work was supported by the Department of Veterans Affairs and by the USPHS Grant CA-28015 from the National Cancer Institute through the National Bladder Cancer Project.

## REFERENCES

1. L. Rehn, Blasengeschwulste bei fuchsinarbeitern, Arch. Klin. Chir. 50: 588 (1895).
2. D.B. Clayson, and R.C. Garner, Carcinogenic aromatic amines and related compounds, in:"Chemical Carcinogens," C.E. Searle, ed., American Chemical Society, Washington, D.C. (1976).
3. M.R. Guerin, and M.V. Buchanan, Environmental exposure to N-aryl compounds, in:"Carcinogenic and Mutagenic Responses to Aromatic Amines and Nitroarenes," C.M. King, L.J. Romano, and D. Schuetzle, eds., Elsevier Science Publishing Co., Inc., New York (1988).
4. G.T. Bryan, "Carcinogenesis: A Comprehensive Survey," Raven Press, New York (1978).
5. S.M. Cohen, Toxicity and carcinogenicity of nitrofurans, in:"Carcinogenesis: A Comprehensive Survey," G.T. Bryan, ed., Raven Press, New York (1978).
6. J.M. Klemencic, and C.Y. Wang, Mutagenicity of nitrofurans, in:"Carcinogenesis: A Comprehensive Survey," G.T. Bryan, ed., Raven Press, New York (1978).
7. C.Y. Wang, Y. Kamiryo, and W.A. Croft, Carcinogenicity 2-amino-4-(5-nitro-2-furyl)thiazole in rats by oral and subcutaneous administration, Carcinogenesis. 3: 275 (1982).
8. L. Spry, V.M. Lakshmi, T. Zenser, and B. Davis, Metabolism and excretion of nitrofurothiazole bladder carcinogens, J. Pharmacol. Exp. Ther. 238: 457 (1986).
9. L.A. Spry, T.V. Zenser, S.M. Cohen, and B.B. Davis, Role of renal metabolism and excretion in 5-nitrofuran-induced uroepithelial cancer in the rat, J. Clin. Invest. 76: 1025 (1985).
10. L.A. Spry, J. Rubinstein, C. Rettke, T.V. Zenser, and B.B. Davis, Renal metabolic/excretory coupling, Am. J. Physiol. 254: F145 (1988).
11. K. Tatsumi, S. Kitamura, H. Amano, and K. Ueda, Comparative Study on Metabolic Formation of N-Arylformamides and N-Aryl-acetamides from Carcinogenic Arylamines in Mammalian Species, Cancer Res. 49: 2059 (1989).
12. T.V. Zenser, M.B. Mattammal, M.O. Palmier, and B.B. Davis, Microsomal Nitroreductase Activity of Rabbit Kidney and Bladder: Implications in 5-Nitrofuran-Induced Toxicity, J. Pharmacol. Exp. Ther. 219: 735 (1981).
13. M.B. Mattammal, T.V. Zenser, M.O. Palmier, and B.B. Davis, Renal reduced nicotinamide adenine dinucleotide phosphate: cytochrome c reductase-mediated

metabolism of the carcinogen N-[4-(5-nitro-2-furyl)-2-thiazolyl]acetamide, Cancer Res. 45: 149 (1985).

14.  M.B. Mattammal, T.V. Zenser, and B.B. Davis, Anaerobic metabolism and nuclear binding of the carcinogen 2-amino-4-(5-nitro-2-furyl)thiazole (ANFT), Carcinogenesis (Lond.) 3: 1339 (1982).

15.  M.B. Mattammal, V.M. Lakshmi, T.V. Zenser, and B.B. Davis, Mass Spectral Identification of 2-Amino-4-(5-nitro-2-furyl)thiazole Metabolites, Biomed. Mass Spectrom. 15: 495 (1988).

16.  M.B. Mattammal, E. White, T.V. Zenser, and B.B. Davis, Mass spectrometry of 2-substituted-4-arylthiazoles. 3--Identification of microsomal nitroreduction products by mass spectrometry, Biomed. Mass Spectrom. 11: 149 (1984).

17.  T.V. Zenser, M.O. Palmier, M.B. Mattammal, R.I. Bolla, and B.B. Davis, Comparative effects of prostaglandin H synthase-catalyzed binding of two 5-nitrofuran urinary bladder carcinogens, J. Pharmacol. Exp. Ther. 227: 139 (1983).

18.  R.W. Wise, T.V. Zenser, and B.B. Davis, Peroxidase metabolism of the urinary bladder carcinogen 2-amino-4-(5-nitro-2-furyl)thiazole, Cancer Res. 43: 1518 (1983).

19.  R.W. Wise, T.V. Zenser, F.F. Kadlubar, and B.B. Davis, Metabolic activation of carcinogenic aromatic amines by dog bladder and kidney prostaglandin H synthase, Cancer Res. 44: 1893 (1984).

20.  T.J. Flammang, Y. Yamazoe, R.W. Benson, D.W. Roberts, D.W. Potter, D.Z.J. Chu, N.P. Lang, and F.F. Kadlubar, Arachidonic acid-dependent peroxidative activation of carcinogenic arylamines by extrahepatic human tissue microsomes, Cancer Res. 49: 1977 (1989).

21.  J.R. Rice, L.A. Spry, T.V. Zenser, and B.B. Davis, Effect of peroxidase inhibitors on an in vivo metabolite of the urinary bladder carcinogen N-[4-(5-nitro-2-furyl)-2-thiazolyl]formamide in rats, Cancer Res. 48: 304 (1988).

22.  M.B. Mattammal, V.M. Lakshmi, T.V. Zenser, and B.B. Davis, Lung prostaglandin H synthase and mixed-function oxidase metabolism of nicotine, J. Pharmacol. Exp. Ther. 242: 827 (1987).

23.  J.R. Rice, T.V. Zenser, and B.B. Davis, Formation of thioether conjugates of the bladder carcinogen ANFT catalyzed by prostaglandin H synthase, Carcinogenesis (Lond.) 6: 585 (1985).

24.  J.R. Rice, T.V. Zenser, and B.B. Davis, Prostaglandin synthase-dependent cooxidation and aromatic amine carcinogenesis, in:"Arachidonic Acid Metabolism and Tumor Initiation," L.J. Marnett, ed., Martinus-Nijhoff Publishing, Boston, (1985).

25.  R.M. Davidson, J.R. Rice, V.M. Lakshmi, T.V. Zenser, and B.B. Davis, [1]H and [13]C NMR study of 2-aminothiazole urinary tract carcinogens, Magnetic Resonance in Chemistry 26: 482 (1988).

26.  T.J. Powles, R.S. Bockman, K.V. Honn, and P. Ramwell, "Prostaglandins and Cancer," Alan R. Liss, Inc., New York, (1982).

27.  I.B. Weinstein, The origins of human cancer: Molecular mechanisms of carcinogenesis and their implications for cancer prevention and treatment, Cancer Res. 48: 4135 (1988).

28.  D. Julius, T.J. Livelli, T.M. Jessell, and R. Axel, Ectopic Expression of the Serotonin 1c Receptor and the Triggering of Malignant Transformation, Science 244: 1057 (1989).

29.  Y.H. Wong, T.V. Zenser, and B.B. Davis, Regulation of Dog Urothelial Cell Arachidonic Acid Release and $PGE_2$ Synthesis, Carcinogenesis 10: 1621 (1989).

30.  A. Danon, T.V. Zenser, D.L. Thomasson, and B.B. Davis, Eicosanoid synthesis by cultured human urothelial cells: Potential role in bladder cancer, Cancer Res. 46: 5676 (1986).

31.  T.V. Zenser, D.L. Thomasson, and B.B. Davis, Characteristics of bradykinin and TPA increases in the $PGE_2$ levels of human urothelial cells, Carcinogenesis 9: 1173 (1988).

# MUTAGENIC ARYLAZIDES, ARYLNITRENES, ARYLNITRENIUM IONS

Dieter Wild

Institute of Pharmacology and Toxicology
University of Würzburg
Versbacher Str. 9
D-8700 Würzburg, Federal Republic of Germany

## INTRODUCTION

The studies by the Millers (1970) have established that electrophilic nitrenium ions are the ultimate reactive metabolites formed in the course of the metabolism of mutagenic and carcinogenic aromatic amines. Nitroarenes can likewise form nitrenium ions (Rosenkranz and Mermelstein, 1983). In both cases, arylhydroxylamines are formed first and are further activated by esterification (e.g. sulfation, acetylation). Hydrolysis of the esters yields the nitrenium ions. Such esters have been extensively used for the production of and studies on nitrenium ions (Scribner et al., 1970). In practice, however, their synthesis and use have been hampered by a number of difficulties such as instability of the required hydroxylamines under aerobic conditions and instability of the esters in the presence of water. A more convenient source of nitrenium ions would therefore be welcome.

We have explored the suitability of arylazides for the generation of arylnitrenium ions. Photolysis of arylazides generates arylnitrenes, electron-deficient species which can react as radicals and as electrophiles. Formally, an arylnitrenium ion can be obtained from an arylnitrene by addition of a proton. The extensive literature on the chemistry of arylazides and nitrenes has been summarized recently (Smith, 1984). Most studies in this field have been performed with non-aqueous solvents; arylazides have, however, found application also in biological, aqueous systems, namely for photoaffinity labelling (Bayley and Staros, 1984). Furthermore, in 1967 Rose speculated that arylnitrenes might have carcinogenic properties. This idea remained dormant despite a few mutagenesis studies with irradiated azides (Sarrif et al. 1978; Brown 1980; Dollinger et al. 1980; Fukunaga et al. 1984). In the course of the last two years we have found that the photolysis of aqueous solutions of arylazides by near UV-light (NUV) produces reactive species which bind to DNA and deoxynucleotides and are mutagenic in Salmonella (Wild and Dirr, 1988; Wild et al., 1988; Wild et al., 1989). Here we summarize and discuss these recent findings as well as yet unpublished data and present conclusions regarding the origin of the mutagenic potency of aromatic amino- and nitro-compounds.

*Nitroarenes*, Edited by P. C. Howard *et al.*
Plenum Press, New York, 1990

MATERIALS AND METHODS

Arylazides can in general be most conveniently synthesized from the arylamines by diazotization and subsequent reaction of the diazonium salt in the cold with azide ion (Smith and Brown, 1951). We have reported the syntheses of numerous new heterocyclic aryl-azides (Wild and Dirr, 1989). Arylazides are instable in daylight and are therefore handled in yellow light. Protected from daylight, they are in our experience stable at refrigerator temperatures.

The photoactivation of arylazides in diluted aqueous solutions or in the top agar layer of Salmonella reversion test plates is performed by means of a black light bulb (Osram HQV 125, $\lambda$ = 365-366 nm) at a distance of 10 cm. The immediate plate irradiation technique has been previously described (Wild et al., 1989).

RESULTS AND DISCUSSION

The thermospray mass spectra of the azides (in ammonium acetate solution containing small amounts of methanol) exhibit at most traces of the parent azide (M + H$^+$), the more prominent signals are due to thermolysis of the azide and reactions of the nitrene. Usually, the most prominent mass is due to the amine, formed from the nitrene by H-abstraction. Masses compatible with an arylhydroxyl-amine structure are especially noteworthy in the spectra of azido-IQ (Wild and Dirr, 1988) and some other heterocyclic azides. In all spectra, masses compatible with arylmethoxyamine structures are present. Furthermore, arylacetoxyamine-like masses are found. Very similar mass spectra are obtained with photolyzed azide solutions.

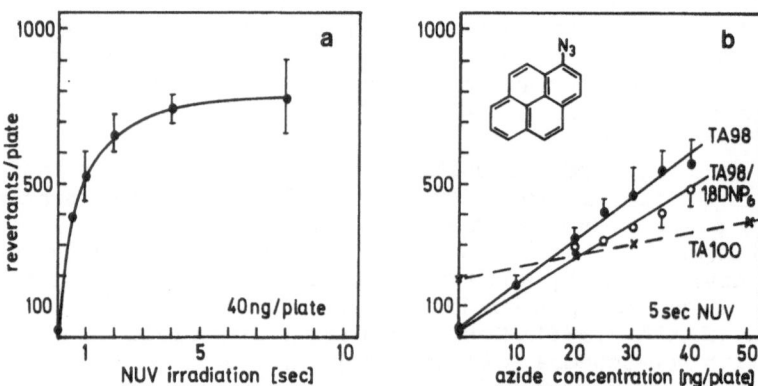

Fig. 1.   Mutagenic activity of 1-azidopyrene + NUV in Salmonella typhimurium TA98, TA98/1,8-DNP$_6$ and TA100.

Salmonella reverse mutation data obtained with 1-azidopyrene + NUV are shown in Fig. 1. 1-Azidopyrene is the most NUV-sensitive azide studied so far: Fig. 1a shows the very steep increase of revertants per plate with increasing NUV irradiation time, the maximal mutagenic response is obtained after 5 sec. Fig. 1b shows the dose response; the slope of the response obtained with Salmo-nella typhimurium TA98 corresponds to a specific mutagenic activity

of 3470 revertants/nmol. Almost the same response (2870 re-
vertants/nmol) was obtained in the acetyltransferase-deficient
strain TA98/1,8-DNP$_6$. Similar findings in the two strains were
obtained also with other arylazides; azido-isoIQ data are presented
in Fig. 2a. For comparison, data obtained with nitro-isoIQ and with
iso-IQ + rat liver S9 are shown in Figs. 2b and 2c. The latter two
compounds owe their mutagenicity to hydroxylamine metabolites;
their lack of esterification in TA98/1,8-DNP$_6$ results in a considerable loss of mutagenic activity. These and additional results concerning the lifetime of the active species (Wild et al., 1989)
suggest that a hydroxylamine is not a major intermediate in azide-
photomutagenesis. We conclude, therefore, that the reactive azide
photolysis product is a nitrene and/or nitrenium ion.

1-Azidopyrene (Fig. 1b) and some other arylazides were also
studied in the Salmonella tester strain TA100. In this strain, NUV
in the absence of an azide was slightly mutagenic and induced 60 re-
vertants/min NUV irradiation. This NUV-effect was subtracted for the
calculation of the mutagenic activities of azides in TA100.

Table 1 gives an overview of the mutation data obtained with a
broad spectrum of arylazides. The activities in TA98 vary over 5
orders of magnitude. The lowest activities are found with the substituted phenyl azides; the activities increase with the size of the
aromatic ring system and reach maximal values for the N-heterocyclic
azido-imidazoaromatic compounds (Wild and Dirr, 1989). The benzene-
and naphthalene-derived azides are more mutagenic in TA100 than in
TA98, 4-azidodiphenyl is equally mutagenic in the two strains, and
the polycyclic azides are considerably less mutagenic in TA100 than
in TA98. Similar relationships between mutagenic activity on the one

Table 1.  Mutagenic activity of carbocyclic and heterocyclic arylazides + NUV in Salmonella typhimurium TA98 and TA100

| Arylazide | Mutagenic activity (revertants/nmol) | | $t_{50}$ (min) |
|---|---|---|---|
| | TA98 | TA100 | |
| Phenyl azide (azidobenzene) | < 0.04 | | − |
| 2,6-Dimethylphenyl azide | < 0.04 | | − |
| 2,4,5-Trimethylphenyl azide | 0.21 | | 1.3 |
| 2,4,6-Trimethylphenyl azide | 0.55 | 6.8 | 1.1 |
| 4,4'-Diazidodiphenylmethane | 3.3 | | 0.3 |
| 1-Azidonaphthalene | 2.4 | 28.7 | 0.25 |
| 2-Azidonaphthalene | 16.1 | 31.4 | 0.43 |
| 4-Azidodiphenyl | 20.3 | 24.6 | 1.0 |
| 4,4'-Diazidodiphenyl | 330 | | 0.1 |
| 2-Azidofluorene | 695 | 511 | 1.0 |
| 6-Azidochrysene | 936 | | 0.17 |
| Azido-Methyl-IQx | 2630 | | 1.6 |
| 1-Azidopyrene | 3470 | 887 | 0.01 |
| Azido-IQ | 13400 | 1950 | 0.5 |
| Azido-Methyl-IQ | 39200 | | 1.2 |
| Azido-isoIQ | 44000 | | 0.5 |

$t_{50}$, period of NUV irradiation required for half-maximal mutation
induction;
data in TA98 and $t_{50}$-values are taken from Wild et al. (1989), Wild
and Dirr (1989), and Wild (1989).

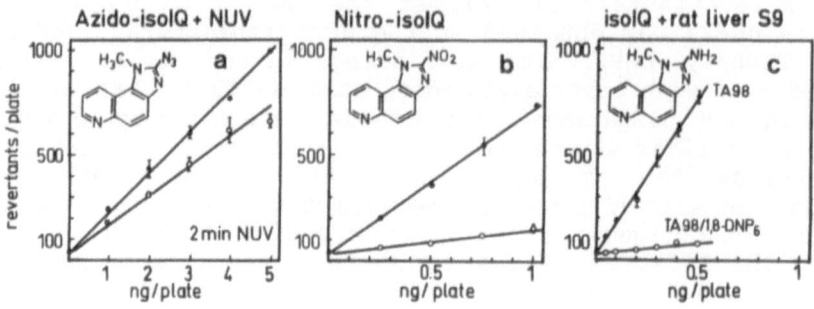

Fig. 2. Mutagenic activity of azido-isoIQ, nitro-isoIQ and isoIQ in Salmonella typhimurium TA98 and TA98/1,8-DNP$_6$.

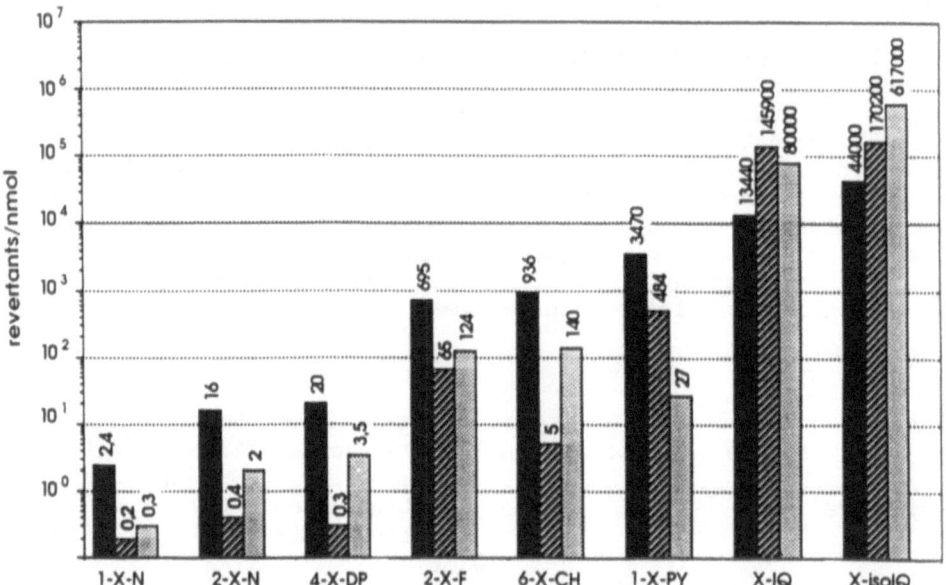

Fig. 3. Synopsis of the mutagenic potencies of matching azido-, nitro- and amino-compounds ("trios") in Salmonella typhimurium TA98. Black columns, X = azido; hatched columns, X = nitro; grey columns, X = amino. Abbreviations used: **N**, naphthalene; **DP**, diphenyl; **F**, fluorene; **CH**, chrysene; **PY**, pyrene. Published data used: Connor et al.(1983); Dirr and Wild (1988); El-Bayoumi et al. (1981, 1989); Kaiser et al. (1986); Later et al.(1984); Mermelstein et al. (1981).

hand and aromatic ring size and tester strain on the other hand have been observed repeatedly in studies on arylamines and nitroarenes (e.g. Later et al., 1984; Rosenkranz and Mermelstein, 1983).

If our assumption is correct and the mutagenic azide photolysis product is (or is closely related to) a nitrenium ion, one may expect a fundamental relationship between the mutagenic activities of azido-, nitro- and amino-compounds. We have therefore compared the mutagenic activities of such compounds. Fig. 2 demonstrates that azido-isoIQ, nitro-isoIQ and isoIQ are extremely potent mutagens in the Salmonella tester strain TA98. Analogous data for azido-nitro-amino "trios" with naphthalene, diphenyl, fluorene, chrysene, pyrene, IQ and isoIQ ring systems have been compiled from the literature and our own work and are shown in Fig. 3. In the carbocyclic trios, the azides are most mutagenic whereas in the heterocyclic trios (including trios not shown here; Wild and Dirr, 1989) the nitro and amino compounds are most mutagenic. These differences within trios can possibly be attributed to differences of compound-specific properties such as partition coefficients, rate of permeation through cell membranes, sites and rates of activation (extra- and intra-cellular activation of the amines, intracellular activation of nitro- and azido-compounds), rates of concurrent inactivation and to different experimental conditions.

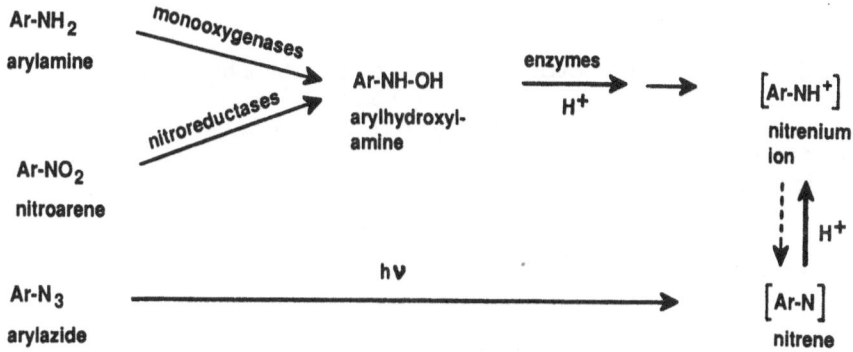

Fig. 4. Formation of electrophilic nitrenes/nitrenium ions

However, the most remarkable feature arising in this comparison is the parallelism of the mutagenic potencies of azido-, nitro- and amino-compounds. This reflects a trinity of structure-activity relationships; it suggests that the ultimate species from azido-, nitro- and amino-compounds are identical, nitrenes/nitrenium ions (Fig. 4), and that their electrophilic reactivity is a main determinant of mutagenic potency in Salmonella for all the members of a trio. The differences of the mutagenic potencies within trios which have above been ascribed to compound-specific properties can now be reconsidered in this light: they are probably due to different intracellular yields of the short-lived nitrene/nitrenium ion which can bind to DNA.

The potency differences between trios can be understood as an expression of corresponding differences in the electrophilic character of the ultimate nitrene/nitrenium ion. The relationship between the electrophilic character and the chemical structure of nitrenes/-nitrenium ions can be explained on the basis of the delocalization concept which assumes that delocalization of the positive charge stabilizes the nitrenium ion and, at the same time, increases its reactivity with DNA. This argumentation can be extended to the electron-deficiency of the nitrene. Parameters relevant to the efficiency of delocalization are: number of conjugated or condensed aromatic rings, position of the nitrenium substituent on the aromatic ring system (compare 1-azido- and 2-azido-naphthalene), substitution by methyl and by amino groups. The extraordinary mutagenic potency of the nitrenes/nitrenium ions derived from IQ and related heterocyclic compounds can be explained by an extraordinarily efficient delocalization of the positive nitrenium charge into the aromatic ring: we propose that the main contributing resonance structures are the imidazole-nitrenium ion and the imidazole-ammonium ion structures (with the positive charge at N1 and N3 of the imidazole ring). Carbenium ion resonance structures can also contribute to the total delocalization.

It may be added that the conclusions drawn from the Salmonella mutagenesis data have been confirmed by $^{32}$P-postlabelling studies on the DNA and nucleotide adducts obtained from azido-, nitro- and amino-aromatic compounds (Wild et al., 1989). Preliminary data suggest a correlation between the dGp-binding and mutation-inducing potencies of the nitrenes/nitrenium ions.


CONCLUSIONS

1. Arylazides can be photoactivated to electrophilic species which might be nitrenes or nitrenium ions and react like the aryl-nitrenium ions formed metabolically from arylamines and nitroarenes. Therefore arylazides can be used as a convenient source of nitrenium ions.

2. The mutagenic potency of azido-, nitro- and amino-aromatic compounds in Salmonella is governed mainly by the <u>electrophilic nature</u> of the trio-specific nitrene/nitrenium ions. This can be seen from the essential similarity of the potencies within the azido-nitro-amino-trios. Compound-specific properties can affect the <u>yield of formation</u> of the nitrenes/nitrenium ions and can thus modify the mutagenic potency; they do, however, not override the governing role of the nitrene/nitrenium ion.

ACKNOWLEDGEMENT

This work is supported by the Deutsche Forschungsgemeinschaft, Sonderforschungsbereich 172 "Molekulare Mechanismen kanzerogener Primärveränderungen".

REFERENCES

Bayley, H. and Staros, J.V., 1984, Photoaffinity labeling and related techniques, in "Azides and Nitrenes", E.F.V. Scriven, ed., Academic Press, Orlando, p.433.
Brown, B.R., 1980, Mutagenesis by photoaffinity labelling using selected azidofluorenes, Mutat. Res., 70:17.
Connor, T.H., Ramanujam, V.M.S., Rinkus, S.J., Legator, M.S. and Trieff, N.M., 1983, The evaluation of mutagenicities of 19 structurally related aromatic amines and acetamides in Salmonella typhimurium TA98 and TA100, Mutat. Res., 118:49.
Dirr, A. and Wild, D., 1988, Synthesis and mutagenic activity of nitro-imidazoarenes. A study on the mechanism of the genotoxicity of heterocyclic arylamines and nitroarenes, Mutagenesis, 3:147.
Dollinger, D.D., Hixon, S.C., Sarrif, A.M. and White, W.E., Jr.,

1980, Mutations and cell transformation with 2-azido-9-fluoren-one oxime, In Vitro, 16:541.

El-Bayoumi, K., Lavoie, E.J., Hecht, S.S., Fow, E.A. and Hoffmann, D., 1981, The influence of methyl substitution on the mutagenicity of nitronaphthalenes and nitrobiphenyls, Mutat. Res., 81:143.

El-Bayoumi, K., Delclos, K.B., Heflich, R.H., Walker, R., Shiue, G.-H. and Hecht, S.S., 1989, Mutagenicity, metabolism and DNA adduct formation of 6-nitrochrysene in Salmonella typhimurium, Mutagenesis, 4:235.

Fukunaga, M., Cox, B.A., von Sprecken, R.S. and Yielding, L.W., 1984, Production of frameshift mutations in Salmonella typhimurium py phenanthridinium derivatives: enzymatic activation and photoaffinity labeling, Mutat. Res., 127:31.

Kaiser, G., Harnasch, D., King, M.-T. and Wild, D., 1986, Chemical structure and mutagenic activity of aminoimidazoquinolines and aminonaphthimidazoles related to IQ, Chem.-Biol. Interact., 57:97.

Later, D.W., Pelroy, R.A., Stewart, D.L., McFall, T., Booth, G.M., Lee, M.L., Tedjamulia, M. and Castle, R.N., 1984, Microbial mutagenicity of isomeric two-, three-, and four-ring amino polycyclic aromatic hydrocarbons, Environ. Mutag., 6:497.

Mermelstein, R., Kiriazides, D.K., Butler, M., McCoy, E.C. and Rosenkranz, H.S., 1981, The extraordinary mutagenicity of nitropyrenes in bacteria, Mutat. Res., 89:187.

Miller, J.A., 1970, Carcinogenesis by chemicals: an overwiew - G.H.A. Clowes Memorial Lecture, Cancer Res., 30:559.

Rose, F.C., 1967, Possible cytotoxic role of nitrenes, Nature, 215:1492.

Rosenkranz, H.S. and Mermelstein, R., 1983, Mutagenicity and genotoxicity of nitroarenes - All nitro-containing chemicals were not created equal, Mutat. Res., 114:217.

Sarrif, A.M., White, W.E., Jr. and DiVito, N., 1978, Photolysis of 2-azidofluorene in situ as a probe in chemical carcinogenesis: bypass of requirement for metabolic activation, Biochem. Biophys. Res. Commun., 83:506.

Scribner, J.D., Miller, J.A. and Miller, E.C., 1970, Nucleophilic substitution on carcinogenic N-acetoxy-N-arylacetamides, Cancer Res., 30:1570.

Smith, P.A.S., 1984, Aryl and heteroaryl azides and nitrenes, in "Azides and Nitrenes", E.F.V. Scriven, ed., Academic Press, Orlando, p. 95.

Smith, P.A.S. and Brown, B.B., 1951, The reaction of aryl azides with hydrogen halides, J. Am. Chem. Soc., 73:2438.

Wild, D., 1989, A novel pathway to the ultimate mutagens of aromatic amino and nitro compounds, Environ. Health Persp., in press.

Wild, D. and Dirr, A., 1988, Synthesis of 2-azido-3-methyl-imidazo[4,5-f]quinoline and photolytic generation of a highly reactive and mutagenic IQ derivative, Carcinogenesis, 9:869.

Wild, D. and Dirr, A., 1989, Mutagenic nitrenes/nitrenium ions from azido-imidazoarenes and their structure-activity relationships, Mutagenesis, 4: in press.

Wild, D., Asan, E., Dirr, A. Fasshauer, I. and Henschler, D., 1988, DNA-adducts of aminoimidazoarenes and structurally analogous nitro- and azido-imidazoarenes, in "Carcinogenic and Mutagenic Responses to Aromatic Amines and Nitroarenes", C.M. King, L.J. Romano and D. Schuetzle, eds., Elsevier, New York, p. 271.

Wild, D., Dirr, A., Fasshauer, I. and Henschler, D., 1989, Photolysis of arylazides and generation of highly electrophilic DNA-binding and mutagenic intermediates, Carcinogenesis, 10:335.

# PRODUCTS OBTAINED BY IN VITRO REACTION OF 4,5-EPOXY-4,5-DIHYDRO-1-NITROPYRENE WITH DNA

Karam El-Bayoumy, Ajit K. Roy, and
Stephen S. Hecht

Section of Biological Chemistry
Division of Chemical Carcinogenesis
American Health Foundation, Valhalla, N.Y. 10595

## INTRODUCTION

A number of nitropolycyclic aromatic hydrocarbons ($NO^2$-PAH) have been identified in environmental sources(1-3). 1-Nitropyrene (1-NP), a representative example, is a bacterial mutagen and an animal tumorigen and has been reported to be the most abundant $NO^2$-PAH in sources such as diesel engine emission (4). 1-NP is activated in bacterial systems (5,6) *via* nitro-reduction to yield a major DNA adduct, N-(deoxyguanosin-8-yl)-1-aminopyrene (N-dG-1-AP). However, studies carried out in our laboratory as well as in others indicate that simple nitro-reduction of 1-NP cannot account for the observed DNA adducts *in vivo* and *in vitro* (7-12). In order to more clearly define the metabolic activation of 1-NP, additional DNA adduct markers are required. Therefore, in this paper we present our initial results on the identification of a major adduct formed from the incubation of DNA with 4,5-epoxy-4,5-dihydro-1-nitropyrene (1-NP-4,5-oxide), a mutagenic ring oxidized metabolite of 1-NP. We have also established the $^{32}$P-fingerprints of the DNA adducts derived from 1-NP-4,5-oxide.

## MATERIALS AND METHODS

1-NP was obtained commercially and purified as described (13). [4,5,9,10-$^3$H]1-NP (12.9 Ci/mmol, >99% radiochemical purity) was acquired from Chemsyn Science Laboratories, Lenexa, KS. 1-NP-4,5-oxide and its radiolabelled analog (23.6 mCi/mmol) were synthesized by treatment of 1-NP and [4,5,9,10-$^3$H]1-NP with m-chloroperbenzoic acid (Aldrich Chem. Co., Milwaukee, WI) according to a published procedure (14). *Cis*- and *trans*-4,5-dihydro-4,5-dihydroxy-1-nitropyrene (*cis*- and *trans*-4,5-DHD-1-NP) were synthesized as previously described (15). Calf thymus DNA (type I), enzymes, and all other biochemical reagents were purchased from Sigma Chemical Co., St. Louis, MO.

The incubations were carried out as previously described (7). Briefly, calf thymus DNA was dissolved (2 mg/ml) in 50 mM citrate buffer, pH 5.8, and the solution was purged with nitrogen gas (10 min). An aliquot of [4,5,9,10-$^3$H]1-NP-4,5-oxide (23.6 mCi/mmol) or 1-NP-4,5-oxide (from a 4.0 mM stock solution in nitrogen purged DMSO) was added to the DNA solution to a final concentration of 20 to 40 $\mu$M. The solution was incubated for 3 h at 37°C. The DNA was purified by solvent extractions (12) and was enzymatically hydrolyzed by a previously published method (16). The adducts were analyzed by reverse phase HPLC; see figure legends for HPLC conditions.

The proton magnetic resonance (PMR) analyses were performed with a Bruker AM-360 MHz spectrometer. Radioactivity was determined on a Beckman liquid scintillation counter.

The DNA which had been modified with 1-NP-4,5-oxide was hydrolyzed to deoxyribonucleoside-3'-monophosphates and the adducts were enriched by extraction with n-butanol (17). The adducts were converted to [5'-$^{32}$P]3',5'-bisphosphates, analyzed by four directional PEI-cellulose TLC, and then located by screen-enhanced autoradiography according to published procedures (17,18).

RESULTS AND DISCUSSION

HPLC analysis of the hydrolysate of DNA which had been modified with 1-NP-4,5-oxide yielded 3 major peaks (Figure 1); other minor peaks were also observed. The yield of the 3 peaks, based on [$^3$H]1-NP-4,5-oxide, was 4.2±0.5%. The formation of multiple adducts was not surprising. For any reactive center in DNA, 4 pairs of diastereomers resulting from *cis*- and *trans*- addition and substitution at the 4 or 5 positions could form.

The PMR spectrum of peak 3 in DMSO-d$^6$ is shown in Figure 2A; the expanded spectrum of the aromatic region of peak 3 in DMSO-d$^6$ is shown in Figure 2B; the corresponding spectra in the presence of D$^2$O are shown in Figures 3A and 3B.

The chemical shifts of the protons in the adduct and in *cis*- and *trans*-1-NP-4,5-DHD (Table 1) are similar. However, in the adduct, additional downfield signals were observed at 7.23, 7.95, and at 10.65 ppm, representing N$^2$-H, C8-H, and N1-H of guanine respectively; the signals at 7.23 and 10.65 ppm disappeared upon adding D$^2$O. These results demonstrate that the adduct is a deoxyguanosine derivative. Several nucleophilic centers in deoxyguanosine are available to react with 1-NP-4,5-oxide. Attachment of the C8 position is ruled out by the presence of the proton at 7.9 ppm. The observation of the amide proton (N1-H) at 10.65 ppm eliminated the possibility of substitution at O$^6$ or N1 (19). If the substitution had occurred at the 7-position of guanine we would have expected the chemical shift of the proton attached to C8 to be further downfield because of the positive charge in the imidazole ring.

The pH partition coefficient pattern of the adduct between aqueous buffers and n-butanol: ether (10:90), calculated (Figure 4) according to the method of Moore and Koreeda, showed that the adduct had basic and acidic pH values (20). This result eliminates the possibility of the adduct being an N1-, $O^6$-, or 7- substituted guanine and supports the $N^2$ substitution of guanine. The adduct was subjected to acid hydrolysis under conditions (0.1 M HCl, 37°C, 14 h) which are known to cleave the deoxyribose moiety (21). The PMR spectra of the adduct after acid hydrolysis (Table 1) further supports the $N^2$-substitution of guanine.

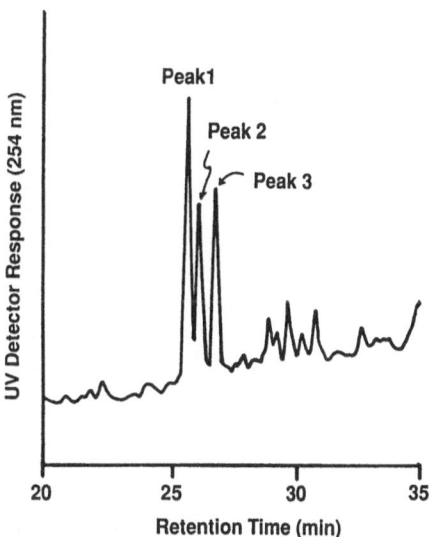

Figure 1. HPLC profile of the hydrolysate of calf thymus DNA incubated with 1-NP-4,5-oxide. The analysis was performed using a Vydac $C_{18}$-10$\mu$ reverse phase HPLC column (1.0 x 25 cm) with a linear gradient from 0% to 75% $CH_3OH$ in $H_2O$ over 30 min at a flow rate of 5 ml/min.

The *cis*- or *trans*- configuration of the hydroxyl and guanine moieties could not be established in this study due to the small coupling constant between the benzylic hydrogens (3.4 Hz). This small coupling indicates that the benzylic protons must adopt either the [ea]- or [ee]- orientations. $N^2$-substitution has also been observed in the reaction of the K-region epoxide of 7,12-dimethylbenz[a]anthracene with poly G *in vitro* (22); however, on the basis of the small coupling constant between the benzylic hydrogens, it was also not possible to distinguish between the *cis*- and *trans*- configuration. It

TABLE 1. Proton magnetic resonance spectral parameters for one of the major products (peak 3) of the reaction of 1-NP-4,5-oxide with calf thymus DNA[a]

| Assignment | Chemical shift (multiplicity, J [Hz]) | | | | |
|---|---|---|---|---|---|
| | Peak 3[b]/DMSO-$d_6$ | Peak 3/DMSO-$d_6$-$D_2O$ | Peak 3 after acid hydrolysis | trans-4,5-DHD-1-NP/DMSO-$d_6$ | cis-4,5-DHD-1NP/DMSO-$d_6$ |
| $H_2$ | 8.42(d,8.1) | 8.38(d7.9) | 8.41(d,8.1) | 8.39(d,8.1) | 8.39(d,8.0) |
| $H_{10}$ | 8.30(d,9.1) | 8.27(d,9.3) | 8.29(D,9.2) | 8.27(d,9.3) | 8.29(d,9.4) |
| $H_9$ | 8.21(d,9.2) | 8.16(d,9.3) | 8.20(d,9.2) | 8.15(d,9.3) | 8.16(d,9.4) |
| $H_7$ | 8.05(m) | 8.03(m) | 8.04(m) | 7.78(dd,7.6,7.5) | 7.80(m) |
| $H_3$ | 7.93(d,8.0) | 7.89(d,7.9) | 7.90(d,8.0) | 7.98[*] | 7.96(dd,8.0,0.9) |
| $H_6$ | 7.78(m) | 7.76(m) | 7.77(d,5.1) | 7.90(d,6.9) | 7.80(m) |
| $H_8$ | 7.78(m) | 7.76(m) | 7.77(d,5.1) | 7.90(d,6.9) | 8.03(dd,7.6,1.7) |
| $H_5$ | 5.70(bs) | 5.75(d,3.9) | 5.72(bs) | 4.97(bs) | 5.04(dd,5.1,4.5) |
| $H_4$ | 5.20(bs) | 5.21(d,3.9) | 5.20(bs) | 4.97(bs) | 5.11(dd,6.2,4.5) |
| -$OH_4$ | 5.70[c] | - | - | 6.14(bs) | 5.80(d,6.2) |
| -$OH_1$ | 5.70[c] | - | - | 5.98(bs) | 5.43(d,5.1) |
| $N^1$-H | 10.65(bs) | - | - | - | - |
| $C_8$-H | 7.78[d] | 7.91[d] | 8.55(s) | - | - |
| $N^2$-H | 7.25(bs) | - | - | - | - |
| -1'-H | 6.10(bs) | 6.09(t,3.8) | - | - | - |
| -2'-$H_a$ | c | c | - | - | - |
| -2'-$H_b$ | c | c | - | - | - |
| -3'-H | 4.25(bs) | c | - | - | - |
| -4'-H | 3.72(bs) | c | - | - | - |
| -5'-$H_a$ | c | c | - | - | - |
| -5'-$H_b$ | c | c | - | - | - |
| -3'-OH | c | - | - | - | - |
| -5'-OH | c | - | - | - | - |
| $N^9$-H | - | - | 7.65[e] | - | - |

[a]Spectra were recorded at 300 MHz and chemical shifts are reported in ppm downfield from tetramethylsilane. m, multiplet; t, triplet; d, doublet; s, singlet; J, coupling constant in Hz. cis- and trans- were included for comparison.

[b]The adduct exhibited a doubling of resonances presumably due to two conformational isomers; one set of resonances disappeared upon adding $D_2O$.

[c]These resonances were not resolved and were tentatively assigned due to either line broadening or to overlap with solvent resonances.

[d]These resonances appear as a multiplet containing the resonance of C8-H.

[e]These resonances appear as a multiplet containing the resonance of H3.

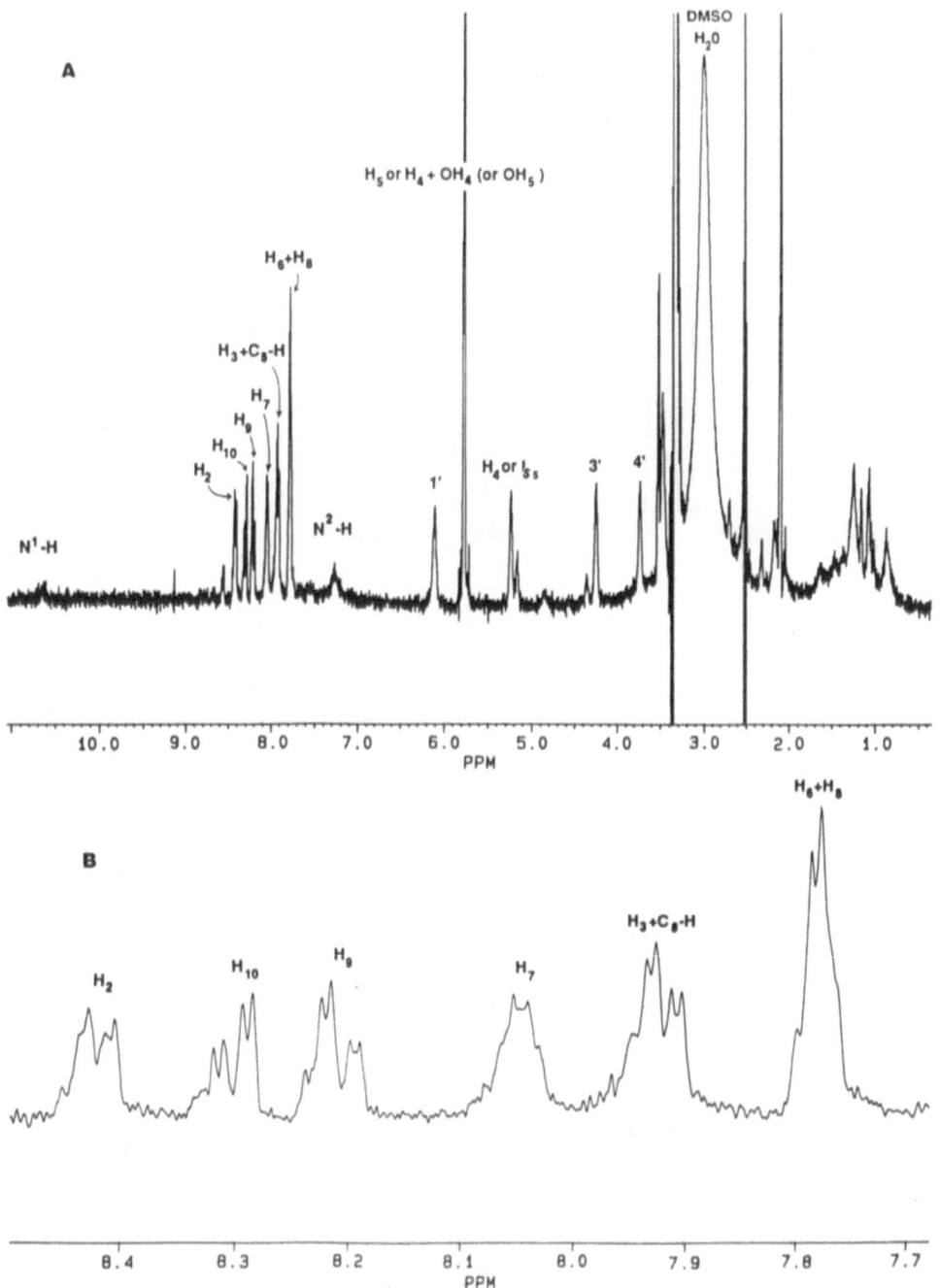

Figure 2. PMR spectrum of the adduct (peak 3) in DMSO-$d_6$ [A] and the expanded portion of the aromatic region from 7.7 to 8.5 ppm [B].

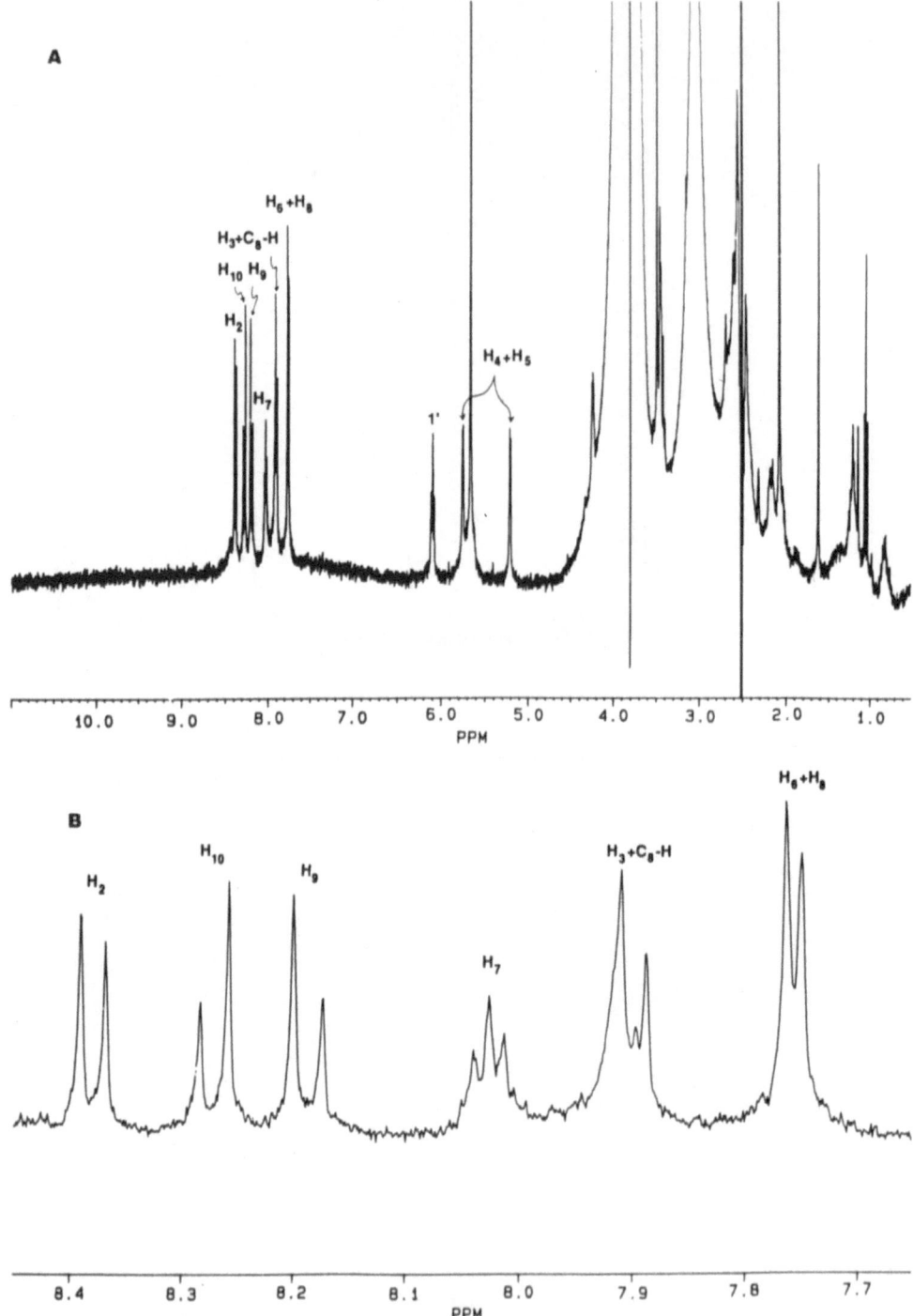

Figure 3. PMR spectrum of the adduct (peak 3) in DMSO-d$_6$/D$_2$O [A] and the expanded portion of the aromatic region from 7.65 to 8.45 ppm [B].

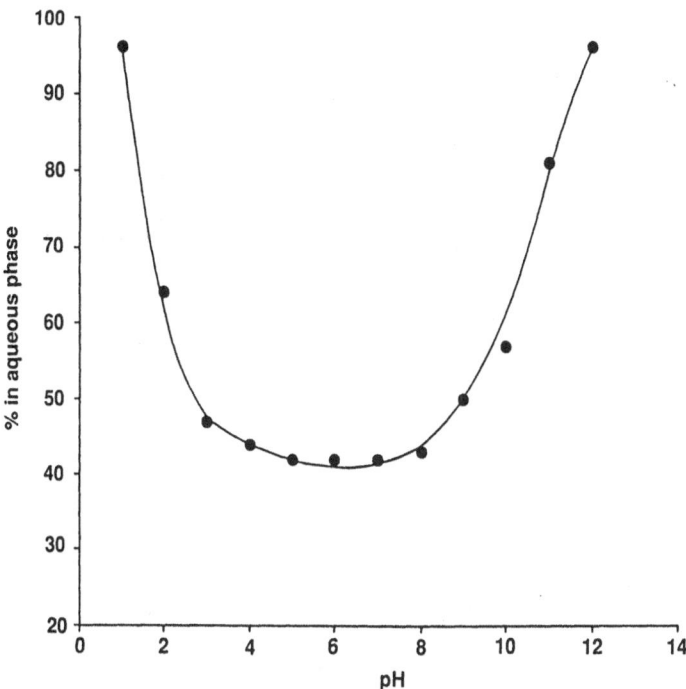

Figure 4. Partition of the adduct (peak 3) between aqueous
buffers, pH 1 to 12, and 10% n-butanol in diethyl
ether.

needs to be confirmed whether the substitution occurred at the
4 or 5 position.  However, an $S_N1$ mechanism is more likely to
occur based on the slightly acidic condition of the incubation
mixture and on electronic factors.  Accordingly we propose that
substitution at the 5-position may have occurred (Figure 5).

     The [32]P-fingerprints of DNA which had been modified with
1-NP-4,5-oxide are shown in Figure 6A.  The pattern of DNA
which had been modified with 1-NP in the presence of xanthine
oxidase *in vitro* (12) is included for comparison in Figure 6C.
Since the [32]P-postlabelling assays were conducted under identical conditions it can be concluded that adducts derived from 1-
NP-4,5-oxide are chromatographically different from the major
adduct derived from nitroreduction of 1-NP.  On the other hand,
the adducts derived from 1-NP-4,5-oxide were chromatographically similar to some of the adduct spots which had been
observed in our previous study in rat liver and rat mammary
glands following the administration of 1-NP (12); however, this
needs to be confirmed.

     In summary, we propose the structures outlined in Figure 7
for peak 3 formed by reaction of 1-NP-4,5-oxide with DNA.  We
also hypothesize that some of the putative 1-NP-DNA adducts
observed *in vivo* may have been formed from 1-NP-4,5-oxide.
Currently, experiments are being conducted to test our
hypothesis.

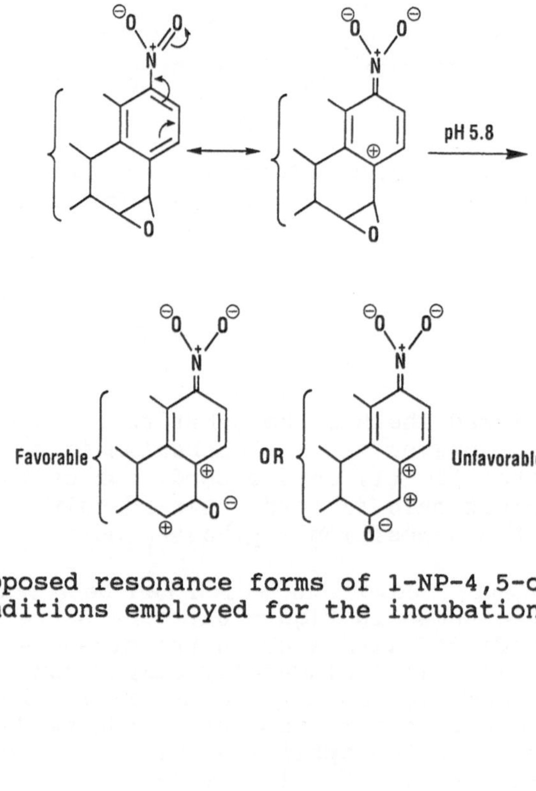

Figure 5. Proposed resonance forms of 1-NP-4,5-oxide under the conditions employed for the incubations.

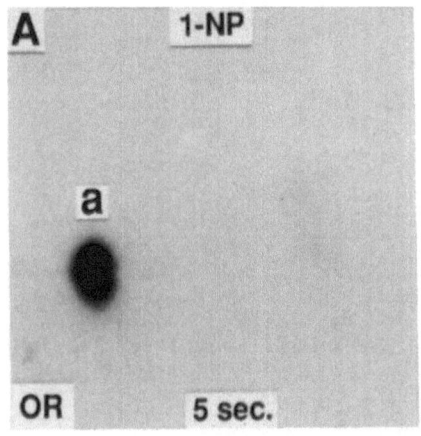

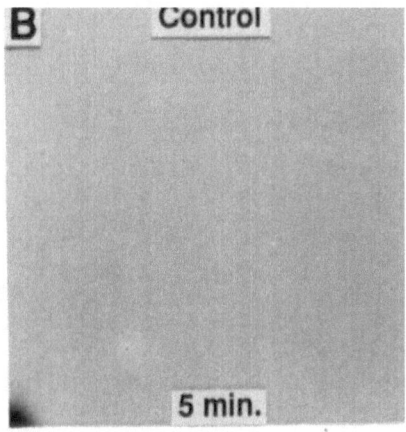

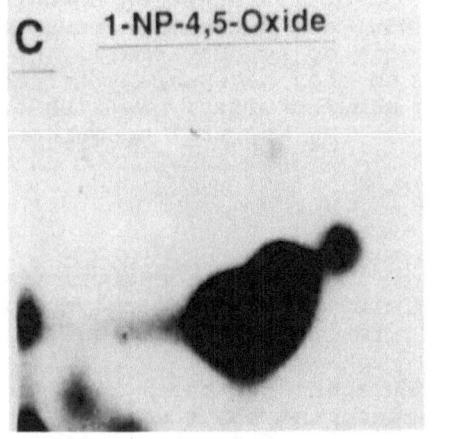

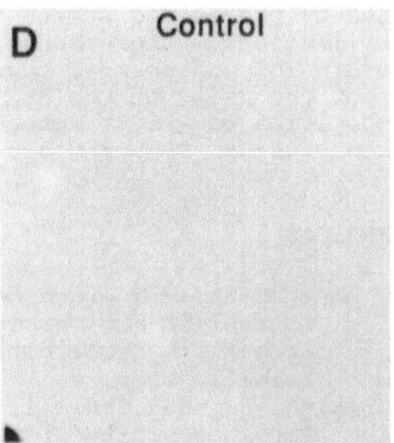

Figure 6. [32]P-Maps of DNA modified with 1-NP in the presence of xanthine oxidase [A], and with 1-NP-4,5-oxide [C]; [B] and [D] represent the corresponding control experiments, in which no xanthine oxidase or 1-NP-4,5-oxide was included in the incubation.

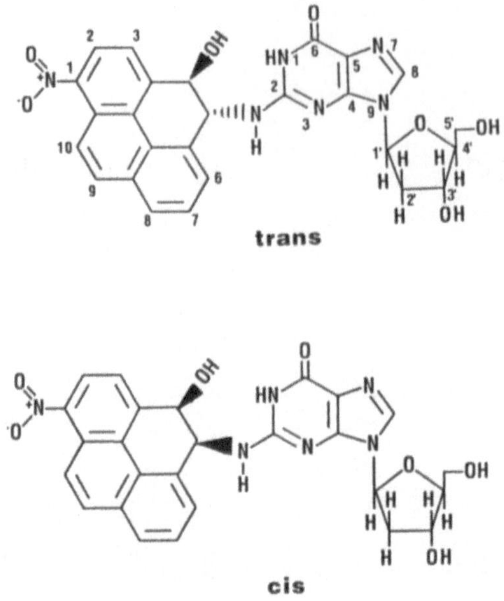

Figure 7. Proposed structures of adduct (peak 3) obtained by reaction of 1-NP-4,5-oxide with calf thymus DNA in vitro.

ACKNOWLEDGEMENTS

This study was supported by National Cancer Institute Grant CA 35519. Part of this work was supported under contract to the Health Effects Institute (HEI), an organization jointly funded by the United States Environmental Protection Agency (EPA) (Assistance Agreement X-812059) and automotive manufacturers. The contents of this article do not necessarily reflect the views of the HEI, nor do they necessarily reflect the policies of EPA or automotive manufacturers. We wish to acknowledge Dr. Bijaya Misra for the PMR spectral measurements.

REFERENCES

1.  Rosenkranz, H.S., and Mermelstein, R.  The genotoxicity, metabolism and carcinogenicity of nitrated polycyclic aromatic hydrocarbons.  J. Environ. Sci. Health, C3: 221-272, 1985.
2.  Tokiwa, H., and Ohnishi, Y.  Mutagenicity and carcinogenicity of nitroarenes and their sources in the environment.  Crit. Rev. Toxicol., 17: 23-60, 1986.
3.  Fu, P. P., Chou, M. W., and Beland, F.A.  Effects of nitro substitution on the in vitro metabolic activation of polycyclic aromatic hydrocarbons.  In: S. K. Yang and B. D. Silverman (eds.), Polycyclic Aromatic Hydrocarbon Carcinogenesis:Structure Activity Relationships, Vol. 2, pp. 37-65.  Boca Raton: CRC Press, Inc., 1988.
4.  Schuetzle, D.  Sampling of vehicle emissions for chemical analysis and biological testing.  Environ. Health Perspect., 47: 65-80, 1983.

5.  Howard, P. C., Heflich, R. H., Evans, F. E., and Beland, F. A. Formation of DNA adducts in vitro and in Salmonella typhimurium upon metabolic reduction of the environmental mutagen 1-nitropyrene. Cancer Res., 43: 2052-2058, 1983.
6.  Messier, F., Lu, C., Andrews, P., McCarry, B. E., Quilliam, M. A., and McCalla, D. R. Metabolism of 1-nitropyrene and formation of DNA adducts in Salmonella typhimurium. Carcinogenesis, 2: 1007-1011, 1981.
7.  Djuric, Z., Fifer, D. K., Howard, P.C., and Beland, F.A. Oxidative microsomal metabolism of 1-nitropyrene and DNA-binding of oxidized metabolites following nitroreduction. Carcinogenesis, 7: 1073-1079, 1986.
8.  Jackson, M. A., King, L. C., Ball, L. M., Ghayourmanesh, S., Jeffrey, A. M., and Lewtas, J. Nitropyrene: DNA binding and adduct formation in respiratory tissues. Environ. Health Perspect., 62: 203-207, 1985.
9.  Mitchell, C. E. Formation of DNA adducts in mouse tissues after intratracheal instillation of 1-nitropyrene. Carcinogenesis, 9: 857-860, 1988.
10. El-Bayoumy, K. E., Shiue, G.-H., and Hecht, S. S. Metabolism and DNA binding of 1-nitropyrene and 1-nitrosopyrene in newborn mice. Chem. Res. Tox., 1: 243-247, 1988.
11. Stanton, C. A., Chow, F. L., Phillips, D. H., Grover, P. L., Garner, R. C., and Martin, C. N. Evidence for N-(deoxyguanosin-8-yl)-1-aminopyrene as a major DNA adduct in female rats treated with 1-nitropyrene. Carcinogenesis, 6: 535-538,1985.
12. Roy, A. K., El-Bayoumy, K., and Hecht, S. S. [32]P-postlabeling analysis of 1-nitropyrene-DNA adducts in female Sprague-Dawley rats. Carcinogenesis, 10: 195-198, 1989.
13. El-Bayoumy, K., Rivenson, A., Johnson, B., DiBello, J., Little, P., and Hecht, S. S. Comparative tumorigenicity of 1-nitropyrene, 1-nitrosopyrene, and 1-aminopyrene administered by gavage to Sprague-Dawley rats. Cancer Res., 48, 4526-4260, 1988.
14. Fifer, E.K., Howard, P.C., Heflich, R.H. and Beland, F.A. Synthesis and mutagenicity of 1-nitropyrene-4,5-oxide and 1-nitropyrene-9,10-oxide, microsomal metabolites of 1-nitropyrene. Mutagenesis, 1: 433-438, 1986.
15. El-Bayoumy, K., Villucci, P., Roy, A. K., and Hecht, S. S. Synthesis of K-region derivatives of the carcinogen 1-nitropyrene. Carcinogenesis, 7: 1577-1580, 1986.
16. Martin, C. N., Beland, F. A., Roth, R. W., Kadlubar, F. F. Covalent binding of benzidine and N-acetylbenzidine to DNA at the C-8 atom of deoxyguanosine in vivo and in vitro. Cancer Res., 42: 2678-2696, 1982.
17. Gupta, R. C., Reddy, M. V., and Randerath, K. [32]P-Postlabelling analysis of non-radioactive aromatic carcinogen-DNA adducts. Carcinogenesis, 3: 1081-1092, 1982.
18. Gupta, R. C. Enhanced sensitivity of [32]P-postlabelling analysis of aromatic carcinogen: DNA adducts. Cancer Res., 45: 5656-5662, 1985.
19. Loveless, A. Possible relevance of O-6 alkylation of deoxyguanosine to the mutagenicity and carcinogenicity of nitrosamines and nitrosamides. Nature, 223: 206-207, 1969.
20. Moore, P. D., and Koreeda, M. Application of the change

in partition coefficient with pH to the structure
determination of alkyl substituted guanosines.
Biochem. Biophys. Res. Commun., 73: 459-464, 1976.

21.  Preuss-Schwartz, D., and Baird, W. M.  Benzo[a]pyrene: DNA
     adduct formation in early-passage wistar rat embryo
     cell culture: evidence for multiple pathways of
     activation of benzo[a]pyrene.  Cancer Res., 46: 545-
     552, 1986.

22.  Jeffrey, A. M., Blobstein, S. H., Weinstein, I. B.,
     Beland, F. A., Harvey, R. G., Kasai, H., and Nakanishi,
     K.  Structure of 7, 12-dimethylbenz[a]anthracene-
     guanosine adducts.  Proc. Natl. Acad. Sci., 73:
     2311-2315, 1976.

# OXIDATIVE METABOLISM OF NITRATED POLYCYCLIC AROMATIC

# HYDROCARBONS IN MAMMALIAN CELL FRACTIONS

Paul C. Howard and Scott F. Purvis

Department of Environmental Health Sciences
Case Western Reserve University School of Medicine
Cleveland, Ohio USA 44106

## INTRODUCTION

The nitrated polycyclic aromatic hydrocarbons (nitro-PAH) are ubiquitous pollutants in the environment, whose genotoxicity is dependent on biological activation. Two pathways for the metabolism of nitro-PAH have been noted: *(1)* cytochrome P450-mediated C-oxidation; and *(2)* nitroreduction of the exocyclic nitro group via cytosolic and/or microsomal enzymes to reactive arylhydroxylamines (1-4). Typically, the cytochrome P450-mediated C-oxidation leads to the formation of phenol and dihydrodiol metabolites that are quickly detoxified by sulfate or glucuronide conjugation (5-7). Alternatively, epoxide intermediates are conjugated to glutathione with glutathione transferases (8).

This has led to the hypothesis in our laboratory that the carcinogenicity of nitro-PAH is dependent on the balance of C-oxidation and nitroreduction [the exceptions to this rule are 6-nitrobenzo(a)pyrene (9-10), and 6-nitrochrysene (11-12)]. This hypothesis is supported by several observations in the literature. First, although DNA-adducts of 1-nitropyrene were detected by Stanton *et al.* (13), in other studies, Beland *et al.* (14) and ourselves (unpublished) have been unable to detect C8-deoxyguanosine adducts in adult rats when dinitropyrene-free 1-nitropyrene was administered. Secondly, several tumor studies have shown that the tumorigenicity of 1-nitropyrene is more pronounced in young animals than in adult animals (15-16). The principal detoxification pathway for 1-nitropyrene is through cytochrome P450 mediated C-oxidation. The cytochromes P450 are differentially expressed through development, and in general, are expressed at much lower levels in neonates than in adult animals (17). This then would suggest that the principal detoxification pathway might be limiting in young animals. Thirdly, the nitro-PAH 1,8-dinitropyrene is highly mutagenic in *in vitro* tests and highly carcinogenic in animal studies (16,18). If 1,8-dinitropyrene is like the mononitro-PAH, the principal detoxification pathway for 1,8-dinitropyrene would be through cytochrome P450 C-oxidation, while the bioactivation would be through nitroreduction. While nitroreduction followed

by O-esterification has been shown to be the activation pathway
for 1,8-dinitropyrene (19-21), we have been unable to detect
any cytochrome P450 mediated C-oxidation of 1,8-dinitropyrene
using rat or rabbit liver microsomes (unpublished).  Therefore,
1,8-dinitropyrene administered to animals is metabolized only
by nitroreduction which leads to obligatory formation of active
intermediates.  This could explain the high carcinogenicity
noted for this compound.

Experiments have been underway in our laboratory to test
the hypothesis that the carcinogenicity of nitro-PAH are the
result of the balance of nitroreduction and C-oxidation.  This
manuscript describes the preliminary results of the kinetics of
the C-oxidative metabolism of 1-nitropyrene using rabbit liver
cellular subfractions.

MATERIALS AND METHODS

Liver microsomes were isolated from healthy adult male New
Zealand White rabbits (Hazelton Laboratories).  Following $CO_2$
asphyxiation, the livers were removed, minced, and allowed to
stand on ice for 10 minutes in 0.25 M sucrose, 0.1 M potassium
phosphate, pH 7.4 (sucrose-$PO_4$ buffer).  The livers were then
weighed and 3 volumes of sucrose-$PO_4$ buffer were added.  The
livers were homogenized using a Brinkman Polytron PT30 probe,
and centrifuged at 12,000xg for 20 min at 2-4°C.  The
supernatant was then centrifuged at 100,000xg for 1 hr at 2-
4°C.  The microsomal pellets were resuspended in 0.15 M KCl
using Potter-Elvehjam homogenizers, and centrifuged at

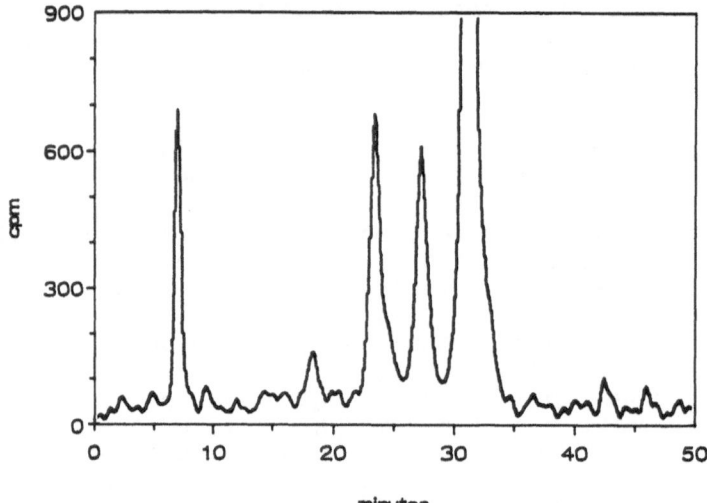

Fig. 1      High performance liquid chromatography
            analysis of the microsomal metabolites of
            [4,5,9,10-³H]1-nitropyrene.  The
            chromatographic conditions are essentially
            the same as those described in (22).  The
            metabolites were identified by
            cochromatography with authentic standards:
            1-nitropyrene-4,5-dihydro-4,5-diol, 7 min;
            1-nitropyren-6-ol and 1-nitropyren-8-ol, 23
            min; 1-nitropyren-3-ol, 27 min; 1-
            nitropyrene, 31 min.

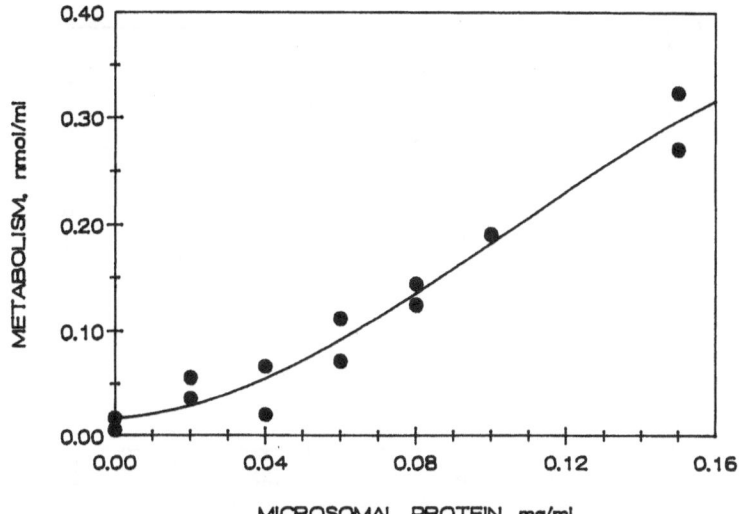

Fig. 2    Dependence of metabolism on microsomal
          protein concentration.  The metabolism of
          [4,5,9,10-³H]1-nitropyrene was determined
          by varying the protein concentration
          between 0 and 150 μg/ml.  Incubations were
          for 10 min under condition described in the
          text.

100,000xg for 1 hr at 2-4°C.  The microsomal pellets were
resuspended in sucrose-$PO_4$ buffer using Potter-Elvehjam
homogenizers, and stored in aliquots at -70°C.

    For the metabolic studies, [4,5,9,10-³H]1-nitropyrene
([³H]1-nitropyrene; 1 Ci/mmol, radiochemical purity >98%,
Chemsyn, Inc., Lenexa, KS) was incubated at 0.21-40 μM in 25 ml
Erlenmeyer flasks in a total of 2 ml containing: 1 mM NADPH, 4
mM glucose-6-phosphate, 1 mM $MgCl_2$, 0.1 units/ml glucose-6-
phosphate dehydrogenase (from *Crotalux atrox*, type XII, Sigma
Chem Co.), 0.05 M potassium phosphate, pH 7.4, and the
microsomal protein.  The amounts of microsomal protein and time
of incubation were adjusted to ensure zero-order kinetics, and
typically were 0.08 mg/ml microsomal protein and 10 minute
incubations.  Following incubation, the substrate and
metabolites were extracted using an equal volume of
chloroform:methanol (2:1).  After centrifugation (200xg, 10
min) the chloroform layer was removed, and the aqueous layer
was re-extracted with 2 ml of chloroform.  The organic layers
were combined, and the substrate and metabolites were separated
by high performance liquid chromatography (hplc), and
quantitated using a flow-through scintillation counter (Flo-One
β, Radiomatic Instr.).  Typically, 0.05 μCi was analyzed for
each sample.  Radiolabeled metabolites were identified through
cochromatography with authentic standards.

    Kinetic analysis of the microsomal metabolism of [³H]1-
nitropyrene was accomplished as described above using 0.21-40
μM [³H]1-nitropyrene.  The rate constants were determined by
non-linear regression analysis of the velocity versus substrate
concentration curve (Enzfitter, Biosoft, Inc.).

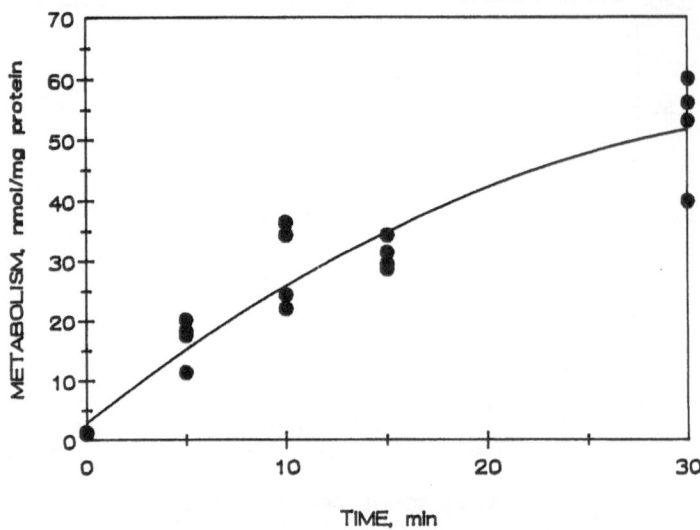

Fig. 3    Dependence of metabolism on time of
          incubations.  The metabolism of [4,5,9,10-
          $^3$H]1-nitropyrene was determined following
          incubation with 80 µg/ml microsomal protein
          under conditions described in the text.

The solubility limit of 1-nitropyrene was determined by
adding increasing amounts of 1-nitropyrene to 0.05 M potassium
phosphate, pH 7.4, and monitoring the absorbance at 390 nm.

RESULTS

As shown in Figure 1, 1-nitropyren-3-ol, a mixture of 1-
nitropyren-6-ol and 1-nitropyren-8-ol, 1-nitropyrene-4,5-
dihydro-4,5-diol, and unidentified epoxides were detected
following the incubation of [$^3$H]1-nitropyrene with rabbit liver
microsomes.  The formation of these metabolites is consistent
with previous reports for rabbits (22), and other species (4-
7,23-24).

In order to determine the conditions for zero-order
kinetics in the metabolism studies, the dependence of
metabolism on microsomal protein concentration and time of
incubation was investigated.  As shown in Figures 2 and 3, the
metabolism was linear at concentrations of microsomal protein
between 40 and 120 µg/ml, and for up to 10 minutes.

Our initial kinetic studies with 0.2-40 µM $^3$H-1NP
demonstrated kinetics consistent with substrate inhibition of
the cytochromes P450 (not shown; a decrease in the rate of the
metabolism at higher substrate concentrations).  However, in
order to rule out insolubility of the substrate, the solubility
limit of 1-nitropyrene at pH 7.4 with 0.05 M potassium
phosphate buffer was determined to be between 20 and 25 µM
(Figure 4).

The kinetics of the rabbit liver microsomal metabolism
were redetermined using 0.2-13 µM [$^3$H]1-nitropyrene.  The
results from a representative experiment is shown in Figure 5,

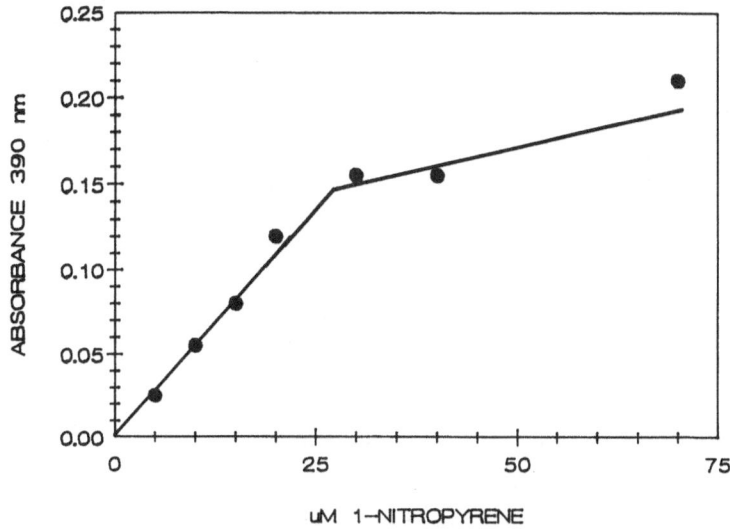

Fig. 4    Solubility of 1-nitropyrene was determined
         by monitoring the absorbance at 390 nm,
         following the addition of indicated amounts
         to 0.05 M potassium phosphate, pH 7.4.

where the $K_m$ and $V_{max}$ were 0.8 µM and 1.76 nmol/min/mg
microsomal protein, respectively.


DISCUSSION

One of the problems we face in determining the health risk
of nitro-PAH to the human population is understanding the *in
vivo* metabolism of the nitro-PAH, and the relationship of *in
vitro* studies to these *in vivo* processes.  Our efforts in this
laboratory have focused on quantitation of the relationship
between the bioactivation and detoxification of nitro-PAH both
*in vitro* and *in vivo*.

In these studies we have shown that the $K_m$ for the
metabolism of [³H]1-nitropyrene *in vitro* by rabbit liver
microsomes is 0.8 µM.  Many *in vitro* studies on the metabolism
of 1-nitropyrene have been performed using rabbit, mouse,
guinea pig, hamster, and rat microsomes and concentrations of
1-nitropyrene ranging from 5-80 µM.  Our results are the first
to quantitate the $K_m$ of the microsomal cytochromes P450 for 1-
nitropyrene, and demonstrate that in the metabolic studies ·
reported in the literature, most investigators were
using 1-nitropyrene at enzyme saturating concentrations (at
least 2x the $K_m$).  However, the use of 1-nitropyrene at
concentrations higher than 20-25 µM results in "apparent"
substrate inhibition of the enzymes due to the insolubility of
the 1-nitropyrene at concentrations above 20-25 µM.  This could
result in an underestimation of the rates of metabolism of 1-
nitropyrene when greater than 20-25 µM 1-nitropyrene is used.

The partitioning of 1-nitropyrene between the aqueous and
microsomal lipid ($K_p = [1NP_{lipid}]/[1NP_{aqueous}]$) was determined to
be approximately $10^4$ (data not shown).  In our *in vitro*
metabolic assays there would be a large partitioning of the 1-
nitropyrene into the microsomal lipid.  If we assume that the

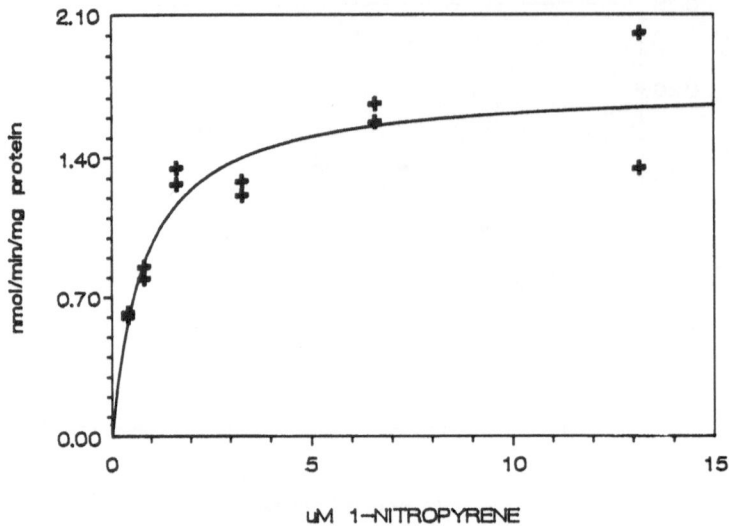

Fig. 5    The $K_m$ and $V_{max}$ of the rabbit liver
microsomal metabolism of [4,5,9,10-${}_3$H]1-
nitropyrene was determined at 80 µg/ml
microsomal protein and 10 min incubation.
The kinetic values were determined by nonlinear least squares analysis: $K_m$, 0.86 µM;
$V_{max}$, 1.76 nmol/min/mg protein.


total 1-nitropyrene ($1NP_t$) equals the amount adsorbed into the
lipid ($1NP_{lipid}$) plus that left in the aqueous fraction
($1NP_{aqueous}$), or $[1NP_t] = [1NP_{lipid}] + [1NP_{aqueous}]$, and that the
volume for the *in vitro* incubation ($V_t$) is divided into lipid
($V_{lipid}$) and aqueous ($V_{aqueous}$) components where $V_t = V_{lipid} + V_{aqueous}$, then:

$$1NP_t V_t = [1NP_{lipid}]V_{lipid} + [1NP_{aqueous}]V_{aqueous} \qquad \text{(equation 1)}$$

In the case of microsomal incubations, the volume of the lipid
is several orders of magnitude less than the aqueous volume.
Therefore, equation 1 would reduce to equations 2 and 3.

$$[1NP_{aqueous}] = [1NP_t]/(K_p(V_{lipid}/V_t)+1) \qquad \text{(equation 2)}$$

$$[1NP_{lipid}] = [1NP_t]/(V_{lipid}/V_t) + (1/K_p) \qquad \text{(equation 3)}$$

Using the assumptions of Parry *et al.* (25) concerning the
specific density of microsomal lipid, and our conditions where
the microsomal protein concentrations were 80 µg/ml the
addition of 13.1 µM [${}^3$H]1-nitropyrene to the incubations
results in a remaining concentration of 10 µM 1-nitropyrene in
the aqueous fraction, and 103 mM 1-nitropyrene in the lipid
phase.  At present, it is unclear whether the aqueous soluble
substrate or lipid bound substrate is important for cytochrome
P450 binding and metabolism of lipophilic xenobiotics.  Due to
this, in our determinations of the $K_m$ and $V_{max}$ of 1-nitropyrene
metabolism, we have not corrected for the true concentration of
$1NP_{aqueous}$.

When 1-nitropyrene is administered to animals, cytochrome
P450-mediated C-oxidation metabolites and nitroreduced

metabolites are detected (4-7). Studies by El-Bayoumy *et al.* (26) using germ-free animals have shown that the nitroreduction of 1-nitropyrene occurs in the intestinal tract via the microflora. The lack of nitroreduction of 1-nitropyrene in germ-free whole animals can be the result of several factors. *(1)* The animals could lack the enzymes capable of nitroreduction of 1-nitropyrene; however, xanthine oxidase (1), cytochrome P450 (2,27), and aldehyde oxidase (28) have been shown to catalyze the nitroreduction of 1-nitropyrene in several species. *(2)* Certain kinetic parameters could explain this phenomenon. The rate of C-oxidation could be considerably higher that the rates of nitroreduction at the concentrations of 1-nitropyrene *in vivo* due to low $K_m$'s of cytochromes P450 and high $K_m$'s of the nitroreductases for the 1-nitropyrene, or the nitroreduced 1-nitropyrene could reoxidize in the presence of physiological $O_2$ concentrations. *(3)* Alternatively, our results on the partitioning of 1-nitropyrene between the aqueous and lipid phases suggests that *in vivo* very little of the 1-nitropyrene would partition into the lipid bilayer of the cell, and then into the cytosol where most of the nitroreductases are located.

Further studies are in progress to determine which of these possibilities explain the lack of nitroreduction of 1-nitropyrene in germ-free animals, and if this also explains the carcinogenicity of 1-nitropyrene in newborn but not adult rats and mice.

ACKNOWLEDGEMENTS

This work was supported by grant ES03648 from the NIH.

REFERENCES

1.    Howard, P.C., and Beland, F.A (1982) Xanthine oxidase catalyzed binding of 1-nitropyrene to DNA. Biochemical Biophys. Res. Commun., 104, 727-732.

2.    Djuric,Z., Potter,D.W., Heflich,R.H., and Beland,F.A. (1986) Aerobic and anaerobic reduction of nitrated pyrenes *in vitro*. Chem.-Biol. Interactions, 59, 309-324.

3.    Beland, F.A., Heflich, R.H., Howard, P.C., and Fu, P.P. (1985) The *in vitro* metabolic activation of nitro polycyclic aromatic hydrocarbons. IN: Polycyclic Hydrocarbons and Carcinogenesis (Harvey,R.G., ed.) pp. 371-396. American Chemical Society, Washington, DC.

4.    Tokiwa,H., and Ohnishi,Y. (1986) Mutagenicity and carcinogenicity of nitroarenes and their sources in the environment. Critical Reviews in Toxicology, 17, 23-60.

5.    Howard,P.C., Flammang,T., and Beland,F.A. (1985) Comparison of the *in vitro* and *in vivo* hepatic metabolism of the carcinogen 1-nitropyrene. Carcinogenesis, 6, 243-249.

6.    El-Bayoumy,K., and Hecht,S.S. (1983) Identification and mutagenicity of metabolites of 1-nitropyrene formed by rat liver. Cancer Res., 43, 3132-3137.

7.   Ball,L.M., Kohan,M.J., Inmon,J.P., Claxton,L.D., and Lewtas,J. (1984) Metabolism of 1-nitro[$^{14}$C]pyrene *in vivo* in the rat and mutagenicity of urinary metabolites. Carcinogenesis, 5, 1557-1564.

8.   Djuric,Z., Coles,B., Fifer, E.K., Ketterer,B., and Beland,F.A. (1987) *In vivo* and *in vitro* formation of glutathione conjugates from the K-region epoxides of 1-nitropyrene. Carcinogenesis, 8, 1781-1786.

9.   Fu,P.P., Chou,M.W., Yang,S.K., Beland,F.A., Kadlubar,F.F., Casciano,D.A., Heflich,R.H., and Evans,F.E. (1982) Metabolism of the mutagenic environmental pollutant, 6-nitrobenzo[a]pyrene: metabolic activation via ring oxidation. Biochem. Biophys. Res. Commun., 105, 1037-1043.

10.  Fu,P.P., Chou,M.W., Miller,D.W., White,G.L., Heflich,R.H., and Beland,F.A. (1985) The orientation of the nitro substituent predicts the direct-acting bacterial mutagenicity of nitrated polycyclic aromatic hydrocarbons. Mutation Res., 143, 173-181.

11.  El-Bayoumy,K., Delclos,K.B., Heflich,R.H., Walker,R., Shiue,G.H., and Hecht,S.S. (1989) Mutagenicity, metabolism and DNA adduct formation of 6-nitrochrysene in *Salmonella typhimurium*. Mutagenesis, 4, 235-240.

12.  Delclos,K.B., El-Bayoumy,K., Casciano,D.A., Walker,R.P., Kadlubar,F.F., Hecht,S.S., Shivapurkar,N., Mandal,S., and Stoner,G.D. (1989) Metabolic activation of 6-nitrochrysene in explants of human bronchus and in isolated rat hepatocytes. Cancer Res., 49, 2909-2913.

13.  Stanton,C.A., Chow,F.L., Phillips,D.H., Grover,P.L., Garner,R.C., and Martin,C.N. (1985) Evidence for N-(deoxyguanosin-8-yl)-1-aminopyrene as a major adduct in female rats treated with 1-nitropyrene. Carcinogenesis, 6, 535-538.

14.  Smith,B.A., and Beland,F.A. (1989) DNA adduct formation in target tissues of Sprague-Dawley rats, CD-1 mice, and A/J mice following tumorigenic doses of 1-nitropyrene. Proc. Amer. Assoc. Cancer Res., 30, 525.

15.  Hirose,M., Lee,M.S., Wang,C.Y., and King,C.M. (1984) Induction of rat mammary gland tumors by 1-nitropyrene, a recently recognized environmental mutagen. Cancer Res., 44, 1158-1162.

16.  Wislocki,P.G., Bagan,E.S., Lu,A.Y.H., Dooley,K.L., Fu,P.P., Han-Hsu,H., Beland,F.A., and Kadlubar,F.F. (1986) Tumorigenicity of nitrated derivatives of pyrene, benz[a]anthracene, chrysene and benzo[a]pyrene in the newborn mouse assay. Carcinogenesis, 7, 1317-1322.

17.  Waxman,D.J., Dannon,G.A., and Guengerich,F.P. (1985) Regulation of rat hepatic cytochrome P-450: age-dependent expression, hormonal imprinitng, and xenobiotic inducibility of sex-specific isoenzymes. Biochemistry, 24, 4408-4417.

18. Ohgaki,H., Negishi,C., Wakabayashi,K., Kusama,K., Sato,S., and Sugimura,T. (1984) Inductions of sarcomas in rats by subcutaneous injection of dinitropyrenes. Carcinogenesis, 5, 583-585.

19. Orr,J.C., Bryant,D.W., McCalla,D.R., and Quilliam,M.A. (1985) Dinitropyrene-resistant *Salmonella typhimurium* are deficient in an acetyl-CoA acetyltransferase. Chem.-Biol. Interactions, 54, 281-288.

20. Saito,K., Shinohara,A., Kamataki,T., and Kato,R. (1985) Metabolic activation of mutagenic N-hydroxylamines by O-acetyltransferase in *Salmonella typhimurium* TA98. Arch. Biochem. Biophys., 239, 286-295.

21. Djuric,Z., Fifer,E.K., and Beland,F.A. (1985) Acetyl coenzyme A-dependent binding of carcinogenic and mutagenic dinitropyrenes to DNA. Carcinogenesis, 6, 941-944.

22. Howard, P.C., Reed, K.A., and Koop, D.R. (1988) Oxidative metabolism of 1-nitropyrene by rabbit liver microsomes and purified microsomal cytochrome P450 isozymes. Cancer Res., 48, 4261-4269.

23. Howard,P.C., DeMarco,G.J., Consolo,M.C., and McCoy,G.D. (1987) Differing effects of chronic ethanol consumption by mice on liver microsomal metabolism of xenobiotics: 1-nitropyrene, nicotine, aniline, and *N*-nitrosopyrrolidine. Molecular Toxicology, 1, 177-189.

24. Rosenkranz,H.S., and Howard,P.C. (1986) Structural basis of the activity of nitrated polycyclic aromatic hydrocarbons. IN: "Carcinogenic and Mutagenic Effects of Diesel Engine Exhaust", (N. Ishinishi, A. Koizumi, R.O. McClellan, and W. Stöber, eds.), Elsevier Science Pub., Amsterdam, pp.141-148.

25. Parry,G., Palmer,D.N., and Williams,D.J. (1976) Ligand partitioning into membranes: its significance in determining $K_m$ and $K_s$ values for cytochrome P-450 and other membrane bound receptors and enzymes. FEBS Letters, 67, 123-129.

26. El-Bayoumy, K., Reddy,B., and Hecht,S.S. (1984) Identification of ring oxidized metabolites of 1-nitropyrene in the feces and urine of germfree F344 rats. Carcinogenesis, 5, 1371-1373.

27. Saito,K., Kamataki,T., and Kato,R. (1984) Participation of cytochrome P-450 in reductive metabolism of 1-nitropyrene by rat liver microsomes. Cancer Res., 44, 3169-3173.

28. Tatsumi,K., Kitamura,S., and Narai,N. (1986) Reductive metabolism of aromatic nitro compounds including carcinogens by rabbit liver preparations. Cancer Res., 46, 1089-1093.

METABOLIC ACTIVATION OF 6-NITROCHRYSENE AND 6-AMINOCHRYSENE IN VITRO AND IN VIVO

K. Barry Delclos, Glenn Talaska, and Ralph P. Walker

National Center for Toxicological Research
Jefferson, Arkansas

Christiane Brassinne and Jean Paul Sculier

Institut Jules Bordet
Brussels, Belgium

INTRODUCTION

The well-documented mutagenicities and tumorigenicities of many nitro polycyclic aromatic hydrocarbons (nitro PAHs) in experimental models have raised the issue of the potential hazards of exposure to these compounds to humans. While several nitro PAHs have become objects of study due to their quantitative importance in the environment or their extremely potent mutagenic activities in Salmonella tester strains, 6-nitrochrysene (6-NC) came to be of interest primarily because of its highly potent carcinogenic activity in the preweanling mouse bioassay (1-5). However, when applied to mouse skin prior to repeated doses of the tumor promoter 12-0-tetradecanoylphorbol 13-acetate, 6-NC was found to be a weak initiator and, in contrast to the situation in the preweanling mouse model, was less active than the parent compound, chrysene (6,7). It is thus of considerable interest to determine the reason for the high carcinogenic potency of 6-NC in the preweanling mouse and the meaning of the results of the mouse bioassays in terms of assessing the potential risk of 6-NC and related compounds to humans.

While 6-NC has become a subject of investigation fairly recently, the biological activities of 6-aminochrysene (6-AC) have been under study for decades. The early literature on 6-AC indicated that this compound had low toxicity and lacked tumorigenic activity in the rat (8,9). Later studies indicated that 6-AC was tumorigenic when administered i.p. to preweanling mice (4,10) or applied topically to adult mice (11), although El-Bayoumy et al. have recently demonstrated that 6-AC is less potent than 6-NC in the preweanling mouse (4). In addition, 6-AC has been found to be one of several amino polycyclic aromatic hydrocarbons that are direct-acting mutagens in Salmonella tester strains (12,13). 6-AC was observed to inhibit the induction of mammary tumors by 7,12-dimethyl-benz[a]anthracene (8), to cause atrophy of the spleen in rodents (8), to inhibit the development of and to cause partial regression of spontaneous mammary tumors in mice (8,14,15), and to inhibit the growth of trans-plantable rhabdomyosarcoma in rats (15). Based on the early rodent

Nitroarenes, Edited by P. C. Howard et al.
Plenum Press, New York, 1990

295

studies, 6-AC was used as a trial therapeutic agent in patients with splenomegaly of various origins (16) and in patients with advanced breast cancer (17). The initial trial of 6-AC in breast cancer patients involved the oral administration of 2 g of 6-AC per day. A weak response was noted in this trial (2 objective responses in a total of 32 patients) and it was suggested that poor uptake of the drug from the gut may have been partially responsible for the observed low efficacy (15). A clinical trial of 6-AC in patients with advanced breast cancer has recently been carried out, with 6-AC administered intravenously in a liposomal preparation. We have developed $^{32}$P-postlabeling methodology that allows us to detect DNA adducts in the peripheral blood cells of these patients. The problems encountered in the development of this assay may have implications for the monitoring of human populations for exposure to environmental carcinogens.

## Metabolic Activation Pathways for 6-Nitrochrysene and 6-Aminochrysene

We have been interested over the past several years in determining the metabolic activation pathways of 6-NC and 6-AC in both animal and human model systems. 6-NC and 6-AC can be metabolized to a common N-hydroxylated derivative, N-hydroxy-6-AC, that can then be further activated to an electrophilic nitrenium ion capable of modifying DNA bases (Figure 1). We isolated and characterized the DNA adducts formed

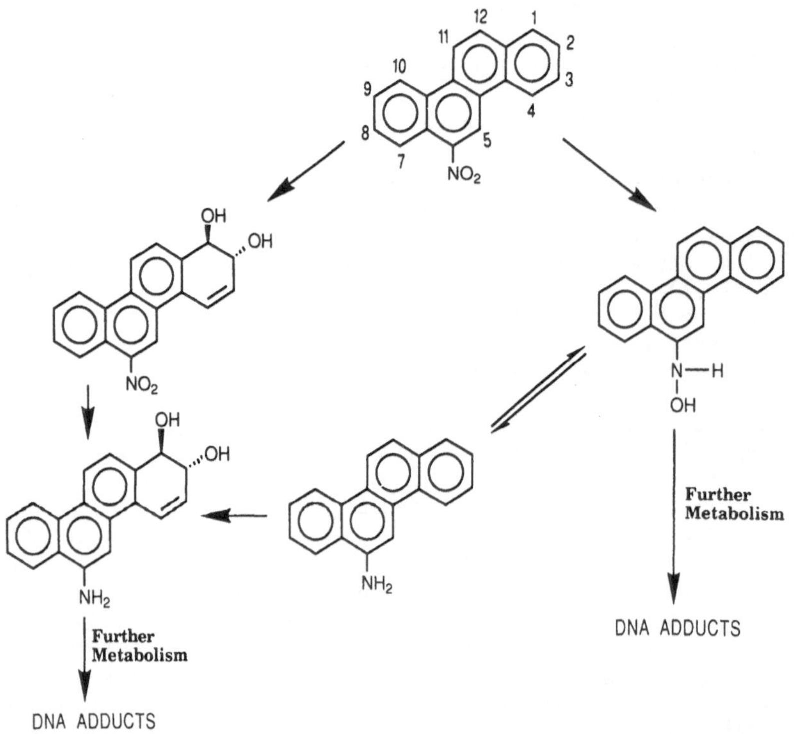

Figure 1. Metabolic activation pathways for 6-NC and 6-AC that lead to the formation of the proximate DNA-binding metabolites 6-AC-1,2-dihydrodiol and N-hydroxy-6-AC.

from N-hydroxy-6-AC and identified DNA adducts with identical HPLC retention times in [$^3$H]6-NC- and [$^3$H]6-AC-treated hepatocytes from uninduced rats (18). In general, an adduct formed between the amine nitrogen of the carcinogen and the C8 position of deoxyguanosine is the major product formed in the reaction between an N-hydroxy-arylamine and DNA, with varying amounts of $N^2$- or $O^6$-modified deoxyguanosine and $N^6$- or C8-modified deoxyadenosine formed depending on the particular carcinogen. In the case of N-hydroxy-6-AC, however, N-(deoxyguanosin-8-yl)-6-AC, 5-(deoxyguanosin-$N^2$-yl)-6-AC and N-(deoxyinosin-8-yl)-6-AC were formed in approximately equal proportions (19). Such a deoxyinosine adduct has not been previously reported and is presumably derived from oxidative deamination of the corresponding deoxyadenosine adduct.

Adducts derived from N-hydroxy-6-AC were not found to be present in the target organs (lungs and livers) of preweanling mice treated with a carcinogenic regimen of 6-NC or 6-AC. Instead, a single major adduct was found in the lungs and livers of these mice (19). In collaboration with Drs. Karam El-Bayoumy and Stephen Hecht of the American Health Foundation, we examined the metabolism of 6-NC in the preweanling mouse and the reaction of the metabolites with DNA (20). As had been previously described by El-Bayoumy and Hecht in rat liver microsomes (12), trans-1,2-dihydroxy-1,2-dihydro-6-nitrochrysene, a mutagenic metabolite of 6-NC, was found to be a major metabolite of 6-NC in the preweanling mouse. However, whereas the further metabolism of 6-NC-1,2-dihydrodiol by rat liver microsomes resulted in the formation of a catechol as the major product (12), further metabolism of 6-NC-1,2-dihydrodiol by the preweanling mouse resulted in formation of the nitro-reduced derivative, trans-1,2-dihydroxy-1,2-dihydro-6-amino-chrysene, as a major metabolite (20). Incubation of 6-AC-1,2-dihydrodiol with calf thymus DNA in the presence of microsomes from 3-MC-pretreated rats and an NADPH-generating system resulted in the formation of a single major adduct with chromatographic and chemical properties identical to those of the adduct formed in 6-NC- and 6-AC- treated preweanling mice. Thus, our studies with isolated hepatocytes from uninduced rats and preweanling mice indicated two major pathways for the metabolic activation of 6-NC and 6-AC to DNA-reactive species (Figure 1).

The biological significance of the formation of 6-AC-1,2-dihydrodiol is underlined by the demonstrations by El-Bayoumy and his colleagues of its potent tumorigenic activity in the preweanling mouse (4) and its mutagenicity in Salmonella (13). The identifications of the ultimate reactive species and the major adduct derived from 6-AC-1,2-dihydrodiol have remained an elusive goal due to difficulties in preparing sufficient quantities of adduct or possible ultimate reactive species for structural studies. Two possibilities for activation are the formation of an N-hydroxy derivative of 6-AC-1,2-dihydrodiol or ring oxidation to a reactive epoxide. The former pathway does not seem to be a major pathway since the reaction of N-hydroxy-6-AC-1,2-dihydrodiol, generated in situ, with calf thymus DNA yielded products that were chromatographically distinct from the adduct formed in preweanling mice treated with 6-NC or 6-AC (data not shown).

## Modulation of the Activation Pathways for 6-NC and 6-AC

Our studies with 6-NC and 6-AC in the preweanling mouse indicated one major activation pathway leading to DNA adduct formation with quantitative differences between adduct levels in 6-NC- and 6-AC-treated mice (2- to 8-fold higher in NC-treated mice) and in lung and liver DNA of mice treated with both compounds (2- to 3-fold higher in hepatic DNA) (19). However, results obtained in systems other than the preweanling mouse indicate that both the formation of N-hydroxy-6-AC and 6-AC-1,2-

dihydrodiol can be important activation pathways in vivo as well as in vitro and that a variety of factors can have an effect on the quantitative importance of these pathways in a given system. The observation in preweanling mice that 6-NC and 6-AC are metabolized to similar ultimate reactive metabolites does not appear to be generally true. The liver DNA of male F344 rats (100 g) treated with a single i.p. dose of [$^3$H]6-NC (0.03 µmol per g body weight) showed a markedly different adduct pattern from the DNA of rats treated with an identical dose of [$^3$H]6-AC (Figure 2). Whereas the liver DNA of the latter animals contained primarily adducts derived from N-hydroxy-6-AC, the liver DNA of 6-NC-treated animals contained adducts derived from both N-hydroxy-6-AC and 6-AC-1,2-dihydrodiol. It is also noteworthy that the liver DNA from the NC-treated animals contains adducts, which account for approximately 10% of the total adducts, that have retention times between those of the adduct derived from 6-AC-1,2-dihydrodiol (Adduct M in Figure 2) and 5-(deoxyguanosin-N$^2$-yl)-6-AC (Adduct I in Figure 2). Lung DNA from the 6-NC treated rats, but not from the 6-AC-treated rats, also contained adducts in this region that accounted for 20-25% of the total adducts (data not shown). Although these adducts have not been identified, hydrolysates of DNA from incubations containing [$^3$H]6-NC-1,2-dihydrodiol, xanthine oxidase and hypoxanthine contain adducts with identical retention times that are presumably derived from N-hydroxy-6-AC-1,2-dihydrodiol (data not shown).

Age, tissue type and intestinal microflora can also influence the activation of 6-NC. We had previously found that the metabolic activation of 6-NC in preweanling male mice of several strains resulted in the formation of a single major adduct derived from a metabolite of 6-AC-1,2-dihydrodiol in both lung and liver (19,20). The lung DNA of 5-week-old male Balb/c mice treated with a single i.p. dose of [$^3$H]6-NC (0.03 µmol per g body weight) showed an adduct profile similar to that of the preweanling mice. However, the hepatic DNA from these animals showed a significant level of adducts derived from N-hydroxy-6-AC (Table 1). Germfree mice treated in an identical fashion excreted 75% less 6-AC in the feces than did the conventional mice and had levels of DNA adducts in lung and liver similar to those in conventional mice. However, the proportion of adducts derived from N-hydroxy-6-AC in the livers of the germfree mice was significantly higher than that in the conventional mice (Table 1). A single experiment with an identical dose of [$^3$H]6-AC indicated no difference in total binding (1,200 vs. 1,150 fmol bound/ mg DNA) or in the distribution of adducts in germfree and conventional animals (ratio of adducts derived from N-hydroxy-6-AC to those derived from 6-AC-1,2-dihydrodiol of 0.75). The basis for the apparent effect of intestinal microflora on the hepatic metabolism of NC could not be determined from the available data. The shift in activation pathways may reflect differences in the levels of 6-NC reaching the livers of germfree and conventional mice or an indirect effect of the microflora on hepatic metabolizing enzymes. Evidence for indirect effects of intestinal microflora on host metabolic enzymes has been presented by Rowland (21).

The data presented above suggest that modification of the expression or activity of enzymes involved in the metabolism of 6-NC can alter the preferred pathway of activation of this compound to DNA binding derivatives. Further evidence for this was obtained in our recently published studies, carried out in collaboration with Dr. Daniel Casciano of NCTR, that indicated isolated hepatocytes could metabolize 6-NC to reactive metabolites by different pathways depending on the pretreatment of the animals with inducers of drug metabolism (22). DNA from 6-NC-treated control hepatocytes or hepatocytes from phenobarbital-treated rats contained adducts derived from N-hydroxy-6-AC whereas DNA from hepatocytes from 3-methylcholanthrene- or Aroclor-pretreated rats contained

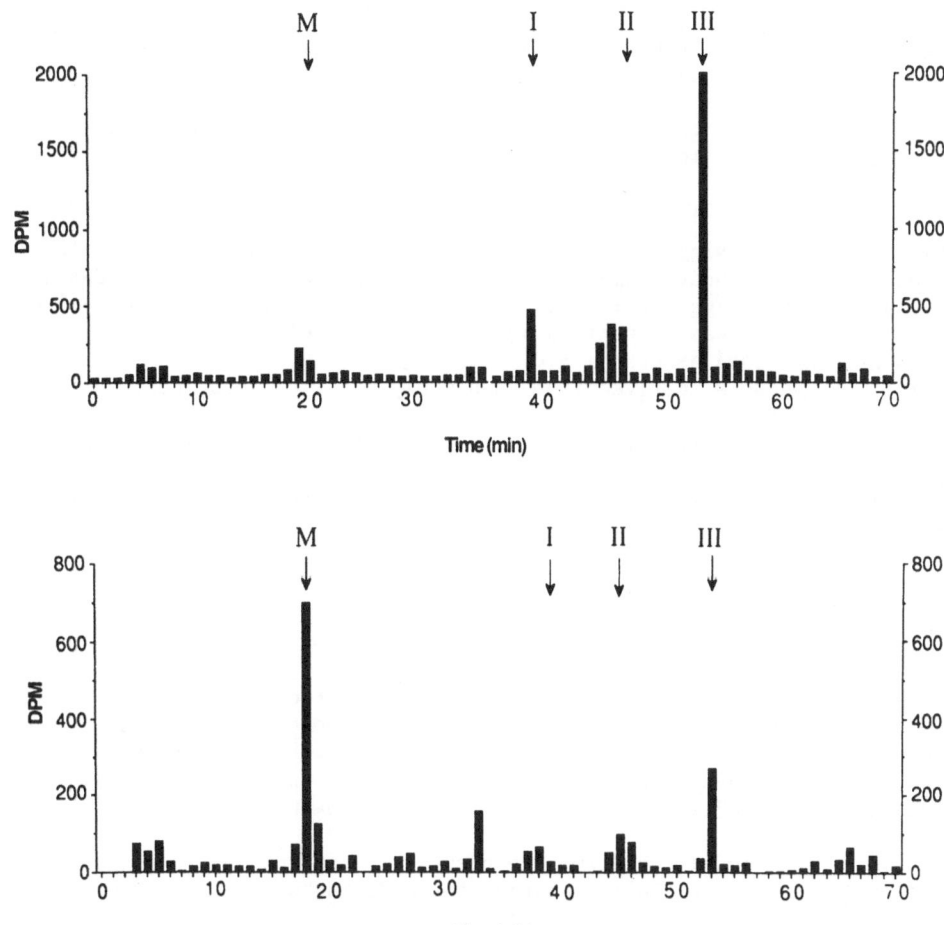

Figure 2. Hplc chromatogram of enzymatic hydrolysates of DNA isolated from the livers of 100 g F344 rats 24 h after a single i.p. treatment with [$^3$H]6-AC (top) or [$^3$H]6-NC (bottom) (0.03 μmol/ 5 μl DMSO/ g body weight). DNA was isolated, hydrolyzed and adducts extracted with n-butanol as described in reference 19. Chromatography was accomplished on a 4.6 mm x 25 cm 5 μm Nucleosil C18 column (Alltech Associates, Deerfield, IL) with a linear gradient from 30% methanol: 70% water to 100% methanol over 70 minutes. Arrows mark the retention times of the major adduct derived from a metabolite of 6-AC-1,2-dihydrodiol (M), 5-(deoxyguanosin-N$^2$-yl)- 6-AC (I), N-(deoxyguanosin-8-yl)-6-AC (II) and N-(deoxyinosin-8-yl)-6-AC (III).

Table 1. Binding of [³H]6-NC to liver DNA of 5-week-old male Balb/c mice[a].

| Gut status | Time of Sacrifice | Total Binding[b] (fmol/mg DNA) | Adduct Ratio[c] |
|---|---|---|---|
| Conventional | 24 h | 2,650 ± 670 (n=5) | 0.3 ± 0.1[d] |
| Conventional | 48 h | 3,200 ± 1,300 (n=8) | 0.3 ± 0.1[e] |
| Germfree | 24 h | 2,430 ± 970 (n=6) | 1.0 ± 0.8 |
| Germfree | 48 h | 2,200 ± 760 (n=7) | 0.7 ± 0.3 |

[a]Mice received a single i.p. injection of [³H]6-NC in DMSO at a rate of 0.03 µmol/5µl/g body weight and were killed 24 or 48 h later. DNA was isolated from lungs and livers of individual animals and the level of adduct formation was determined as described in references 18 and 19.
[b]Data are expressed as mean ± S.D.
[c]Ratio of adducts derived from N-hydroxy-6-AC to those derived from 6-AC-1,2-dihydrodiol.
[d]Significantly different from ratio for value for germfree 24 h liver, $p < 0.05$.
[e]Significantly different from ratio for value for germfree 48 h liver, $p < 0.001$.

primarily adducts derived from 6-AC-1,2-dihydrodiol. Similarly, DNA adducts formed in <u>Salmonella</u> strain TA 100 exposed to [³H]6-NC in the presence of liver S9 from Aroclor-induced rats were shown to be derived from 6-AC-1,2-dihydrodiol whereas adducts formed in the bacteria in the absence of liver S9 were derived from N-hydroxy-6-AC (13).

That genetic variability or exposure to environmental agents might also influence the metabolic pathways for 6-NC in humans was suggested by data obtained on 6-NC metabolism and DNA adduct formation in explants of human bronchus that were carried out in collaboration with Dr. Gary Stoner of the Medical College of Ohio and Drs. El-Bayoumy and Hecht of the American Health Foundation (22). The explants were capable of activating 6-NC by either nitroreduction or by the combination of nitroreduction and ring oxidation that leads to the reactive metabolite of 6-AC-1,2-dihydrodiol. The predominant pathway of activation differed in explants prepared from different donors.

<u>Detection of DNA Adducts Derived from N-hydroxy-6-AC and 6-AC-1,2-dihydrodiol by ³²P-postlabeling</u>

Analysis of samples of human tissues for adducts derived from metabolites of 6-NC and 6-AC required the development of a sensitive assay that does not rely on radioactive carcinogen. Antibody detection methods are not feasible at this time because of the lack of sufficient quantities of the possible expected adducts for the production of antibodies. We instead turned to ³²P-postlabeling methodology. When standard methods (23) were used to detect adducts in samples of DNA that had been modified <u>in vitro</u> by [³H]N-hydroxy-6-AC or by a microsomal metabolite of [³H]6-AC-1,2-dihydrodiol, we found that the levels of adducts determined by ³²P-postlabeling were more than an order of magnitude lower than the adduct levels determined by analysis of the bound tritium. The retention time of the adduct derived from microsomal metabolite of 6-AC-1,2-dihydrodiol on a reversed-phase hplc column (see Figure 2) suggested that the polarity of the adduct might be resulting in the loss of the adduct in the D1 separation used to remove non-adducted nucleotides. Several alternative D1 solvents were used, and it was found that 0.65 M sodium phosphate, pH 6.0, provided the optimal separation of the adduct and normal nucleotides. In addition, the adduct derived from a metabolite of 6-AC-1,2-dihydrodiol was found to be resistant to hydrolysis by nuclease P1 and thus treatment of modified DNA with this enzyme enhanced the sensitivity of the assay (24).

In the case of N-hydroxy-6-AC-modified DNA, it was found that the digestion of the modified DNA to 3'-mononucleotide phospates was dramatically dependent on pH (Figure 3). Usual conditions call for the digestion to be carried out at pH 6, but relative adduct labeling at pH 6 was approximately 70-fold lower than that found at pH 7. pH did not appear to have an effect on the hydrolysis of the adduct derived from 6-AC-1,2-dihydrodiol, or on N-hydroxy-4-aminobiphenyl-adducted DNA, or DNA modified by N-hydroxy derivatives of 1,6- and 1,8-dinitropyrenes (not shown). Thus, for the analysis of DNA samples in which adducts derived from NC and AC are suspected, it is necessary to carry out the hydrolysis with micrococcal nuclease and spleen phosphodiesterase at pH 7.0.

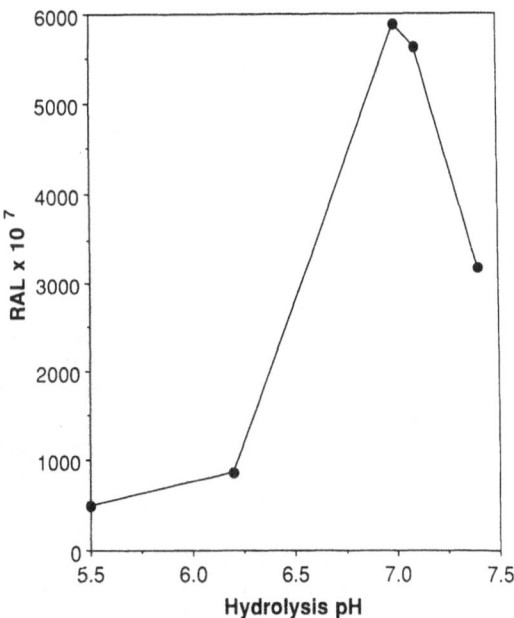

Figure 3. The effect of hydrolysis pH on relative adduct labeling for N-hydroxy-6-AC-modified DNA. N-Hydroxy-6-AC-modified DNA (17) was digested and labeled with [γ-$^{32}$P]ATP as described in reference 23 except that the pH of the digestion with micrococcal nuclease and spleen phosphodiesterase was adjusted as indicated.

Thin layer chromatograms of [32]P-postlabeled calf thymus DNA modified with N-hydroxy-6-AC or a microsome-generated metabolites of 6-AC-1,2-dihydrodiol are shown in Figures 4 and 5. The N-hydroxy-6-AC-modified DNA shows one major treatment-related spot. Two additional spots that run slightly to the right of the major spot are occasionally observed, although they are not visible in the chromatogram shown. Whether the major spot observed contains all three of the N-hydroxy-6-AC-derived adducts [5-(deoxyguanosin-N²-yl)-6-AC, N-(deoxyguanosin-8-yl)-6-AC and N-(deoxyinosin-8-yl)-6-AC] or two of the three adducts are poorly hydrolyzed or labeled remains to be determined. The mobility of the major adduct derived from 6-AC-1,2-dihydrodiol (Figure 5) clearly distinguishes this adduct from the pattern obtained with N-hydroxy-6-AC-modified DNA. A minor spot visible to the right of the major adduct spot in the 6-AC-1,2-dihydrodiol-modified DNA is a possible treatment-related spot observed in these DNA samples.

## DNA Adducts in Peripheral Blood Cells of Patients Treated with 6-AC in a Phase II Clinical Trial

A previous clinical trial, in which 6-AC was administered orally to patients with advanced breast cancer, resulted in an objective response in 2 of 32 patients (17). Gastrointestinal intolerance was observed and poor absorption of 6-AC from the gut was suggested to play a role in limiting the effectiveness of the drug (15). Recently, it was found that 6-AC entrapped in liposomes (composition egg lecithin, cholesterol, stearylamine in a ratio of 4:3:1) could be safely administered to cancer patients and an objective response was observed in one of 13 patients in a phase I trial (25). A phase II clinical trial of this preparation of 6-AC was instituted as salvage therapy in patients with advanced breast cancer metastatic to the liver (J.P. Sculier, M. Piccart et al., manuscript submitted). Liposomal preparations of this type are concentrated in the liver, spleen and bone marrow and thus it was felt that patients with this type of metastatic disease would be the most likely patients to show a response if the drug was effective. Although most of the patients in the trial had received prior treatment with hormones or alkylating drugs, they had not received chemotherapy for the four weeks prior to the initiation of the trial. Fourteen patients were treated with 3 infusions of 6-AC (700 to 1020 ml of liposomes, mean 6-AC concentration of 368 µg/ml) at bi-weekly intervals. Blood samples were taken prior to dosing and at various intervals for up to 2 weeks after the infusions. No therapeutic response was observed in these patients, and the trial was thus terminated. However, portions of blood samples taken from several of the patients before dosing and at various time points up to 2 weeks after infusion were made available for DNA adduct analysis. Samples from only two patients have been examined since the methodological problems with the [32]P-postlabeling of these adducts have been resolved, and similar results have been observed in these cases. An autoradiogram of the chromatogram of blood DNA from a patient 24 hr after the first infusion of 6-AC is shown in Figure 6. Three adduct spots not present in the pretreatment blood sample were observed, and the major adduct spot has the chromatographic mobility of the major adduct derived from the microsomal metabolite of 6-AC-1,2-dihydrodiol. No adducts derived from N-hydroxy-6-AC have been observed in samples analyzed thus far. The origin of the two minor adduct spots remains to be determined. Analyses of the remaining samples will focus on demonstrating that the major adduct spot is the same adduct as the major adduct formed from the microsomal metabolite of 6-AC-1,2-dihydrodiol and examining the possible identities of the minor adducts.

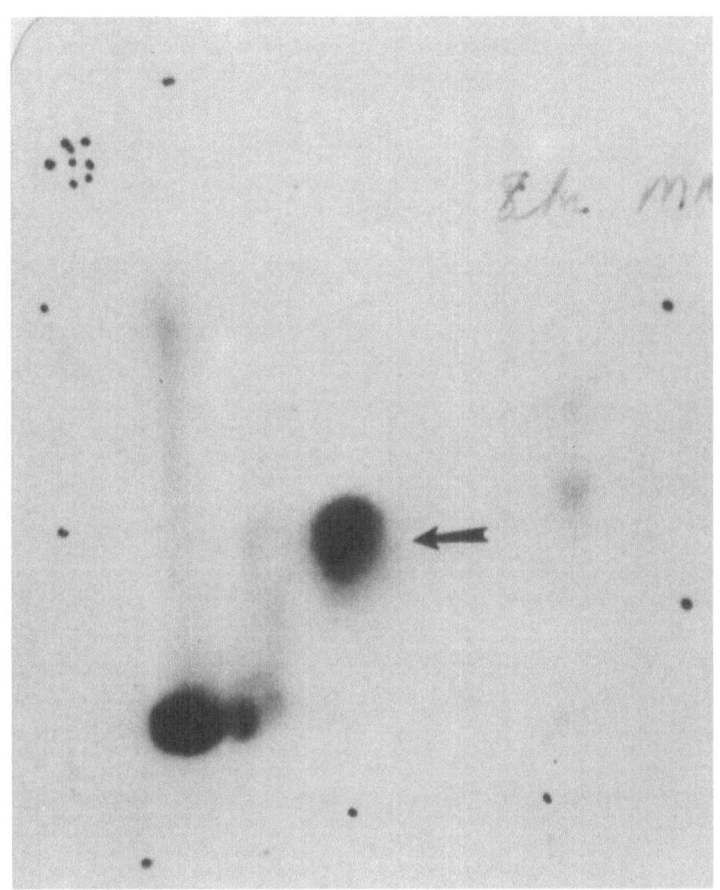

Figure 4.   Autoradiogram of a chromatogram of calf thymus DNA modified
with N-hydroxy-6-AC.  The arrow indicates the major treatment-
related adduct spot referred to in the text.  DNA was modified
as described in reference 18 and digested and labeled with
$[\gamma\text{-}^{32}P]ATP$ as described in reference 23 except that the pH of
the digestion with micrococcal nuclease and spleen phosphodiesterase was 7.0.   Chromatography was carried out on PEI
cellulose plates (Alltech Associates, Deerfield, IL) using the
following solvent conditions: D1, 0.65 M sodium phosphate, pH
6.0;  D3, 3.2 M lithium formate, 7.7 M urea, pH 3.5; D4, 0.75 M
lithium chloride, 0.47 M Tris, 8.0 M urea, pH 8.0; D5, 0.9 M
sodium phosphate, pH 6.8.

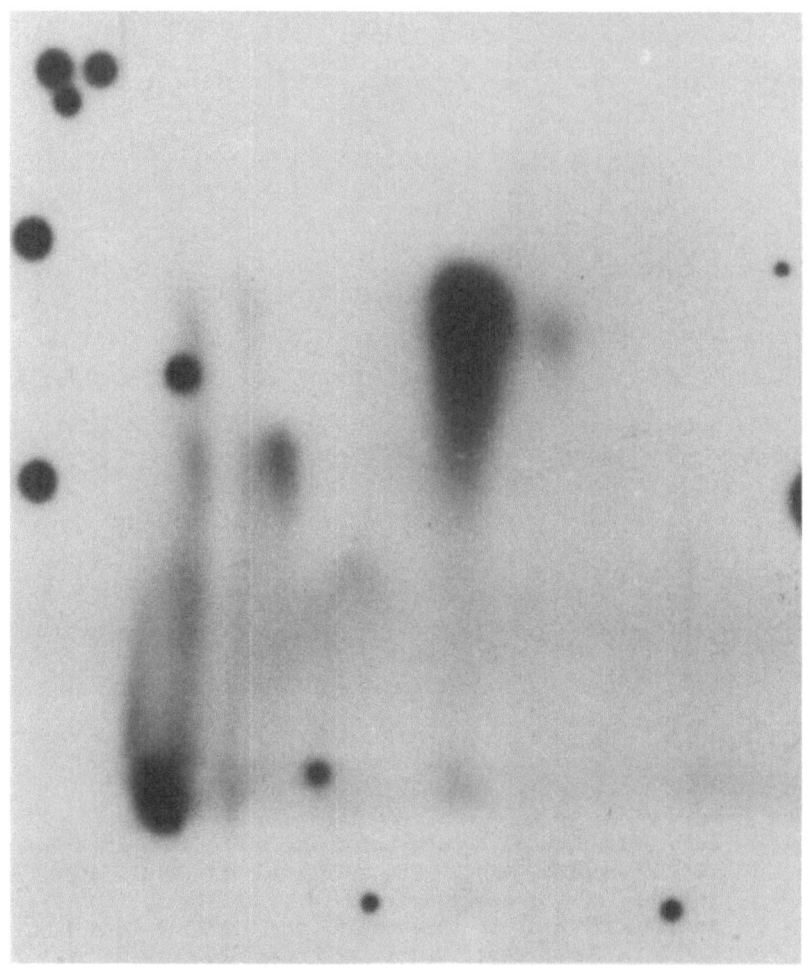

Figure 5.  Autoradiogram  of a chromatogram of calf thymus DNA modified by
a  microsomal  metabolite  of  6-AC-1,2-dihydrodiol.   DNA  was
modified as described in reference 20 and analyzed as described
in the legend for Figure 4.

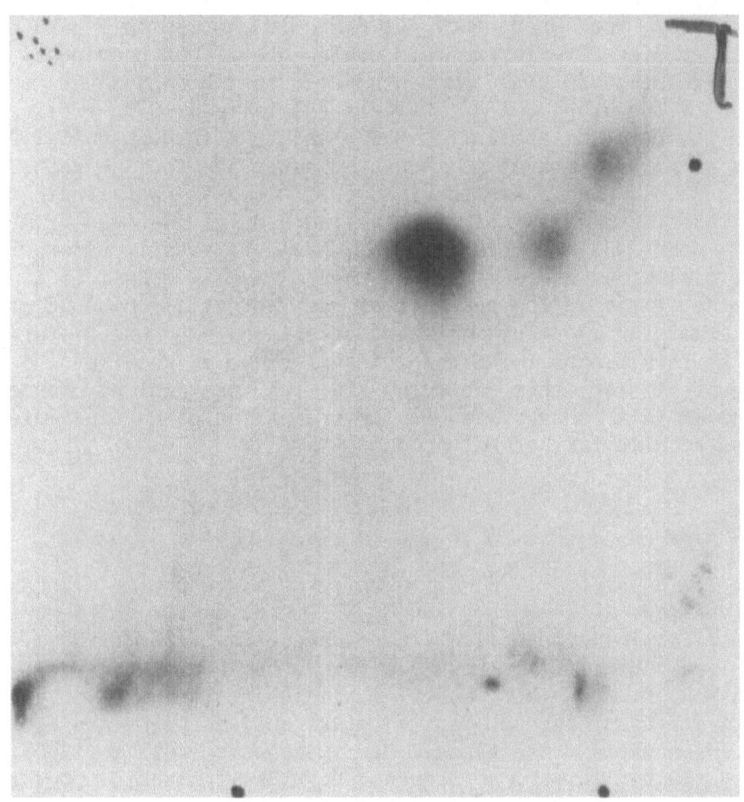

Figure 6. Autoradiogram of a chromatogram of lymphocyte DNA of a patient 24 hr after an infusion of 6-AC encapsulated in liposomes. DNA was isolated from whole blood by a modification of the method of Jeanpierre (26) that included treatment of the DNA with RNase. The DNA was hydrolyzed as described above, treated with nuclease P1 as described by Reddy and Randerath (24) and $^{32}$P-postlabeled as indicated in Figure 4. Chromatography was carried out as described in the legend to Figure 4. Relative adduct labeling values were 144, 44, and 44 per $10^9$ normal nucleotides for the major and two minor adduct spots, respectively.

## CONCLUSIONS

As with other nitro PAHs, 6-NC can be activated by a variety of pathways. 6-NC is unique among the nitrated PAHs thus far examined in that its primary pathway of activation in the system in which it has been shown to be a potent carcinogen, the preweanling mouse, appears to involve the intermediate formation of an amino dihydrodiol. The further activation of this amino dihydrodiol, possibly through a diol epoxide, leads to a single major DNA adduct. As is shown in the data in this manuscript, a variety of factors can influence the predominant metabolic pathway that leads to adduct formation, but the biological consequences of the various activation pathways remain to be determined.

Whether significant human exposure to 6-NC or 6-AC occurs in sufficient quantities to pose a health risk cannot be determined at this time. 6-NC has been detected in diesel exhaust (27 and Scheutzle, this conference) and has been reported in air particulates in West Germany (28), although the occurrence of 6-NC in ambient air has not been a universal finding (Zielinska, this conference). Our previously published results of studies with explants of human bronchus and our limited results with the blood DNA of patients receiving 6-AC in a clinical trial indicate that the metabolism of 6-NC and 6-AC through 6-AC-1,2-dihydro-diol is a major pathway in human tissues and thus exposure to these compounds could pose a health risk if it occurs. Our findings with the $^{32}$P-postlabeling assay indicate that the adducts derived from 6-NC, and possibly other potential carcinogens, could be greatly underestimated if optimal conditions are not used in the analysis. Thus, it is important to examine carefully the behavior of each potential compound of interest before examining a human sample of uncertain exposure history for DNA damage by this method. The role of the unique metabolism of 6-NC in the carcinogenicity of this compound in the preweanling mouse and the relevance of this assay system to the prediction of carcinogenic response in humans require further attention.

REFERENCES

1. W. F. Busby, Jr., R. C. Garner, F. L. Chow, C. N. Martin, E. K. Stevens, P. M. Newberne, and G. N. Wogan, 6-Nitrochrysene is a potent tumorigen in newborn mice. Carcinogenesis 6: 801 (1985).
2. P. G. Wislocki, E. S. Bagan, A. Y. H. Lu, K. L. Dooley, P. P. Fu, H. Han-Hsu, F. A. Beland, and F. F. Kadlubar, Tumorigenicity of nitrated derivatives of pyrene, benz[a]anthracene, chrysene and benzo[a]pyrene in the newborn mouse assay. Carcinogenesis 7: 1317 (1986).
3. W. F. Busby, Jr., E. K. Stevens, E. R. Kellenbach, J. Cornelisse, and J. Lugtenburg, Dose-response relationships of the tumorigenicity of cyclopenta[c,d]pyrene, benzo[a]pyrene and 6-nitrochrysene in a newborn mouse lung adenoma bioassay. Carcinogenesis 9: 741 (1988).
4. K. El-Bayoumy, G.-H. Shiue, and S. S. Hecht, Comparative tumorigenicity of 6-nitrochrysene and its metabolites in newborn mice. Carcinogenesis 10: 369 (1989).
5. W. F. Busby, Jr., E. K. Stevens, C. N. Martin, F. L. Chow, and R. C. Garner, Comparative tumorigenicity of parent and mononitro-poly-nuclear aromatic hydrocarbons in the BLU:Ha newborn mouse assay. Toxicol. Appl. Pharm. 99: 555 (1989).
6. K. El-Bayoumy, S. S. Hecht, and Hoffmann, D., Comparative tumor initiating activity on mouse skin of 6-nitrobenzo[a]pyrene, 6-nitrochrysene, 3-nitroperylene, 1-nitropyrene and their parent hydrocarbons. Cancer Lett. 16: 333, (1982).
7. M. Sala, C. Lasne, Y. P. Lu, and I. Chouroulinkov, Morphological transformation in three mammalian cell systems following treatment with 6-nitrochrysene and 6-nitrobenzo[a]pyrene. Carcinogenesis 8: 503 (1987).
8. G. Rudali, N. P. Buu-Hoi, and A. Lacassagne, Sur quelques effets biologiques du 2-amino-chrysene. C. R. Acad. Sci. (Paris) 236: 2020 (1953).
9. G. Lambelin, G. Mees, and N. P. Buu-Hoi, Chronic toxicity studies of 6-aminochrysene in the rat. Arzneim.-Forsch. 17: 1117 (1967).
10. F. J. C. Roe, R. L. Carter, and S. Adamthwaite, Induction of liver and lung tumours in mice by 6-aminochrysene administered during the first three days of life. Nature 221: 1063 (1969).
11. G. Lambelin, J. Roba, R. Roncucci, and R. Parmentier, Carcinogenicity of 6-aminochrysene in mice. Eur. J. Cancer 11: 327 (1975).
12. K. El-Bayoumy and S. S. Hecht, Identification of trans-1,2-dihydro-1,2-dihydroxy-6-nitrochrysene as a major mutagenic metabolite of 6-nitrochrysene. Cancer Res. 44: 3408 (1984).
13. K. El-Bayoumy, K. B. Delclos, R. H. Heflich, R. Walker, G.-H. Shiue,

and S. S. Hecht, Mutagenicity, metabolism and DNA adduct formation of 6-nitrochrysene in Salmonella typhimurium. Mutagenesis 4: 235 (1989).

14. J. Gelzer and P. Loustalot, Chrysenex in experimental advanced mammary cancer. Eur. J. Cancer 3: 79 (1967).

15. H. J. Tagnon, A. Coune, S. Garattini, R. Rosso, G. Lambelin, M. Gautier, and N.P. Buu-Hoi, The antitumoral activity of some derivatives of 6-aminochrysene. Eur. J. Cancer 6: 81 (1970).

16. N. P. Buu-Hoi, D.-P. Hien, and P. Mabille, Some biological properties of 6-aminochrysene and its derivatives, a family of carcinostatic compounds devoid of cytotoxic action, in: "Cancer Chemotherapy. Proc. Takeda Int. Conf., Osaka", A. Goldin, ed., Maruzen Co., Tokyo (1966).

17. Groupe Europeen du Cancer du Sein, Induction par le 6-aminochrysene de remission du cancer du sein en phase avancee chez la femme. Eur. J. Cancer 3: 75 (1967).

18. K. B. Delclos, D. W. Miller, J. O. Lay, Jr., D. A. Casciano, R. P. Walker, P. P. Fu, and F. F. Kadlubar, Identification of C8-modified deoxyinosine and $N^2$- and C8-modified deoxyguanosine as major products of the in vitro reaction of N-hydroxy-6-aminochrysene with DNA and the formation of these adducts in isolated rat hepatocytes treated with 6-nitrochrysene and 6-aminochrysene. Carcinogenesis 8: 1703 (1987).

19. K. B. Delclos, R. P. Walker, K. L. Dooley, P. P. Fu, and F. F. Kadlubar, Carcinogen-DNA adduct formation in the lungs and livers of preweanling CD-1 male mice following administration of [$^3$H]-6-nitro-chrysene, [$^3$H]-6-aminochrysene, and [$^3$H]-1,6-dinitropyrene. Cancer Res. 47: 6272 (1987).

20. K. B. Delclos, K. El-Bayoumy, S. S. Hecht, R. P. Walker, and F. F. Kadlubar, Metabolism of the carcinogen [$^3$H]6-nitrochrysene in the preweanling mouse: identification of 6-aminochrysene-1,2-dihydrodiol as the probable proximate carcinogenic metabolite. Carcinogenesis 9: 1875 (1988).

21. I. R. Rowland, Interactions of the gut microflora and the host in toxicology. Tox. Path. 16: 147 (1988).

22. K. B. Delclos, K. El-Bayoumy, D. A. Casciano, R. P. Walker, F. F. Kadlubar, S. S. Hecht, N. Shivapurkar, S. Mandal, and G. D. Stoner, Metabolic activation of 6-nitrochrysene in explants of human bronchus an in isolated rat hepatocytes. Cancer Res. 49: 2909 (1989).

23. R. C. Gupta, Non-random binding of the carcinogen N-hydroxy-2-AAF to repetitive sequences of rat liver DNA in vivo. Proc. Natl. Acad. Sci. (USA) 81: 6943 (1984).

24. M. V. Reddy and K. Randerath, Nuclease P1-mediated enhancement of sensitivity of 32P-postlabeling test for structurally diverse DNA adducts. Carcinogenesis 7: 1543 (1986).

25. J. P. Sculier, C. Brassinne, C. Laduron, A. Coune, C. Hollaert, and C. Delcroix, A phase I study, with pharmacokinetic analysis, of intravenous administration of 6-aminochrysene entrapped into sonicated liposomes in patients with advanced cancer. J. Liposome Res., in press.

26. M. Jeanpierre, A rapid method for the purification of DNA from blood. Nucleic Acids Res. 15: 9611 (1987).

27. J. Schilhabel and K. Levsen, Identification of nitrated polycyclic hydrocarbons in diesel particulate extracts by negative ion chemical ionization and tandem mass spectrometry. Fresenius Z. Anal. Chem. 333: 800 (1989).

28. R. C. Garner, C. A. Stanton, C. N. Martin, F. L. Chow, W. Thomas, D. Hubner, and R. Herrmann, Bacterial mutagenicity and chemical analysis of polycyclic aromatic hydrocarbons and some nitro derivatives in environmental samples collected in West Germany. Environ. Mut. 8: 109 (1986).

# THE POSSIBLE ROLE OF NITROARENES IN HUMAN CANCER

Stephen S. Hecht and Karam El-Bayoumy

Division of Chemical Carcinogenesis
American Health Foundation
Valhalla, NY 10595

## INTRODUCTION

In this paper, we consider the possible role of nitroarenes as causative compounds for human cancer. At the outset,
it is important to recognize that adequate data to answer this
question are not available. However, we will evaluate existing
data with respect to three major cancer sites: lung, bladder,
and breast. These sites were chosen because either experimental or epidemiologic studies, as summarized below, suggested
that nitroarenes were possible etiologic agents. Our evaluation of the role of nitroarenes in each of these diseases will
be based on their abilities to induce the relevant tumors in
experimental animals, on their concentrations in the appropriate environments, and on data from analyses of human tissues or
blood. We recognize that this approach is limited by the complex interactions that occur in environmental mixtures to which
humans are exposed. Interactions between carcinogens and
cocarcinogens, inhibitors, or promoters as well as host factors
such as levels of carcinogen metabolizing enzymes tend to con-
found analyses of this type. Nevertheless, we have found it
useful to focus on individual compounds within complex mixtures
as this can lead to new insights regarding mechanisms of cancer
induction.

### Lung Cancer

Large scale epidemiologic studies in several countries
have demonstrated that 80-90% of lung cancer deaths in men and
60-80% in women are attributable to tobacco smoking (1). It
does not appear likely that nitroarenes make a significant contribution to the carcinogenicity of tobacco smoke. Cigarette
smoke condensate was analyzed for 1-nitronaphthalene, 1-nitro-
pyrene, and 6-nitrochrysene; none was detected (detection limit
1-10 ng/cigarette) (2). However, the possible roles of nitroarenes in sidestream smoke or endogenously formed nitroarenes
have not been evaluated.

The contribution of urban air pollution to lung cancer
deaths is uncertain, but it may not exceed 1% (3). Estimates
of occupational causes of lung cancer range from 3-17% (4).

Presently available data do not demonstrate a convincing association between exposure to diesel exhaust and lung cancer incidence (5,6). However, inhalation studies of diesel exhaust have shown that it is tumorigenic to rodent lung (7-10). Since certain nitroarenes are present in diesel exhaust as well as ambient urban air, and have been shown to elicit respiratory tract tumors in rodents, they could play a role in lung cancers which may be associated with exposure to urban air or diesel exhaust. Although the percent of lung cancer deaths attributable to air pollution and occupational exposures may be low, the annual death rate from lung cancer is high—139,000 lung cancer deaths are estimated for the U.S. in 1988 (11).

Several classes of carcinogenic agents are known to be present in ambient air. These include polynuclear aromatic hydrocarbons (PAH) and their heterocyclic analogues, aldehydes, nitroPAH, as well as miscellaneous volatile organic carcinogens such as benzene (12,13). Among these, PAH are the most thoroughly characterized with respect to their concentrations and tumorigenic activities. Several lines of evidence support the role of PAH as major tumorigenic agents in polluted air. Particulate matter collected from urban air, and its PAH-containing subfractions are tumorigenic on mouse skin (13). Subfractions of diesel exhaust condensate enriched in PAH were more tumorigenic than those enriched in nitroPAH when implanted in rat lung (14). A variety of PAH have been shown to induce respiratory tract tumors in experimental animals when administered by lung implantation, intratracheal instillation, or inhalation (15). Among these, benzo[a]pyrene (BaP) is the most thoroughly investigated and one of the most carcinogenic. It can be considered as a representative of this class of carcinogens. To gain further insight on the potential role of nitroPAH in tumorigenesis induced by ambient air particulates or diesel exhaust, it is thus useful to compare the respiratory tract carcinogenic activities and concentrations of BaP with those of representative nitroPAH.

Two nitroPAH have been shown to be highly tumorigenic to rodent lung—1,6-dinitropyrene (1,6-diNP) and 6-nitrochrysene (6-NC). Their structures and that of BaP are illustrated in Figure 1.

Upon implantation in rat lung, 0.5 μmol of 1,6-diNP induced tumors in 82% of the animals (16). Under similar conditions, doses of 0.4-4.0 μmol BaP induced lung tumors in 29-94% of the treated rats (17). In Syrian golden hamsters,

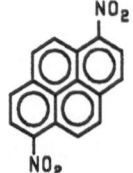
1,6-dinitropyrene

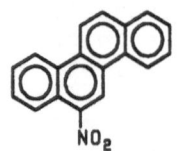
6-nitrochrysene

benzo[a]pyrene

Fig. 1. Structures of the respiratory carcinogens 1,6-dinitro-pyrene (1,6-diNP), 6-nitrochrysene (6-NC), and benzo-[a]pyrene (BaP).

intratracheal instillation of 1.7 μmol 1,6-diNP per week for 26 weeks induced lung tumors in 95% of the animals (18). Intratracheal instillation of 2 μmol BaP per week for 28 weeks induced lung tumors in 31% of the hamsters; the incidence of total respiratory tract tumors was 66% (19). These data indicate that 1,6-diNP is a stronger respiratory carcinogen than is BaP; this was confirmed by the study reported by Tokiwa et al in this book. The tumors induced by the two compounds were also histologically different. 1,6-DiNP gave mostly undifferentiated neoplasms whereas BaP induced well-differentiated squamous cell carcinomas.

6-NC is an exceptionally potent tumorigen in the newborn mouse. Comparisons of 6-NC and BaP have been carried out. The results showed that 6-NC was 10-25 times more tumorigenic toward newborn mouse lung than was BaP (20). It is one of the most potent tumorigens known in the newborn mouse. These data indicate that nitroPAH can be stronger tumorigens than BaP, a representative highly carcinogenic PAH. However, levels of BaP in ambient air (typically 400 pg/m$^3$) are greater than those of 1,6-diNP (typically 1 pg/m$^3$) (21,22). In diesel exhaust particulates, a BaP concentration of 2.2 ppm has been reported, compared to 0.4-1.2 ppm for 1,6-diNP (21,22). The higher concentrations of BaP than of 1,6-diNP in these systems may counteract its lower carcinogenicity.

Perhaps the best way to assess the relationship between exposure to nitroarenes and risk for lung cancer is to deter-mine levels of nitroarene-DNA adducts in human lung tissue. Only three studies have been published in which lung tissue from living subjects has been analyzed for "aromatic DNA ad-ducts." One measured putative PAH-DNA adducts by immunoassay while the other two used $^{32}$P-postlabelling (23-25). In all three cases, adduct levels appeared to depend on exposure to tobacco smoke. The adducts have not been identified. They could arise from PAH, nitroPAH, or other carcinogens present in tobacco smoke. In a recent study, levels of 4-aminobiphenyl-DNA adducts in human lung were quantified by immunoassay. There was no relationship to smoking status (26). Subcellular fractions from the same lung tissue were capable of reducing 4-nitrobiphenyl to the DNA binding species but were not able to oxidize 4-aminobiphenyl. These results suggested that exposure to 4-nitrobiphenyl may have accounted for the detection of these DNA adducts in lung. It will be important to carry out further studies of this type, particularly with respect to exposures to 1,6-diNP and 6-NC.

In summary, presently available data are insufficient to evaluate the role of nitroarenes in human lung cancer. It is unclear whether occupational or environmental exposures cause lung cancer, and to what extent PAH, nitroPAH, or other car-cinogens might be involved as causative factors. However, it is clear that 1,6-diNP and 6-NC are potent respiratory carcino-gens in rodents. These compounds present a potential hazard to exposed humans. Measures should be taken to eliminate them from the human environment.

Bladder Cancer

An epidemiologic study of motor exhaust-related occupa-tions has suggested a possible enhanced risk for bladder cancer (27). Two nitroPAH—2-nitronaphthalene and 4-nitrobiphenyl(Fig 2)—have been shown to induce bladder tumors, although the bio-assay data are limited. 2-Nitronaphthalene produced bladder

tumors in 1 of 3 treated monkeys, and 4-nitrobiphenyl induced
bladder tumors in 2 of 4 dogs (28,29). Thus, by comparison to
aromatic amines such as 2-naphthylamine and 4-aminobiphenyl,
which are established human bladder carcinogens, the data sup-
porting a role for nitroarenes in human bladder cancer are
quite limited. Nevertheless, the metabolic connection between
nitroarenes and aromatic amines supports further investigation
of their possible role as human bladder carcinogens.

2-nitronaphthalene                    4-nitrobiphenyl

Fig. 2. Structure of nitroarenes which induce bladder tumors.

        4-Aminobiphenyl-hemoglobin adducts have been detected in
humans and experimental animals. Their levels are higher in
smokers than in non-smokers. Some of the adducts detected in
non-smokers probably result from environmental exposures. Adducts have also been detected in untreated rats and dogs (30).
The extent to which these adducts might have arisen by exposure
to 4-nitrobiphenyl is not known, but is worth investigating. A
recent study has indicated the presence of 4-aminobiphenyl-DNA
adducts in human bladder biopsy samples, according to analyses
performed by $^{32}$P-postlabelling (31). Aromatic amines have also
been detected in human urine, independent of smoking status
(32). These data indicate that human exposure to nitroarenes
leading to adduct formation in the urinary bladder is possible.
Further studies are required to confirm this and to determine
the potential significance of these adducts as initiators of
bladder cancer in humans.

### Breast Cancer

        The initiator of human breast cancer is not known. Poten-
tial candidates include PAH such as BaP (33), aromatic amines
such as 4-aminobiphenyl (34), 2-amino-3-methylimidazo[4,5-f]-
quinoline (35), and nitroPAH. Four nitroPAH—1-nitropyrene, 4-
nitropyrene, 5-nitroacenaphthene, and 1,8-diNP—induce mammary
tumors in female rats (Fig 3) (36-39). Among these, 1,8-diNP
is the most potent, inducing a statistically significant
incidence of mammary adenocarcinoma after i.p. injection of 16
$\mu$mol to weanling CD rats. 1-Nitropyrene, the most predominant
of these nitroPAH in the environment, is a relatively weak
mammary carcinogen.
        One study has reported the detection of putative aromatic
DNA adducts in human mammary epithelial cells (40). Adducts
were detected in 3 of 10 samples analyzed by $^{32}$P-postlabelling.
The TLC retention times of these spots were different from that
of the major diol epoxide adduct of BaP. The identification of
such adducts present in human breast tissue would be a major
advance in understanding the etiology of breast cancer.

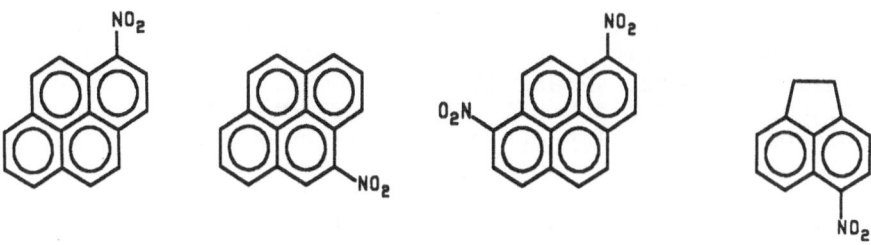

1-nitropyrene      4-nitropyrene      1,8-dinitropyrene      5-nitroacenaphthene

Fig. 3.   Structures of nitroarenes which induce mammary tumors
in laboratory animals.

Perspectives

The data presented in this review suggest that nitroarenes
might be involved in the etiology of some human cancers.  However, insufficient evidence is available to strongly support
this suggestion.  The most reasonable approach to assessing the
role of nitroarenes in cancer etiology would appear to be development of methods for detection and quantitation of their
metabolites and adducts in humans.  Such methods could be used
first to establish the extent of an individual's exposure and
uptake of nitroarenes and second in combination with epidemiologic investigations to determine a possible relationship between adduct levels and risk for particular cancers.  Ideally,
measurements of other carcinogen adducts could be performed on
the same samples resulting in a characteristic adduct or metabolite spectrum.  As discussed in this review and in other
chapters of this book, methods are being developed to measure
nitroarene metabolites and adducts and initial results
suggesting human uptake of these carcinogens are already
available.  Quantitation of adducts in human DNA or protein
samples requires exceedingly sensitive methods.  Presently
available technology seems capable of reaching the required
detection limits and provides exciting new approaches toward
realistically assessing cancer risks associated with environmental exposures.

REFERENCES

1.   International Agency for Research on Cancer, IARC Mono-
      graphs on the Evaluation of the Carcinogenic Risk of
      Chemicals to Humans, Volume 38, Tobacco Smoking, IARC,
      Lyon, France (1986) 228.
2.   K. El-Bayoumy, M. O'Donnell, S. S. Hecht, and D. Hoffmann,
      On the analysis of 1-nitronaphthalene, 1-nitropyrene,
      and 6-nitrochrysene in cigarette smoke, Carcinogenesis,
      6:505 (1974).
3.   E. L. Wynder and G. B. Gori, Contribution of the environ-
      ment to cancer incidence: an epidemiologic exercise, J.
      Natl. Cancer Inst., 58:825 (1977).
4.   P. Vineis, T. Thomas, R. B. Hayes, W. J. Blot,
      T. J. Mason, L. W. Pickle, P. Correa, E. T. H. Fontham,

and J. Schoenberg, Proportion of lung cancers in males, due to occupation, in different areas of the U.S.A., Int. J. Cancer, 42:851 (1988).

5. National Research Council, Health Effects of Exposure to Diesel Exhaust: Impacts of Diesel-Powered Light-Duty Vehicles, National Academy Press, Washington, D.C. (1981), 137.

6. K. Steenland, Lung cancer and diesel exhaust: a review, Am. J. Ind. Med., 10:177 (1986).

7. K. Iwai, T. Udagawa, M. Yagamishi, and H. Yamada, Long-term inhalation studies of diesel exhaust on F344 SPF rats. Incidence of lung cancer and lymphoma, in: "Carcinogenic and Mutagenic Effects of Diesel Engine Exhaust," N. Ishinishi, A. Koizumi, R. O. McClellan, and W. Stöber, eds., Elsevier Science Publishers, New York (1986), 349.

8. J. L. Mauderly, R. K. Jones, R. O. McClellan, R. F. Henderson, and W. C. Griffith, Carcinogenicity of diesel exhaust inhaled chronically by rats, in: "Carcinogenic and Mutagenic Effects of Diesel Engine Exhaust," N. Ishinishi, A. Koizumi, R. O. McClellan, and W. Stöber, eds., Elsevier Science Publishers, New York (1986), 397.

9. W. Stöber, Experimental induction of tumors in hamsters, mice and rats after long-term inhalation of filtered and unfiltered diesel engine exhaust, in: "Carcinogenic and Mutagenic Effects of Diesel Engine Exhaust," N. Ishinishi, A. Koizumi, R. O. McClellan, and W. Stöber, eds., Elsevier Science Publishers, New York (1986), 421.

10. J. J. Vostal, Factors limiting the evidence for chemical carcinogenicity of diesel emissions in long-term inhalation experiments, in: "Carcinogenic and Mutagenic Effects of Diesel Engine Exhaust," N. Ishinishi, A. Koizumi, R. O. McClellan, and W. Stöber, eds., Elsevier Science Publishers, New York (1986), 381.

11. American Cancer Society, Cancer Statistics, 1988, Ca-A Cancer Journal for Clinicians, 38:5 (1988).

12. T. E. Graedel, Ambient levels of anthropogenic emissions and their atmospheric transformation products, in: "Air Pollution, The Automobile, and Public Health," A. Y. Watson, R. R. Bates, and D. Kennedy, eds., National Academy Press, Washington, D.C. (1988), 133.

13. D. Hoffmann and E. L. Wynder, Organic particulate pollutants—chemical analysis and bioassays for carcinogenicity, in: "Air Pollution, Third Edition," A. C. Stern, ed., Academic Press, New York (1977), 361.

14. G. Grimmer, H. Brune, R. Deutsch-Wenzel, G. Dettbarn, J. Jacob, K.-W. Naujack, U. Mohr, and H. Ernst, Contribution of polycyclic aromatic hydrocarbons and nitro-derivatives to the carcinogenic impact of diesel engine exhaust condensate evaluated by implantation into the lungs of rats, Cancer Lett., 37:173 (1987).

15. S. S. Hecht, Potential carcinogenic effects of polynuclear aromatic hydrocarbons and nitroaromatics in mobile source emissions, in: "Air Pollution, The Automobile, and Public Health," A. Y. Watson, R. R. Bates, and D. Kennedy, eds., Washington, D.C., National Academy Press, (1988), 555.

16. T. Maeda, K. Izumi, H. Otsuka, Y. Manabe, T. Kinouchi, and Y. Ohnishi, Induction of squamous cell carcinoma in rat

lung by 1,6-dinitropyrene, J. Natl. Cancer Inst., 76:693 (1986).

17. R. P. Deutsch-Wenzel, H. Brune, G. Grimmer, G. Dettbarn, and J. Misfeld, Experimental studies in rat lungs on the carcinogenicity and dose-response relationships of eight frequently occurring environmental polycyclic aromatic hydrocarbons, J. Natl. Cancer Inst., 71:539 (1983).

18. S. Takayama, T. Ishikawa, H. Nakajima, and S. Sato, Lung carcinoma induction in Syrian golden hamsters by intra-tracheal instillation of 1,6-dinitropyrene, Jpn. J. Cancer Res., (Gann) 76:457 (1985).

19. M. Ketkar, U. Green, P. Schneider, and U. Mohr, Investigations on the carcinogenic burden by air pollution in man. Intratracheal instillation studies with benzo[a]pyrene in a mixture of Tris buffer and saline in Syrian golden hamsters, Cancer Lett., 6:279 (1979).

20. W. F. Busby, Jr., E. K. Stevens, E. R. Kellenbach, J. Correlisse, and J. Lugtenburg, Dose-response relationships of the tumorigenicity of cyclopenta[c,d]-pyrene, benzo[a]pyrene, and 6-nitrochrysene in a new-born mouse lung adenoma bioassay, Carcinogenesis, 9:741 (1988).

21. H. Tokiwa and Y. Ohnishi, Mutagenicity and carcinogenicity of nitroarenes and their sources in the environment, CRC Critical Reviews in Toxicology, 17:23 (1986).

22. T. L. Gibson, Nitro derivatives of polynuclear aromatic hydrocarbons in airborne and source particulate matter, Atmos. Env., 16:2037 (1982).

23. F. P. Perera, M. C. Poirier, S. H. Yuspa, J. Nakayama, A. Jaretzki, M. M. Curnen, D. N. Knowles, and I. B. Weinstein, A pilot project in molecular cancer epidemiology: determination of benzo[a]pyrene-DNA adducts in animal and human tissues by immunoassays, Carcinogenesis, 3:1405 (1982).

24. E. Randerath, R. H. Miller, D. Mittal, T. A. Avitts, H. A. Dunsford, and K. Randerath. Covalent DNA damage in tissues of smokers as determined by [32]P-postlabelling assay, J. Natl. Cancer Inst., 81:341 (1989).

25. D. H. Phillips, A. Hewer, C. N. Martin, R. C. Garner, and M. M. King, Correlation of DNA adduct levels in human lung and cigarette smoking, Nature, 336:790 (1988).

26. D. W. Roberts, S. J. Culp, P. P. Fu, C. H. Teitel, R. W. Benson, G. Talaska, J. Massengill, N. P. Lang, and F. F. Kadlubar, Proc. Amer. Assoc. Cancer Res., 30:310 (1989).

27. D. T. Silverman, R. N. Hoover, T. J. Mason, and G. M. Swanson, Motor exhaust-related occupations and bladder cancer, Cancer Res., 46:2113 (1986).

28. G. M. Conzelman, Jr., J. E. Moulton, and L. E. Flanders, III, Tumors in the urinary bladder of a monkey: induction with 2-nitronaphthalene, Jpn. J. Cancer Res. (Gann), 61:79 (1970).

29. W. B. Deichmann, W. M. MacDonald, M. M. Coplan, F. M. Woods, and W. A. D. Anderson, Para nitrobiphenyl, a new bladder carcinogen in the dog, Industr. Med. Surg., 27:634 (1958).

30. M. S. Bryant, P. L. Skipper, S. R. Tannenbaum, and M. Maclure, Hemoglobin adducts of 4-aminobiphenyl in smokers and non-smokers, Cancer Res., 47:602 (1987).

31. G. Talaska, J. Massengill, A. Z. S. S. Al-Juburi, and

F. F. Kadlubar, Carcinogen-DNA adducts in human urinary bladder biopsy samples in relation to cigarette smoking, Proc. Amer. Assoc. Cancer Res., 30:1265 (1989).

32. K. El-Bayoumy, J. Donahue, S. S. Hecht and D. Hoffmann, Identification and quantitative determination of aniline and toluidines in human urine, Cancer Res., 46:6064 (1986).

33. D. L. McCormick, F. J. Burns, and R. E. Albert, Inhibition of benzo[a]pyrene-induced mammary carcinogenesis by retinyl acetate, J. Natl. Cancer Inst., 66:559 (1981).

34. J. C. Arcos and J. Simon, Effect of 4'-substituents on the carcinogenic activity of 4-aminoazobenzene derivatives, Arzneim. Forsch, 12:270 (1962).

35. T. Tanaka, W. S. Barnes, G. M. Williams, and J. H. Weisburger, Multipotential carcinogenicity of the fried food mutagen 2-amino-3-methylimidazo[4,5-f]-quinoline in rats, Jpn. J. Cancer Res., (Gann) 76:570 (1985).

36. M. Hirose, M. S. Lee, C. Y. Wang, and C. M. King, Induction of rat mammary gland tumors by 1-nitropyrene, a recently recognized environmental mutagen, Cancer Res., 44:1158 (1984).

37. K. El-Bayoumy, A. Rivenson, B. Johnson, J. DiBello, P. Little, and S. S. Hecht, Comparative tumorigenicity of 1-nitropyrene, 1-nitrosopyrene, and 1-aminopyrene administered by gavage to Sprague-Dawley rats, Cancer Res., 48:4256 (1988).

38. C. M. King, Metabolism and biological effects of nitropyrene and related compounds, Health Effects Institute Research Report Number 16, HEI, Cambridge, Mass (1988).

39. N. Takemura, C. Hashida, and M. Terasawa, Carcinogenic action of 5-nitroacenaphthene, Br. J. Cancer, 30:481 (1974).

40. L. A. Seidman, C. J. Moore, and M. N. Gould, $^{32}$P-Postlabelling analysis of DNA adducts in human and rat epithelial cells, Carcinogenesis, 9:1071 (1988).

REACTION OF GLUTATHIONE WITH 4-NITROSOBIPHENYL.  John S. Wheeler, Bradford W. Manning, Fred F. Kadlubar, Jack O. Lay, Dwight W. Miller, and Jack A. Hinson, National Center for Toxicological Research, Jefferson, Arkansas, 72079-9502

It has been shown that exposure to 4-aminobiphenyl (4-ABP) can be monitored by measurement of covalent hemoglobin (HB) adducts.  The mechanism of HB binding is believed to be hepatic N-hydroxylation followed by HB-catalyzed oxidation of N-OH-4-ABP to 4-nitrosobiphenyl (4-NBP) which then binds to a cysteine group on HB.  Since the reaction of 4-NBP with endogenous thiols such as glutathione (GSH) may be important in relating HB adducts as an index of 4-ABP exposure, we have investigated the reaction of GSH with 4-NBP.  [$^3$H]4-NBP (15 $\mu$M) was incubated with 10 mM GSH (ph 7.3) at 37° C for up to 1 hr.  HPLC analysis of the reaction indicated that multiple products were formed. The major product was isolated and identified by a combination of MS and $^1$H-NMR spectroscopy to be N-(glutathion-S-yl)-4-ABP S-oxide.  A time course indicated that disappearance of 4-NBP was rapid (app. 80% at 1 min).  Concomitant with this disappearance was formation of the GSH conjugate (38.5%) and of N-OH-4-ABP (20.2%).  Reaction with $^3$H-GSH also indicated a minor GSH conjugate was formed; however, insufficient amounts were formed for conclusive identification.  At 1 hr, the amount of the N-OH-4-ABP and NBP decreased and N-(glutathion-S-yl)-4-ABP S-oxide increased to 70.2% of the total reaction products.  Also, the amount of the minor conjugate decreased and 4-ABP was detected as a significant product.  Thus, individual variations in GSH levels may modulate the amount of 4-NBP binding to HB.

DETERMINATION OF CARCINOGENIC ARYLAMINE N-OXIDATION PHENOTYPE IN HUMANS BY ANALYSIS OF CAFFEINE URINARY METABOLITES. M.A. Butler, N.P. Lang, J.F. Young, G. Talaska, J. Massengill, C. Teitel, and F.F. Kadlubar. The National Center for Toxicological Research, Jefferson, AR 72079 and The John A. McClellan Memorial Veterans Hospital, Little Rock, AR 72205.

Epidemiological studies have shown wide variation in human urinary bladder and colo-rectal cancer incidences that may arise in part from genetic differences in susceptibility to carcinogenic arylamines. The hepatic N-oxidation of several primary arylamines, regarded as an important activation step leading to carcinogenesis, is catalyzed selectively by human liver cytochrome $P-450_{PA}$; and several studies have indicated that considerable inter-individual variability in $P-450_{PA}$ exists in human populations. Recently, we have shown that hepatic microsomal caffeine 3-demethylation, the initial major step in caffeine disposition in humans, is also selectively catalyzed by human $P-450_{PA}$. Thus, caffeine 3-demethylation activity in humans may be used as an indirect measure of carcinogenic arylamine N-oxidation activity. In this study, we have developed a metabolic phenotyping procedure to assess caffeine-3-demethylation proficiency. A 200-µl urine sample, obtained between 4-5 hours after an individual has consumed 50-100 mg caffeine (as coffee or soft drink), is analyzed by an HPLC method that quantifies caffeine and its 14 metabolites. Pharmacokinetic studies indicate that a molar ratio of 1,7-dimethylxanthine/caffeine in this urine sample reflects quasi-steady state blood levels and closely approximates hepatic caffeine 3-demethylation activity. In addition, probit analysis shows a bimodal distribution of caffeine 3-demethylation activity (n = 30), which suggests the existence of phenotypic slow and rapid metabolizers. This method is being applied to determine the role of the human N-oxidation phenotype in inter-individual differences in susceptibility to arylamine-induced cancers.

ASSESSING THE BIOLOGICAL SIGNIFICANCE OF NITRACRINE ANALOGUES:
THE RELATIONSHIP BETWEEN MICROBIAL AND MAMMALIAN MUTAGENICITY.

Lynnette R. Ferguson, and William A. Denny, Cancer Research Laboratory,
University of Auckland Medical School, Private Bag, Auckland, New Zealand

Because of an ongoing interest in nitracrine (1-nitro-9-(dimethyl-aminopropylamino)-acridine) and analogues as potential radiosensitisers and hypoxia-selective anticancer drugs, it is important to assess their potential mutagenic effects in humans.  Although there is obvious value in using mammalian assays for such evaluations, the number of compounds under consideration precludes such experiments as a routine.  Therefore, we have assessed mammalian mutagenesis for a limited group of such compounds (1-, 2-, 3-, 4-nitracrine, and the des-nitro analogue), and then considered whether any microbial assay or combination of such could have predicted the mammalian results. The pattern of mutagenic specificity and of nitroreductase requirements for mutagenicity in a range of strains of Salmonella typhimurium showed no relationship to the mammalian effects.  However, the ability of the compounds for recombination in Saccharomyces cerevisiae strain D5 correlated completely with their clastogenic and mutagenic properties in mammalian cells.  We conclude that assays for recombination in yeast may be the prescreen of choice for nitracrine analogues and possibly also other nitroarenes of environmental importance.

THE DIRECT-ACTING MUTAGENICITY OF NITROTHIOPHENES DERIVATIVES IN
SALMONELLA TYPHIMURIUM . Hrelia P., Morotti M., Scotti M., Paolini
M., Spinelli D.*, Cantelli Forti G. - Institute of Pharmacology and
*Department of Organic Chemistry, University of Bologna, Italy.

Nitroheterocyclic drugs have in recent years become of great
interest for their mutagenicity and carcinogenicity. A series of
nitrocompounds (structurally-related to the 5-nitro-3-
thiophenecarboxylic acid) were synthesized as chemotherapeutic agents
and assayed for their direct-acting and S9 mediated mutagenicity in
Salmonella typhimurium TA100 and TA98 strains. Effects of different
substituents, such as NO2, Cl, Br, F, CH3 and OCH3 groups, were
studied to evaluate the structural feature that affects the
metabolism and the bacterial mutagenic potency of nitrothiophenes.
All the derivatives were found to be mutagenic, TA100 was the most
sensitive strain. Increased activity was shown by substituted
compounds; the Br derivatives were the weakest mutagens, Cl and NO2
compounds exhibited the most dramatic mutagenicity. The mutagenic
activity did not require the S9 fraction but was largely dependent on
bacterial nitroreductase. The primary basis of nitrothiophene
direct-acting mutagenicity appears to be a reduction of the
nitrofunction to the corresponding hydroxylamine, via diamagnetic and
free radical intermediates. The determination of physico-chemical
parameters, as expression of the liphophilic character of the
molecules, will contribute to carry out structure-activity studies in
order to recognize the base structure for the receptor-drug binding.

AN EVALUATION OF THE METABOLISM OF 1-NITRO[$^{14}$C]PYRENE BY RABBIT TRACHEAL EPITHELIAL CELLS: KINETIC ANALYSIS.   Leon C. King[1],[2] Ernest Hodgson[2] and Joellen Lewtas[1]   [1]U. S. Environmental Protection Agency, Research Triangle Park, NC;   [2]Toxicology Program, North Carolina State University, Raleigh, NC

The metabolism of 1-nitro[$^{14}$C]pyrene ($^{14}$C-1-NP) by freshly isolated rabbit tracheal cells has been investigated in order to determine the kinetic parameters $K_M$ amd $V_{Max}$.   Experiments to optimize the cell number and the incubation time yielding maximum metabolite formation following incubations with $^{14}$C-1-NP were also performed. Metabolites from the incubation medium and cell lysates at each stage of this investigation were extracted, analyzed and quantitated by HPLC and liquid scintillation spectrophotometry. Maximum rate of metabolite production was attained at $2 \times 10^6$ cells and 4 hours incubation, with no significant increase at 20 hours. Using these optimized conditions, experiments were performed to assess the kinetics of the metabolism of $^{14}$C-1-NP over a concentration range of (0.1 to 50$\mu$M). The apparent $K_M$ ranged from $2.43 \pm 1.81 \times 10^{-1}\mu$M (S.E.) to $3.89 \pm 7.31 \times 10^{-1}\mu$M (S.E.). $V_{Max}$ was determined to range from $5.62 \times 10^{-3} \pm 3.9 \times 10^{-4}\mu$moles/mg/hr (S.E.) to $7.65 \times 10^{-3} \pm 8.7 \times 10^{-4}$ $\mu$moles/mg/hr (S.E.).   The apparent $K_M$ and $V_{Max}$ were determined using three linear transformation models (Lineweaver-Burk, Hofstee, Woolf and a nonlinear model). Scatchard analysis of the kinetic data resulted in a curvilinear plot indicating the presence of two enzymes, one with a low affinity ($K_M$ = 0.584$\mu$M) and high capacity (Vmax = $2.38 \times 10^{-3}\mu$moles/mg/hr) and another with a high affinity ($K_M$ = 0.062$\mu$M) and low capacity ($V_{Max}$ = $8.1 \times 10^{-4}\mu$moles/mg/hr).   The results of this study  provide information needed to optimize experimental conditions in evaluating the comparative metabolism and genotoxicty of $^{14}$C-1-NP by respiratory tract cells from various species.

This abstract does not necessarily reflect EPA policy.

Comparison of Anti-genotoxic Activities of Dibenzoylmethane, 1,3-Indandione and 2-(Methylmercapto)benzimidazole in Bacteria and Human Cells. H. Nagase, and C. Y. Wang, Michigan Cancer Foundation, Detroit, Michigan 48201.

The nucleophilic compounds, diacylmethanes and thioethers, are potential scavengers of reactive electrophilic carcinogens. We have previously reported that some diacylmethanes inhibit the mutagenicities in Salmonella, and the *in vitro* nucleic acid-binding of some carcinogens. The present study extended the investigation to a mammalian system using unscheduled DNA synthesis (UDS) as an end-point. The amount of dibenzoylmethane (DBM), 1,3-indandione (IDD) and 2-(methylmercapto)benzimidazole (TE) were 0.1, 0.5 and 1 $\mu$mol/plate in the Ames' test and were 0.1 and 1 mM in the UDS test. In the Ames' test, DBM inhibited the mutagenicity of 2-nitrofluorene (2-NF) and 4-nitroquinoline-N-oxide (4-NQO), and IDD inhibited only that of 2-NF. TE did not inhibit the mutagenicity of either 2-NF or 4-NQO. The induction of UDS in a human urothelial cell line, HCV-29, with 4-NQO or *N*-hydroxy-2-acetylaminofluorene (N-OH-AAF) was inhibited by IDD and TE. DBM slightly inhibited the UDS-induction with N-OH-AAF, but not with 4-NQO. The inhibition in the Ames' test and UDS system was dose-related with respect to the inhibitory agents. These results suggest that, in addition to scavenging of reactive carcinogens, other mechanisms, such as metabolism, may also be involved in the inhibition of genotoxicity. This may account for the differences in the inhibitory activities in bacteria and mammalian cells. (Supported by NIH grant CA23800)

DIFFERENTIAL CYTOTOXICITY AND GENOTOXICITY OF 1,3- and 1,6-DINITROPYRENE IN MAMMILIAN CELL LINES DERIVED FROM VARIOUS TISSUES AND SPECIES. Eike Roscher, Olga Cumpelik, and Friedrich J. Wiebel, GSF-Institute of Toxicology, GSF Research Center, D-8042 Neuherberg/München, F.R.G.

We have assessed the cytotoxicity and genotoxicity of 1,3-dinitropyrene (1,3-DNP) and 1,6-dinitropyrene (1,6-DNP) in a panel of cell lines derived from various tissues of mouse, rat, hamster or man. The cell lines used were defined for their expression of xenobiotic metabolizing enzymes, notably cytochrome P-450-containing monooxygenases. As biological endpoints we used inhibition of growth and induction of micronuclei. In addition, micronuclei were analysed for the presence of kinetochores.

1,6-DNP was highly cytotoxic to cell lines such as V79, H4IIEC3 or 208F whithout apparent correlation to their cytochrome P450 activity. 1,3-DNP was selectively toxic to other cell lines, e.g. HepG2, BWIJ or 5L, all of which contain cytochrome P450. Some cell lines were sensitive to both test compounds, others were entirely resistant. - The cytotoxicity of 1,6-DNP was positively correlated with the induction of micronuclei in the cell lines tested. In contrast, 1,3-DNP did not induce micronuclei in cell lines which were highly sensitive to the cytotoxic effects of the compound. Analysis of micronuclei for the presence of kinotochores indicated that 1,6-DNP can cause chromosomal damage by breakage as well as abnormal distribution depending on the test cell line. - The patterns of 1,3- and 1,6-DNP metabolites formed by various sensitive and insensitive cell lines showed that the two compounds are activated by different pathways.

The results suggest that 1,6-DNP and 1,3-DNP differ in their metabolic activation and mechanism of action. The data furthermore show that the responses to the agents are highly cell specific.

COOKED-MEAT DERIVED AROMATIC AMINE MUTAGENS AND THEIR
IMMUNOASSAY. Martin Vanderlaan, Bruce E. Watkins, Mona Hwang, Mark G. Knize
and James S. Felton, Biomedical Sciences Division, Lawrence Livermore National
Laboratory, Livermore, CA 94550

Typical household cooking of meat produces a family of aromatic amine mutagens
termed aminoimizodazaarenes (AIAs). This family contains among other members
PhIP(2-amino-1-methyl-6-phenylimidazo[4,5-b]pyridine), IQ (2-amino-3-
methylimidazo[4,5-f]quinoline), and MeIQx (2-amino-3,8-dimethylimidazo[4,5-f]
quinoxaline). Along with the nitro-aromatics, these are some of the most mutagenic
compounds ever tested in bacterial mutagenesis assays. Where tested, the AIAs are
also carcinogens. Typical levels of AIAs in well-done cooked beef are 0.1, 1.0, and
10 ppb for IQ, MeIQx, and PhIP, respectively. This low level of presence in foods
hampers analysis of AIAs in the diet. To facilitate AIA assay we have developed a set
of monoclonal antibodies to the AIAs. Individual antibodies have been selected that
react with IQ, MeIQx and PhIP. These antibodies are being used to quantify the levels
of individual AIAs in various cooked meats. In addition, analysis of well-done beef
shows the presence of other, currently unknown, structurally related compounds that
immunochemically cross-react with the antibodies. The antibodies also recognize
compounds present in the urine of people on diets of well-done beef, but not people
eating vegetarian diets. These findings suggest that the antibodies may be useful as
biochemical markers of human exposure to dietary meat mutagens. The antibodies
also may provide a means of concentrating and purifying human metabolites of AIAs
from urine.

Research supported by NCI grants RO1 CA40811-03 and CA48446-02 and
performed at LLNL under contract W-7405-ENG-48 with the Department of Energy.

ESTABLISHMENT OF NEW STRAINS OF <u>S. TYPHIMURIUM</u> HIGHLY SENSITIVE TO NITROARENES AND AROMATIC AMINES: TA98 AND TA100 SUBSTRAINS WITH RICH NITROREDUCTASE OR ACETYLTRANSFERASE ACTIVITIES. Masahiko Watanabe, Motoi Ishidate, Jr., and Takehiko Nohmi, Division of Mutagenesis, National Institute of Hygienic Sciences, Kamiyoga, Setagaya-ku, Tokyo 158, Japan

Nitroreductase and acetyltransferase genes of <u>S. typhimurium</u> TA1538 are cloned into pBR322 (see Watanabe et al., Biochem. Biophys. Res. Commun., **147** 974-979 (1987)). The plasmids (pYG216, which contains a nitroreductase gene, and pYG219, which contains an acetyltransferase gene) were introduced into TA98 and TA100. TA98(pYG216), TA98(pYG219), TA100(pYG216) and TA100(pYG219) were named as YG1021, YG1024, YG1026 and YG1029, respectively. YG1021 and YG1026 had nitrofurazone-reductase activity about 50 times higher than the original strain, TA1538(pBR322), and were highly sensitive to 2-nitrofluorene (2-NF), 1-nitropyrene (1-NP) and 2-nitronaphthalene (2-NN)(Mutat. Res., in press (1989)). YG1024 and YG1029 had isoniazid-, 2-aminofluorene-<u>N</u>-acetyltransferase and <u>N</u>-hydroxy-Glu-P-1-<u>O</u>-acetyltransferase activities 100 times higher than TA1538(pBR322), and were highly sensitive to 2-NF, 1-NP, 1,8-dinitropyrene, 2-NN, Glu-P-1(+S9) and 2-aminoanthracene(+S9). These newly established strains are recommended to use for the detection of a small amount of nitroarenes and aromatic amines in the environment, and also for the estimation of metabolic pathway of chemical mutagens.

This work was supported by a Grant-in-Aid from the Japan Health Sciences Foundation.

# CONTRIBUTOR INDEX

# SUBJECT INDEX

Plasmid (continued)
  pS40A, 157-165
  pSM14, 106-107, 110, 111
  pZ189, 149-155
Polycyclic aromatic hydrocarbons,
    16,17
Postlabeling,
  $^{32}$P detection, 127, 169, 181-185,
      191, 201-208, 279-282, 295-300,
      312
Pyrene,
  1-amino, 158
  azido, 267-270
  benzo[a], 2, 6, 7, 29, 31, 32, 203
  1,3-dinitro, 35, 323
  1,6-dinitro, 29-32, 310-311, 323
  1,8-dinitro, 167-179
  1-nitro, 4, 35, 67, 80, 85-92, 157-
      165, 201-208, 273-274, 285-291,
      312, 321
    biliary metabolites, 88
    cysteine conjugates, 88-90
    4,5-dihydroxy-4,5-diol, 87, 88
    9,10-dihydroxy-9,10-diol, 88
    glutathione conjugates, 88-90
    3-hydroxy, 85
    6-hydroxy, 85
    8-hydroxy, 85
    metabolites, 285-291
    4,5-oxide, 87, 88, 92, 181-185, 273-
      275
    9,10-oxide, 87, 88, 92, 181-185

Pyrene, (continued)
  2-nitro, 67
  4-nitro, 67
  1-nitroso, 150, 158
  1-nitroso-8-nitro, 168

Rabbits,
  New Zealand White, 203, 286
Rat,
  carcinogenicity studies, 2,3
  strain,
    F344, 2-5, 9-11, 30, 125-132, 190
    Wistar, 220
Respiratory burst,
  human granulocyte, 249-253

*Salmonella typhimurium*, 167, 325
Silica,
  quartz, 9
*supF* gene, 149-155

Thiophene,
  nitro, 320
  5-nitro-3-carboxylic acid, 320
Toluenes,
  nitro, 66
Tracheobronchial clearance, 17
Tumors,
  lung, 2-9, 31, 32

XAD-2 resin, 66
Xanthine oxidase, 181-185, 202, 203